Lecture Notes in Mathematics

Edited by A. Dold and B. Eckmann

616

Abelian Group Theory

Proceedings of the
2nd New Mexico State University Conference,
Held at Las Cruces,
New Mexico, December 9–12, 1976

Edited by
D. Arnold, R. Hunter, and E. Walker

Springer-Verlag
Berlin Heidelberg New York 1977

Editors
David M. Arnold
Roger H. Hunter
Elbert A. Walker
Department of Mathematics
New Mexico State University
Las Cruces, NM 88003/USA

AMS Subject Classifications (1970): 02H15, 02K05, 13C05, 13L05, 16A18, 18E05, 18E10, 18E25, 18G05, 18G10, 18G15, 18G20, 18G25, 20J05, 20K10, 20K15, 20K20, 20K25, 20K30, 20K35, 20K40, 20K45, 20K99

ISBN 3-540-08447-9 Springer-Verlag Berlin Heidelberg New York
ISBN 0-387-08447-9 Springer-Verlag New York Heidelberg Berlin

Printing and binding: Beltz Offsetdruck, Hemsbach/Bergstr.
2141/3140-543210

TABLE OF CONTENTS

PREFACE

There have been a number of exciting developments in Abelian group theory in the last few years. The solution of the Whitehead problem, using independence results of set theory, has given impetus to the investigation of meta-mathematical techniques. Breakthroughs have been made in the study of finite rank torsion free groups. The theory of simply presented groups and their summands, in both the local and global cases, has been advancing rapidly. Valuated groups have emerged in a variety of contexts and show promise of becoming the key structural concept of the future. New insights have been gained into what makes a group a direct sum of cyclics. Categorical techniques have been employed with ever increasing effectiveness. The entire subject has been revitalized, and the time seemed ripe for a conference. The suggestion that one be held in late 1976 met with overwhelming response from the Abelian groups community. Funding was obtained from the New Mexico State University Mathematics Department, the New Mexico State University College of Arts and Sciences Research Center, and the National Science Foundation. The Conference was held December 9-12, 1976 at the Holy Cross Retreat just South of Las Cruces, New Mexico. This volume contains all the papers presented at the Conference, together with a few other papers submitted by participants.

Several people deserve special mention. Fred Richman was one of the organizers of the Conference. John DePree, Chairman of the New Mexico State University Mathematics Department, and Jack Monagle, Associate Dean of the College of Arts and Sciences, were instrumental in providing financial support for the Conference. Sofora Davis, Evelyn Fox, Cathy Granger and Marnie Solomon performed cheerfully and conscientiously the secretarial duties, and typed a good share of this manuscript. Mona Sailer provided invaluable service during the conference itself. Brother Sean Carr and the staff at the Holy Cross Retreat did their usual superb job of feeding, watering, and bedding down the participants.

David Arnold
Roger Hunter
Elbert Walker

June, 1977
Las Cruces, New Mexico

LIST OF PARTICIPANTS

David Arnold	New Mexico State University, Las Cruces, New Mexico
Khalid Benabdallah	University of Montreal, Montreal, Quebec
Dennis Bertholf	Oklahoma State University, Stillwater, Oklahoma
Eddie Boyd	Oklahoma State University, Stillwater, Oklahoma
Del Boyer	University of Texas, El Paso, Texas
Willy Brandal	University of Tennessee, Knoxville, Tennessee
Don Cook	Albany Junior College, Albany, Georgia
Yonina Cooper	University of Kansas, Lawrence, Kansas
Doyle Cutler	University of California, Davis, California
John DePree	New Mexico State University, Las Cruces, New Mexico
Tom Dixon	Lander College, Greenwood, South Carolina
Vlasta Dlab	Carleton University, Ottawa, Ontario
Don Dubois	University of New Mexico, Albuquerque, New Mexico
Paul Eklof	University of California, Irvine, California
Ronald Ensey	Appalachian State University, Boone, North Carolina
H. K. Farahat	University of Calgary, Calgary, Alberta
Ed Fisher	University of Wisconsin, Madison, Wisconsin
Don Fitzgerald	Oklahoma State University, Stillwater, Oklahoma
Laszlo Fuchs	Tulane University, New Orleans, Louisiana
John Giever	New Mexico State University, Las Cruces, New Mexico
Ralph Grimaldi	Rose-Hulman, Terre Haute, Indiana
William Gustafson	Texas Tech University, Lubbock, Texas
Alfred Hales	University of California, Los Angeles, California
Dave Harrison	University of Oregon, Eugene, Oregon
Neal Hart	Sam Houston State University, Huntsville, Texas
Jutta Hausen	University of Houston, Houston, Texas
Paul Hill	Auburn University, Auburn, Alabama
Billy Hobbs	Point Loma College, San Diego, California
Ed Howard	San Diego State University, San Diego, California

LIST OF PARTICIPANTS con't.

Roger Hunter	New Mexico State University, Las Cruces, New Mexico
John Irwin	Wayne State University, Detroit, Michigan
Catarina Kiefe	University of New Mexico, Albuquerque, New Mexico
Art Knoebel	New Mexico State University, Las Cruces, New Mexico
Art Kruse	New Mexico State University, Las Cruces, New Mexico
Lee Lady	University of Kansas, Lawrence, Kansas
Michel LeBorgne	New Mexico State University, Las Cruces, New Mexico
Wolfgang Liebert	Technische Universität, Munich, Germany
Warren May	University of Arizona, Tucson, Arizona
Adolf Mader	University of Hawaii, Honolulu, Hawaii
Alan Mekler	Carleton University, Ottawa, Ontario
Linda Miller	New Mexico State University, Las Cruces, New Mexico
Ray Mines	New Mexico State University, Las Cruces, New Mexico
Judy Moore	New Mexico State University, Las Cruces, New Mexico
Charles Murley	New Mexico State University, Las Cruces, New Mexico
Ron Nunke	University of Washington, Seattle, Washington
Barbara O'Brien	Texas Tech University, Lubbock, Texas
Robin O'Callaghan	University of Texas at Permian Basin, Odessa, Texas
John O'Neill	University of Detroit, Detroit, Michigan
Ed Oxford	University of Southern Mississippi, Hattiesburg, Miss.
James Parr	Illinois State University, Normal, Illinois
Dick Pierce	University of Arizona, Tucson, Arizona
Jim Reid	Wesleyan University, Middletown, Connecticut
Fred Richman	Princeton University, Princeton, New Jersey
Laurel Rogers	University of Colorado, Colorado Springs, Colorado
Phillip Schultz	University of Washington, Seattle, Washington
Clayton Sherman	New Mexico State University, Las Cruces, New Mexico
Otis Solomon	New Mexico State University, Las Cruces, New Mexico
Bob Stanton	St. John's University, Jamaica, New York
David Tabor	University of Texas at San Antonio, San Antonio, Texas

LIST OF PARTICIPANTS con't.

Sharon Theleman	University of New Mexico, Albuquerque, New Mexico
Elias Toubassi	Tulane University, New Orleans, Louisiana
Mary Turgi	University of Illinois, Urbana, Illinois
Charles Vinsonhaler	University of Connecticut, Storrs, Connecticut
Carol Walker	New Mexico State University, Las Cruces, New Mexico
Elbert Walker	New Mexico State University, Las Cruces, New Mexico
Gary Walls	Oklahoma State University, Stillwater, Oklahoma
Stuart Wang	Texas Tech University, Lubbock, Texas
Bob Warfield	University of Leeds, Leeds, England
Roger Wiegand	University of Nebraska, Lincoln, Nebraska
Sylvia Wiegand	University of Nebraska, Lincoln, Nebraska
Robert Wilson	University of California, Long Beach, California
Julius Zelmanowitz	University of California, Santa Barbara, California

THE STRUCTURE OF MIXED ABELIAN GROUPS

Robert B. Warfield, Jr.[1]

1. Introduction. It is only within the last few years that research on mixed Abelian groups has begun to be a major part of Abelian group theory. This paper is intended as an introduction to the theory of mixed Abelian groups, emphasizing the new techniques which have made the recent progress possible. In addition to a survey of basic techniques, the paper contains some new cancellation theorems, a review of recent work on the classification theory of mixed groups, and a collection of open problems.

For an area of mathematics to remain viable, it must have more than some interesting open questions and occasional good theorems. There must be a variety of questions and methods, so that the life of the subject can go on between the high points. It is for this reason, I think, that the theory of p-groups has always occupied a central place in Abelian group theory. There have always been a variety of methods and open questions available. There have been more high points as well, the most recent being the theory and classification of totally projective groups. Work continues in the theory of p-groups today, but it does not now occupy the central position in research that it usually has.

The theory of torsion-free groups has a rather more sporadic history than that of p-groups, and there have been considerable periods in which there has been very little activity. The difficulty has been a paucity of usable methods and approachable problems. Today, on the other hand, we are seeing a great deal of activity on two fronts: the theory of "big" torsion-free groups and the theory of groups of finite rank. In each case, the change has been caused by the invention of new methods, which have solved old problems while suggesting entirely new kinds of problems.

1. This research was partially supported by a grant from the National Science Foundation.

The theory of mixed groups has a history which is closer to vacuous than to sporadic. Until recently, the theory had not really gotten off of the ground. How serious the situation has been can be seen by noticing that the most frequently studied problem has been to determine when a mixed group isn't really mixed at all--when does it split? ([3,17,38]) I do not mean to disparage the important work that has been done on that problem, but only to point out that the prominence of that problem reflected the total lack of suitable tools to study groups that were really mixed. Mixed groups have also turned up coincidentally in the study of special problems and special classes of groups (e.g. direct sum decompositions [44], cotorsion groups [13,18,29]). In 1951, [24], Kaplansky had an idea which looked as if it should be the beginning of a general theory, when he extended ideas from Mackey's proof of Ulm's theorem to classify countably generated modules of rank one over a complete discrete valuation ring. However, despite the fact that there have been a number of interesting sequels to the Kaplansky-Mackey paper, notably the papers of Rotman and Megibben, these ideas did not immediately lead to the development of a general theory. I think this was partly because the necessary techniques were not well developed, and partly because there was not yet a natural setting for "looking at a mixed group while ignoring the torsion".

It is the thesis of this paper that general techniques are now available for working with mixed groups, and that a large variety of problems can now be attacked with reasonable hope of success. The techniques developed in the last few years have not only solved some old problems, but have also led to a great variety of new problems.

Since this paper is intended as propaganda for the subject of mixed Abelian groups, we will begin with a brief review of conventions and terminology (section 2), followed by a discussion of some of the recently developed techniques (section 3). The survey of techniques is by no means complete, and, in particular, omits the theory of valuated groups, which have been very important in recent work on mixed groups by Arnold, Hunter, Richman and E. Walker. In section 4 we apply these techniques to prove some cancellation theorems for mixed groups, which are the analogues of recent results (not yet published) in the theory of torsion-free groups of finite rank. In sections 5 and 6 we give an exposition of the main results now available in the classification theory of mixed Abelian groups, surveying results which generalize the older classification theorems for totally projective p-groups, direct sums of torsion-free groups of rank one, and countable mixed groups of rank one. Section 7 is a list of open problems in the theory of mixed groups.

In closing, I want to express my gratitude to the faculty and staff of New Mexico State University, for organizing an extremely successful conference. I owe special thanks to Elbert Walker, the guiding spirit of this conference, for the special efforts needed to arrange for me to come to this conference from abroad, and for the encouragement he has given me for my work over the years.

2. Basic Definitions and Conventions. We collect in this section the basic definitions, notations, and conventions that we will use. All are familiar to any expert. The standard reference is volume II of Fuchs's book, to which we refer for more examples and discussion. This subject, like all others, is plagued with a variety of inconsistent notations and terms. Therefore, whenever we introduce a term below, we have tried to include after it any other frequently used names for the same concept, especially if our terminology differs from that in Fuchs [15].

A p-local group (for a fixed prime p) is a group such that for all primes q, $q \neq p$, multiplication by q is an automorphism. We let $Z_{(p)}$ be the subring of the ring Q of rationals consisting of rational numbers which can be written as fractions with denominators prime to p. It is clear that any p-local group can be given the structure of a $Z_{(p)}$-module in a unique way, and that the submodules are precisely the p-local subgroups. It is usually convenient to treat these groups as $Z_{(p)}$-modules. When we refer to a local problem, therefore, we refer to a problem involving p-local groups, while by a global problem, or a global method, we mean a problem or method without this restriction. More generally, if π is any set of prime numbers, we can call a group π-local if for every prime q, $q \notin \pi$, multiplication by q is an automorphism, and these groups may be regarded as modules over the ring Z_π, consisting of rational numbers which can be written as fractions in which the denominators are prime to the elements of π. For any set of primes π and group G, there is a corresponding localization of G, $G_\pi = G \otimes Z_\pi$. It is easy to verify that if

$$\phi_\pi : G \to G_\pi$$

is the natural map (taking g to $g \otimes 1$), then $\mathrm{Ker}(\phi_\pi) = \{x \in G: nx=0$ for some integer $n \neq 0$ all of whose prime factors are not in $\pi\}$. The map ϕ_π is surjective if and only if $G = pG$ for all primes p, $p \notin \pi$. When π consists of one element, $\pi = \{p\}$, we write $G_\pi = G_{(p)}$. If H is a subgroup of G, we may identify H_π with a subgroup of G_π (because of the exact sequence of the tensor product, or, if you like that sort of thing, because Z_π is flat).

If G is an Abelian group, p a prime, and α an ordinal, we define $p^\alpha G$ in the obvious inductive way, and we define $p^\infty G = \cap p^\alpha G$ where the intersection is over all ordinals α. If $x \in G$, the p-height of x, $h_p(x)$, is that ordinal α such that $x \in p^\alpha G$ and $x \notin p^{\alpha+1}G$, if such an α exists, and otherwise $h_p(x) = \infty$. In the global situation, we have to consider the p-height for all primes p, and so we define a height to be a formal product $\Pi_p\, p^{v(p)}$ where each $v(p)$ is an ordinal or ∞, and we define

$$hG = \cap_p\, p^{v(p)}G.$$

If $x \in G$, we define $h(x)$ in the obvious way, and note that $h(x) = \Pi_p\, p^{h_p(x)}$. The multiplicative notation suggests obvious definitions of hk and g.c.d. (h,k) for heights h and k. In particular, if n is a positive integer and h is a height, then nh makes perfectly good sense. Two heights h and k are equivalent if for some positive integers n and m, $nh = mk$. This notion was introduced by Baer in 1937 [4] to study torsion-free groups, where he noted that $h(nx) = nh(x)$ (in a torsion-free group) and that an equivalence class of heights was an invariant for a torsion-free group of rank one. (Incidentally, he also used the multiplicative notation we are using.)

In groups which are not torsion-free it is no longer true that $h(nx) = nh(x)$. Kaplansky discovered how to replace Baer's notion, and introduced the Ulm sequence of an element in a p-local group. More generally, the p-Ulm sequence $u_p(x)$ of an element x in a group is the sequence $\{h_p(p^n x), n \geq 0\}$. An Ulm sequence is a sequence of ordinals or symbols ∞, $\{\alpha_0, \alpha_1, \ldots, \alpha_n, \ldots\}$ such that $\alpha_{i+1} \geq \alpha_i$ and $\alpha_{i+1} = \alpha_i$ only if $\alpha_i = \infty$. If $u = \{u_n : n \geq 0\}$ is an Ulm sequence and G a p-local group, then $uG = \{x \in G: u_p(x) \geq u\}$ (where the last inequality means $h_p(p^n x) \geq u_n$). Kaplansky defines two Ulm sequences $\{\alpha_i\}$ and $\{\beta_i\}$ to be equivalent if for some integers n and m, $\alpha_{n+k} = \beta_{m+k}$ for all $k \geq 0$. He noted that if G is a p-local group of rank one, then any two elements of infinite order have equivalent Ulm sequences, and thus these equivalence classes define an invariant of the group. In the global situation, we define the Ulm matrix (or "height matrix") of an element x to be $\{u(p,n,x)\}$, where $u(p,n,x) = h_p(p^n x)$. We usually think of this Ulm matrix as a collection of Ulm sequences, one for each prime. If we have a group G of rank one, and x and y are two elements of infinite order in G, then for some positive integers n and m, $mx = ny$, from which it follows easily that for all but a finite number of primes (those dividing n or m) the p-Ulm sequences of x and y are the same, while for the other primes, the p-Ulm sequences are equivalent in the previous sense. In general, if u and v are Ulm matrices (defined in the obvious way as functions of two variables, $u(p,n)$, where for each prime p, $u(p, \ )$

is an Ulm sequence), then u and v are equivalent if (i) for all but a finite number of primes p, $u(p,n) = v(p,n)$ for all non-negative integers n, and (ii) for each of the remaining primes p, the Ulm sequences $u(p,\ )$ and $v(p,\ )$ are equivalent. The previous discussion shows that a group G of rank one determines a unique equivalence class of Ulm matrices, which we denote $U(G)$.

If v is an Ulm matrix and G a group, we define vG to be $\{x \in G: u(p,n,x) \geq v(p,n)\}$. These subgroups are characteristic and fully invariant subgroups. In the local case they were introduced by Kaplansky (see [23]). (Kaplansky showed that for countable p-groups, all fully invariant subgroups are of this form.)

We recall that the p-adic integers $Z^*_{(p)}$ are defined to be the completion of Z (or of $Z_{(p)}$) in the p-adic topology. $Z^*_{(p)}$ is complete (and, in fact, compact) in its p-adic topology, and is frequently useful for this reason. For any group G, we define $G^*_{(p)}$ to be $G \otimes Z^*_{(p)}$. (We remark that this is not the p-adic completion of G, unless G is finitely generated.) Since we will need to talk about $Z^*_{(p)}$-modules (such as $G^*_{(p)}$) as well as $Z_{(p)}$-modules in the following work, we will frequently discuss local considerations in the context of modules over an arbitrary discrete valuation ring (with prime p), even though more general discrete valuation rings are of little interest to us as far as the questions considered here are concerned. (For most purposes it makes very little difference when we consider a group which has a module structure as a module rather than as a group. We note that $Z_{(p)}/Z$ is a torsion divisible group with no p-torsion, and that if a group can be given the structure of a $Z_{(p)}$-module, then all of its torsion is p-torsion. Clearly, if G is a group with no torsion elements of order prime to p, then a homomorphism from Z into G can extend to a homomorphism from $Z_{(p)}$ into G in at most one way. It follows that if a group can be given a $Z_{(p)}$-module structure, then this can be done in only one way. It also follows that if A and B

are $Z_{(p)}$-modules, then all group homomorphisms between them are actually module homomorphisms, so that

$$\mathrm{Hom}_Z(A,B) = \mathrm{Hom}_{Z_{(p)}}(A,B).$$

The same analysis does not hold for $Z^*_{(p)}$-modules, but it almost does. In this case we notice that $Z^*_{(p)}/Z$ is divisible. From this it immediately follows that if G is a <u>reduced</u> group which can be given a $Z^*_{(p)}$-module structure, then this can be done in only one way, and that if A and B are reduced $Z^*_{(p)}$-modules, then

$$\mathrm{Hom}_Z(A,B) = \mathrm{Hom}_{Z^*_{(p)}}(A,B).$$

It is not in general true, of course, that if G is a reduced group then $G^*_{(p)}$ is reduced, and it is frequently important to treat $G^*_{(p)}$ as a module, without getting bogged down in its group structure.)

Let B be a module over a discrete valuation ring with prime p, (e.g., $Z_{(p)}$ or $Z^*_{(p)}$), A a submodule, and $\phi: B \to B/A$ the natural map. A is a <u>nice</u> submodule if for all ordinals α, $\phi(p^\alpha B) = p^\alpha(B/A)$. (This terminology is due to Paul Hill. The importance of this notion was first observed by Rotman [34], who pointed out that it is equivalent to saying that every coset of A in B contains an element of maximal height.) A is <u>isotype</u> if for all ordinals α, $p^\alpha A = A \cap p^\alpha B$. A is <u>balanced</u> ([15,p.77]) if it is both nice and isotype, or, in other words, if for all α, the sequence $0 \to p^\alpha A \to p^\alpha B \to p^\alpha(B/A) \to 0$ is exact. Finitely generated torsion submodules are always nice. If the valuation ring is complete (e.g., $Z^*_{(p)}$ -- the ring of p-adic integers), then every finitely generated submodule is nice.

A module M over a discrete valuation ring <u>satisfies Hill's condition</u> ("satisfies the third axiom of countability", "has a nice system" ([15, p.84]) if it has a family $\underline{C}$ of submodules such that (i) $0 \in \underline{C}$, (ii) every element of $\underline{C}$ is nice, (iii) if $N_i (i \in I)$ are in $\underline{C}$ then so is $\sum N_i$, (iv) if $N \in \underline{C}$ and X is a countable subset of M, there is an $N' \in \underline{C}$ containing N and X and such that N'/N is countably generated. A p-group is <u>totally projective</u> if it satisfies Hill's condition. (This differs from some definitions in that we do not require a totally projective group to be reduced.) Hill's famous theorem [15, 41] is that these groups are determined by their Ulm invariants.

3. Basic Techniques. We will be mostly concerned with problems which involve the existence or extension of a homomorphism or isomorphism between mixed groups. That includes such problems as classification problems, for example, where you want the existence of an isomorphism, given certain data. The general approach I wish to emphasize involves treating problems for mixed groups in two stages. We first treat the groups like torsion-free groups, getting conclusions which ignore the torsion in some sense. We then localize in order to treat mixed groups as if they were torsion groups.

We discuss the second technique first. One thing that makes the theory of torsion groups work is the presence of only one prime. The corresponding thing in the case of mixed groups is to look at p-local groups. Various arguments in the theory of p-groups depend heavily on the finiteness of finitely generated subgroups. Sometimes this can be adequately replaced by compactness, in which case we have a chance by passing not to $Z_{(p)}$-modules, but to modules over the ring $Z^*_{(p)}$ of p-adic integers. Hence, given a group G, we look at $G^*_{(p)} = G \otimes Z^*_{(p)}$ as a $Z^*_{(p)}$-module. In this paper we will concentrate on results which reduce to $Z^*_{(p)}$-module questions. We have the following local-to-global lemma, which can be thought of as an Abelian group theorists version of the "Hasse principle".

Lemma 1. Let G and H be groups and S and T subgroups such that G/S and H/T are torsion, and for every prime p, identify $S^*_{(p)}$ and $T^*_{(p)}$ with submodules of $G^*_{(p)}$ and $H^*_{(p)}$ respectively. Let $f: S \to T$ be a homomorphism, and $f_p: S^*_{(p)} \to T^*_{(p)}$ the induced map. Then f extends to a homomorphism from G to H if and only if for every prime

p, f_p extends to a homomorphism from $G^*_{(p)}$ to $H^*_{(p)}$, and, similarly, f extends to an isomorphism from G to H if and only each f_p extends to an isomorphism.

Proof. If f extends to a homomorphism or isomorphism, then by tensoring with $Z^*_{(p)}$ we obtain a corresponding extension of f_p. To prove the converse we imbed G into the product over all primes p of the complete localizations $G^*_{(p)}$. We call this product P. If we regard G and S as subgroups of P, then we can identify G as the set of elements $x \in P$ such that for some integer n, $nx \in S$ and $n \neq 0$. (This follows from the fact that P/G is torsion-free, which we now verify. It suffices to show that if x is an element of P and $px \in G$, then $x \in G$. The map $G \to G^*_{(p)}$ preserves p-height, so we know that $px \in pG$. If $px = py$, for some $y \in G$, then $x = (x-y) + y$, where $p(x-y) = 0$. Since all of the torsion of P is in G (an easy exercise) we conclude that $x \in G$.)

If the corresponding product for H is P', then the existence of an extension of each f_p gives us an extended map $g: P \to P'$. Since $g(S) \subseteq T$, the identification we have given of the elements of G and H in P and P' shows that the restriction of g to G is an extension of f, and this restriction is clearly an isomorphism if g is.

The main theorem available for extending homomorphisms locally (to set up a situation for the previous lemma) is an extension result first proved by Hill for p-groups. A more general version with a very smooth proof was given by Elbert Walker in [41]. A version valid for modules over a discrete valuation ring is stated in [50] (the proof being essentially the same, except that the version in [50] is the first not to be restricted to isomorphisms). We state that version.

Lemma 2. Let M and N be modules, S a nice submodule of M such that M/S is torsion and totally projective, and $f: S \to N$ a homomorphism such that for all ordinals α, $f(S \cap p^\alpha M) \subseteq p^\alpha N$. Then f

extends to a homomorphism of M into N. If $f(S) = T$, f is bijective, T is a nice submodule of N, N/T is totally projective and for all ordinals α, $f(S \cap p^{\alpha}M) = T \cap p^{\alpha}N$, the Ulm invariants of M and N are equal, and the relative Ulm invariants of S in M are equal to those of T in N, then f extends to an isomorphism of M onto N.

We remark that our totally projective torsion modules need not be reduced (we defined this term using Hill's condition only). We will not discuss the relative Ulm invariants in this paper. However, we remark that the condition on relative Ulm invariants is automatically satisfied if the other conditions are satisfied and S has finite rank. Hence for most of this paper we can safely ignore the relative Ulm invariants. To give an example of how the lemmas stated so far can be used, we pause here to prove a classification theorem for groups of rank one.

Theorem 1. Let G and H be groups of rank one such that for each prime p, $G^*_{(p)}$ and $H^*_{(p)}$ satisfy Hill's condition. Suppose that $U(G) = U(H)$ and G and H have the same Ulm invariants. Then $G \cong H$.

Proof. We choose elements x and y in G and H of infinite order such that $u(x) = u(y)$. If f is defined on the subgroup $[x]$ by $f(x) = y$, then it is trivial to verify that for each prime p, $f^*_{(p)}: [x]^*_{(p)} \to [y]^*_{(p)}$ is a height preserving isomorphism. Since finitely generated submodules of a module over a complete discrete valuation ring are always nice, the conditions of Lemma 2 are satisfied, and $f^*_{(p)}$ extends to an isomorphism $G^*_{(p)} \to H^*_{(p)}$. By Lemma 1, this shows that f extends to an isomorphism of G onto H.

The class of rank one groups described above has some nice properties. In particular, every equivalence class of Ulm matrices arises from some group in this class. One can show ([42,51]) that in a rank one module over any discrete valuation ring, any finitely generated submodule is nice, so that this class can be described as those rank one groups which locally satisfy Hill's condition. Anticipating some definitions, we remark that these are also precisely those rank one groups which are summands of simply presented groups. These groups include those rank one groups whose torsion subgroups are direct sums of totally projective groups (a class studied by Wallace in [42]), but Wallace's class is not large enough because, in particular, not all equivalence classes of Ulm sequences can occur for groups in his class. It is still an unsolved problem to describe the class of torsion groups which are torsion subgroups of rank one groups of the type described in Theorem 1.

We now turn to the question of how you can temporarily ignore the torsion in a mixed group, without ignoring it so successfully that you throw away all of your data. Clearly to factor out the torsion subgroup is to ignore the torsion too successfully. In [34] and [36], Rotman and Rotman and Yen consider an equivalence relation on modules of finite rank in which G and H are equivalent if there are free submodules S and T, such that G/S and H/T are torsion, and an isomorphism between S and T which preserves heights (in G and H). They refer to this as an "invariant" for the module. It is indeed an invariant, though certainly not the kind of invariant we would like to see in a classification theorem. It can, however, be turned into isomorphism in a suitable category which turns out to be a useful category to work in. We define, therefore, a category $\underline{H}$ in which the objects are Abelian groups, but in which the morphisms are changed. A candidate for a morphism from G to H in the new category will be a homomorphism $f \in \mathrm{Hom}(S,H)$, where $S \subseteq G$, S is torsion-free, G/S is torsion, and for every prime p, $h_p(f(x)) \geq h_p(x)$ for all $x \in S$, where the heights are computed in G and H. For this to make good sense, and, in particular, for composition to be well-defined, we need to put an equivalence relation on such candidates. If f and f' are candidates with domains S and S', we say they are equivalent if there is some subgroup $S'' \subseteq S \cap S'$ such that G/S'' is torsion and the restriction of f and f' to S'' coincide. Under this equivalence relation, the candidates form a group, and we do indeed get a category $\underline{H}$ in which these are the morphism groups. This is an additive category with infinite direct sums, with kernels but without cokernels in general -- in other words, a category resembling in many ways the category of torsion-free groups. This category is studied and used in [46], [50] and [51]. It is easy to see that two groups of rank one are isomorphic in $\underline{H}$ if and only if they correspond to the same equivalence class of Ulm matrices. Classical results which can be rephrased in terms of the category $\underline{H}$ include the results of Rotman and Rotman-Yen [34,35,36], their extensions by Bang [5,6], and a result of Stratton's [39], in which one condition on a mixed group for it to be split can be described as requiring that the group be $\underline{H}$-isomorphic to a torsion-free group.

One way to think of $\underline{H}$ is that the morphisms are homomorphisms defined on full torsion-free subgroups which might have come from honest homomorphisms, at least as far as height considerations are concerned. One can take the point of view that we threw away too much data in doing that, and that we should have restricted ourselves to homomorphisms that really do come from honest homomorphisms. Proceeding along these lines, we obtain another category, first studied by Elbert Walker, and which we will denote as Walk. If G and H are groups, Walk(G,H) $= \mathrm{Hom}(G,H)/t(G,H)$, where $t(G,H)$ is the set of elements f of $\mathrm{Hom}(G,H)$ such that $f(G)$ is torsion. One can verify that G and H are isomorphic in Walk if and only if there are torsion groups T and S with $G \oplus T \cong H \oplus S$. We will call such groups Walker isomorphic. Walk is again an additive category with infinite sums and kernels.

The last result of this section fits these various ideas together in a result which is very useful for the study of groups of finite torsion-free rank. We recall that Rotman and Yen proved in [36] that if G and H are countably generated modules of finite torsion-free rank over a complete discrete valuation ring, then G and H are isomorphic if they are $\underline{H}$-isomorphic and have the same Ulm invariants. Stratton shows in [38] that the completeness is essential, and that the corresponding result fails for p-local groups. A consequence of the Rotman-Yen result is that for countably generated p-adic modules of finite rank, $\underline{H}$-isomorphism and Walker-isomorphism coincide. It turns out that we can prove a global version of the Rotman-Yen theorem, if we replace $\underline{H}$ by Walk.

Theorem 2. If G and H are countable groups of finite torsion-free rank with the same Ulm invariants and which are Walker isomorphic, then $G \cong H$.

Proof. Let T and S be torsion groups and $f: G \oplus T \to H \oplus S$ an isomorphism. Let F be a free subgroup of G such that G/F is torsion and $f(F) \subseteq H$. Let $F' = f(F)$, and let $g: F \to F'$ be the restriction of f to F. We want to extend g to an isomorphism from G to H. To do this, we not only need the fact that g is height preserving (which is clear) but, in view of the earlier discussion, we need the fact that for every prime p, the map $g_p: F^*_{(p)} \to (F')^*_{(p)}$ is a height preserving isomorphism. Now it is clear that

$$f_p: G^*_{(p)} \oplus T^*_{(p)} \to H^*_{(p)} \oplus S^*_{(p)}$$

is an isomorphism, and that g_p is the restriction of f_p to $F^{*'}_{(p)}$. From this it is clear that g_p is height preserving, so by Lemma 2, g_p extends to an isomorphism $G^*_{(p)} \to H^*_{(p)}$. Since this holds for every prime p, Lemma 1 implies that g extends to an isomorphism, as desired.

4. Cancellation theorems. In this section we use the techniques developed in the previous section to prove some cancellation theorems for countable mixed groups of finite torsion-free rank. The proofs

serve to advertise the usefulness of the category Walk, and also to show the importance of some ideas from ring theory.

Lemma 3. Let R be a ring which as an Abelian group is torsion-free of finite rank and which is p-divisible for all but a finite number of primes p, and let J be the Jacobson radical of R. Then R/J is Artinian, and is the product of a finite dimensional algebra over the field Q of rational numbers and a finite ring.

Proof. To show that R/J is Artinian, it is enough to show that there is an upper bound on the length of semi-simple cyclic R-modules, and to show the rest of the statement it suffices to show, in addition, that each of these simples is a divisible torsion-free group or finite. If M is an R-module and p a prime, then pM and $M[p]$ are submodules, so it follows that if M is a simple module then if M has p-torsion, $pM = 0$, and that if M is torsion-free then M is divisible. The finiteness of the rank of R makes it clear that there is a bound on the number of torsion-free simple summands that can appear in a semi-simple cyclic R-module. As far as the other simple modules are concerned, the above considerations show that a simple module M which is not torsion-free is annihilated by some prime p, and the divisibility condition on R implies that only a finite number of primes can appear in this way. For a given prime p, any cyclic module annihilated by p is a homomorphic image of the finite group R/pR. Putting these facts together, we obtain the desired result.

Theorem 3. Let A, B, and C be countable Abelian groups of finite torsion-free rank such that the Ulm invariants of A are finite and $A \oplus B \cong A \oplus C$, and suppose that there is a finite set π of primes such that either (i) $A = pA$ for all primes $p, p \notin \pi$, or (ii) $B = pB$ for all primes $p, p \notin \pi$. Then $B \cong C$.

Proof. We let $d(X)$ denote the maximal divisible subgroup of X.

Among the Ulm invariants we include the "infinite" Ulm invariant $f(\infty,A)$ which is the dimension of the Z/pZ-vector space of elements of order p in $d(A)$. Since $d(A) \oplus d(B) \cong d(A) \oplus d(C)$, the hypothses easily imply that $d(B) \cong d(C)$, and since the divisible subgroups always split off, we may assume $d(A) = d(B) = d(C) = 0$. (For details of the Abelian group theory involved here, see [23].) We let Z_π be the set of rational numbers which can be written as fractions with denominators prime to the primes in π, and, for any Abelian group X, $X_\pi = X \otimes Z_\pi$. If $d(X) = 0$ and $X = pX$ for all primes, p, $p \notin \pi$, then it is well-known that $X \cong X_\pi$. We now assume that (i) has been proved and use it to prove (ii). The hypothesis of (ii) implies that $A_\pi \oplus B_\pi \cong A_\pi \oplus C_\pi$, and (i) implies that $B_\pi \cong C_\pi$. Since we are assuming $d(B) = d(C) = 0$, it follows that $B \cong B_\pi$, so if we can show $C \cong C_\pi$ we will be done. If $M = A \oplus B$ then M is imbedded in M_π, so we may consider

$$M_\pi/M \cong A_\pi/A \cong A_\pi/A \oplus C_\pi/C.$$

Now A_π/A is a torsion divisible group, so if it is isomorphic to a proper summand of itself then so is some p-primary component. However, since A has finite torsion-free rank, each p-primary component of A_π/A is Artinian, so A_π/A is not isomorphic to a proper summand of itself and $C \cong C_\pi$ as required. This shows that (i) implies (ii), so we need only prove (i).

Since the torsion-free rank of A is finite, it is clear that if E is the endomorphism ring of A in the category Walk, then E is a torsion-free ring of finite rank, and is p-divisible for all primes p, $p \notin \pi$. It follows from Lemma 3 that if $J(E)$ is the Jacobson radical of E, then $E/J(E)$ is an Artinian ring. It follows from the results of [52], or by combining [12] and [7, Lemma 6.4], that A has the cancellation property in the category Walk. (More generally, if A is an object in an additive category and the endomorphism ring of A is Artinian modulo its radical, then A has the cancellation property.)

It follows that B and C are Walker-isomorphic. Since the Ulm invariants of A are finite, it follows that B and C have the same Ulm invariants, from which, using Theorem 2, we infer that $B \cong C$.

In the case in which A, B, and C are torsion-free, case (ii) above is a theorem obtained by Lady in [26] using different methods. In [52], it is shown that in the torsion-free case and in both cases (i) and (ii), the finite rank restriction on B and C is unnecessary. Presumably, it is unnecessary in the mixed case also, but the above proof seems to depend on it. In [10], Crawley showed that a p-group with finite Ulm invariants has the cancellation property in general.

Lemma 4. Let $\underline{A}$ be an additive category in which idempotents split, L an object of $\underline{A}$, and E the endomorphism ring of L. Then the functor taking X to Hom(L,X) gives an equivalence of categories between the category of summands of finite direct sums of copies of L in $\underline{A}$ and the category of finitely generated projective E-modules. If L is a small object, and $\underline{A}$ has infinite sums (coproducts), then the same functor gives an equivalence between the category of summands of direct sums of arbitrarily many copies of L and the category of all projective E-modules.

Remark. We refer to [2] for a variety of applications of this idea in the theory of torsion-free groups. This lemma is slightly more general than the one used by Arnold and Lady in [2] since they are able to construct a specific inverse using a tensor product -- a method not available to us here, and not available for the particular categories $\underline{A}$ to which we wish to apply the lemma.

Proof. The smallness of L in the second part of the lemma is only used to guarantee that in a homomorphism from L to a direct sum of copies of L, all but a finite number of the components are zero. It follows easily that, in every case, the values of the functor are projective E-modules. If we let $\bar{X} = \mathrm{Hom}(L,X)$, then we need to show that for every projective E-module P, there is an X with $\bar{X} \cong P$, and that the map $\mathrm{Hom}(X,Y) \to \mathrm{Hom}(\bar{X},\bar{Y})$ is an isomorphism. Taking direct sums of copies of L, we see that for each free module F there is an X with $\bar{X} \cong F$, and the functor induces an isomorphism $\mathrm{End}(X) \to \mathrm{End}(\bar{X})$. If $Y = e\bar{X}$ is a summand of $\bar{X}$ with idempotent e, and e' is the corresponding idempotent in End(X), then it is easy to calculate $\overline{e'X} = Y$. Finally, the fact that the map $\mathrm{Hom}(X,Y) \to \mathrm{Hom}(\bar{X},\bar{Y})$ is an isomorphism follows easily from the special case in which X and Y are direct sums of copies of L.

Theorem 4. Let A, B, and C be countable Abelian groups of finite torsion-free rank such that the Ulm invariants of A are finite, and such that $A \oplus B \cong A \oplus C$. Then for some positive integer n, $B^n \cong C^n$.

Remark. B^n denotes the direct sum of n copies of B. In the case in which A, B, and C are torsion-free, this was obtained by the author in [52]. Goodearl generalized this in [16] to show that if A is torsion-free of finite rank and B and C are arbitrary groups (not necessarily commutative), and $A \oplus B \cong A \oplus C$, then for some positive integer n, $B^n \cong C^n$. Possibly his methods could be extended to similarly generalize the above result.

Proof. As in the proof of the previous theorem, it suffices to show that B^n and C^n are Walker-isomorphic. We let $L = A \oplus B \oplus C$, and let E = Walk(L,L), the endomorphism ring in Walk of L. The functor taking X to Walk(L,X) yields an equivalence of categories from the category of summands of finite direct sums of copies of L in Walk to the category of finitely generated projective E-modules (Lemma 4). If $\bar{A}$, $\bar{B}$, and $\bar{C}$ are the images of A, B, and C under this functor, then the problem is to show that for some positive integer n, $\bar{B}^n \cong \bar{C}^n$, given that $\bar{A} \oplus \bar{B} \cong \bar{A} \oplus \bar{C}$. We have now reduced the problem to the same problem for finitely generated projective modules over a torsion-free ring of finite rank, and, in effect, it is for these rings that the result is already known, since the result is known for torsion-free groups. (In more detail, we note that to consider projective modules, it is sufficient to reduce everything modulo the nil radical of E, and thus we may assume that $E \otimes Q$ is a semi-simple Q-algebra. If D is the maximal divisible subgroup of E, then D is a two-sided ideal in E, and if we identify E as a subgroup of $E \otimes Q$, then $D = D \otimes Q$, so D is a two-sided ideal of the semi-simple ring $E \otimes Q$, and hence a summand. We therefore have a ring decomposition $E = D \times E'$, where E' is reduced. If a ring decomposes as a product, then so does every module over the ring in a corresponding way, and cancellation certainly holds for finitely generated projective D-modules (since D is semi-simple), so the entire problem is reduced to the corresponding problem for finitely generated projective E'-modules. By Corner's theorem, E' is the endomorphism ring of a torsion-free group of finite rank, so using the category equivalence argument again, we see that the theorem for finitely-generated projective E'-modules follows from the same theorem (proved in [52]) for torsion-free Abelian groups of finite rank.)

5. Balanced Projectives. In both the local and global cases, the balanced projectives are a very nice class of groups. They have a number of natural descriptions, so they cannot be considered an *ad-hoc* class. They include the torsion totally-projective groups and the torsion-free groups which are direct sums of groups of rank one, and they have a classification theorem which extends the known classification theorems in both of those cases. A classification theorem is proved using a naturally defined functorial family of invariants. Finally, there are a great many structural results known for these groups -- in particular they are all direct sums of groups whose torsion-free rank is one. In the following section we will consider a larger class of groups -- the summands of simply presented groups. These will

presumably eventually also have a complete classification theory, though the results are still incomplete in the global case, and will share many of the other nice properties of the balanced projectives. However, even when the theory of these groups is completed, they will be in many respects a less agreeable class than the balanced projectives, and I think that the balanced projectives will still be of considerable independent interest.

In [15, p.77] Fuchs defines a short exact sequence $0 \to A \to B \to C \to 0$ of p-groups to be balanced if A is nice and isotype in B. Equivalently, the sequence is balanced if for every ordinal α, the sequence $0 \to p^{\alpha}A \to p^{\alpha}B \to p^{\alpha}C \to 0$ is exact. For $Z_{(p)}$-modules, we adopt exactly the same definition. If we want a global form of this notion, suitable for application to Abelian groups in general, the best extension is to replace p^{α} by an arbitrary height h, and to look at short exact sequences $0 \to A \to B \to C \to 0$ of Abelian groups such that for all heights h, the sequence $0 \to hA \to hB \to hC \to 0$ is exact. In this section we review the theory of balanced projectives -- groups which are projective with respect to all such sequences. (That is, M is a balanced projective if for every balanced short exact sequence as above, the map $\mathrm{Hom}(M,B) \to \mathrm{Hom}(M,C)$ is surjective.) We refer to [49] and [50] for details, and give only a brief review here, emphasizing the basic methods involved.

We first consider the local case, working with p-local groups ($Z_{(p)}$-modules). It is easy to see that totally projective p-groups are balanced projectives, as are divisible groups. We next construct an additional family of balanced projectives. If λ is a limit ordinal or 0, a module M is a λ-elementary balanced projective (or, in another terminology, a λ-elementary KT-module) if $p^{\lambda}M \cong Z_{(p)}$ and $M/p^{\lambda}M$ is torsion and totally projective. Using these, the totally projective groups, and divisible groups, one can construct for any module a balanced resolution, which proves that every balance projective is a summand of a direct sum of divisible groups, totally-projective groups, and λ-elementary balanced projectives for various limit ordinals λ.

If we want a set of invariants to describe these groups, we will clearly want to use the usual invariants for divisible groups and the Ulm invariants, and something more. We now notice that if M is such a group, and λ a limit ordinal, then $p^{\lambda}M/p^{\lambda+\omega}M$ is a direct sum of cyclic $Z_{(p)}$-modules, and the number of infinite cyclic summands (clearly an invariant of M) is precisely the number of λ-elementary balanced projectives in the given decomposition. Describing this invariant in a way that is functorial and makes sense for any module, we let $h(\lambda,M)$ be the Z/pZ-dimension of $p^{\lambda}M/(p^{\lambda+1}M + t(p^{\lambda}M))$, where $t(X)$ is the torsion subgroup of X. We let $h(\infty,M)$ be the Q-dimension of $p^{\infty}M \otimes Q$. As usual we denote the Ulm invariants by $f(\alpha,M)$ and $f(\infty,M)$.

Theorem 5. If M and N are balanced projective $Z_{(p)}$-modules, then $M \cong N$ if and only if $f(\infty,M) = f(\infty,N)$, $h(\infty,M) = h(\infty,N)$, and for all ordinals α and limit ordinals λ, $f(\alpha,M) = f(\alpha,N)$ and $h(\lambda,M) = h(\lambda,N)$. Furthermore, every balanced projective module is a direct sum of a divisible group, a totally projective p-group, and λ-elementary balanced projectives for various limit ordinals λ.

The proof of this goes roughly as follows. One first notices that in the category $\underline{H}$, a λ-elementary balanced projective is an indecomposable object with endomorphism ring isomorphic to $Z_{(p)}$. Since these endomorphism rings are local, a suitable Krull-Schmidt theorem in the category $\underline{H}$ ([40]) shows that any summand of a direct sum of λ-elementary balanced projectives is at least $\underline{H}$-isomorphic to a direct sum of λ-elementary balanced projectives. This means that any balanced projective M has a K-basis: a subset $X \subseteq M$ which is a basis (the elements are independent and $M/[X]$ is torsion) such that for all sets $\{x_1,\ldots,x_k\}$ of elements of X, $h(n_1x_1+\cdots+n_kx_k) = \text{g.c.d.}\{n_ih(x_i)\}$.

Now if M and N are balanced projectives with the same invariants, then it is easy to see that M and N have K-bases X and Y such that there is a bijective map $f: X \to Y$ satisfying $h(x) = h(f(x))$ for all $x \in X$. If we could extend this to an isomorphism from M to N we would be done. By the basic theorems on extending homomorphisms we could do this by showing that any K-basis generated a nice submodule such that $M/[X]$ was totally projective. When I did this work, I did not know any direct proof of this, though a consequence of the theory was that it is true. Very recently, a proof of this using homological methods and the theory of valuated groups has been given by Hunter, Richman and Walker. The original proof of Theorem 5 got rather less information than this -- it got an isomorphism from M to N but the isomorphism did not agree with f except on a full subgroup of $[X]$. This proved to be inadequate when it came time to do the global theory, so it became necessary to have a better local result. This was proved for modules over a complete discrete valuation ring, and the completeness was used heavily in the proof ([50, Theorem 4.4]). A consequence of the argument is the following characterization of balanced projectives over a complete discrete valuation ring:

Theorem 6. ([50, Theorem 4.4]). A module over a complete discrete valuation ring is a balanced projective if and only if it satisfies Hill's condition and has a K-basis.

I do not know if this is true for modules over an arbitrary discrete valuation ring, or whether there is a global analogue.

To return to the argument for Theorem 5, once one has the classification theorem one also needs a theorem saying what values of the invariants can occur. Once this was done, it turned out to be possible to give an existence proof which showed that any possible set of values of the invariants could be realized by a balanced projective which was in fact the direct sum of a divisible group, a totally projective group, and · λ-elementary balanced projectives for various limit ordinals λ. Since we already have a uniqueness theorem, it follows that all balanced projectives must be of this form. This completes the outline of the proof of Theorem 5. (We remark that is is fairly common for an existence theorem to have a structure theorem built into it in this way. For example, after Hill had proved the classification theorem for totally projective groups, Crawley and Hales proved an existence theorem for simply presented groups which showed as a consequence that all totally projective groups are simply presented. The recent work of Hunter, Richman and Walker [21] is another example of an existence theorem with structural consequences.)

The global theory of balanced projectives develops in a parallel manner. We need for an arbitrary height h a global substitute for the λ-elementary balanced projective. It might appear at first that since we are working over Z rather than over $Z_{(p)}$ we would like a group G with $hG \cong Z$, but this is not generally possible if $h = \Pi p^{v(p)}$ and for some primes p we have $v(p) = \infty$. We therefore let π be the set of primes for which $v(p) \neq \infty$. An h-elementary balanced projective is a group G such that $hG \cong Z_\pi$ and G/hG is a π-torsion, totally projective group.

We now state the resulting theorem.

Theorem 7. If M is a balanced projective group then M is the direct sum of a divisible group, totally projective p-groups (for various primes p) and h-elementary balanced projectives (for various heights h).

The invariants for these groups are obtained by using the Ulm invariants, the invariants of the divisible subgroups, and, for every height h, the number of k-elementary balanced projectives in a decomposition with k equivalent to h. To make these invariants functorial, and easier to state, we give a definition. If h is a height, we let

[h] be the corresponding <u>type</u>: the set of heights equivalent to h. For any group M, we let $[h]M = \sum_{k \in [h]} kM$ and $[h]^*M = \sum_{[k] > [h]} kM$. (Note that $[h]M = \{x \in M: [h(x)] \geq [h]\}$.) We define

$$g([h],M) = \dim[Q \otimes ([h]M/[h]^*M)].$$

<u>Theorem 8</u>. If M and N are balanced projective groups, then $M \cong N$ if and only if M and N have the same Ulm invariants and for every height h, $g([h],M) = g([h],N)$.

We note that, in particular, the number of copies of Q in a direct sum decomposition of the divisible subgroup of M is just $g([h],M)$ where $h = \prod_p p^\infty$.

We give a brief review of how these are proved, omitting all details, and indicating the plan of attack. One first notices that the groups described in Theorem 7 provide a sufficiently large collection of balanced projectives so that one can construct resolutions, and thus conclude that any balanced projective is at least a summand of a group of the type described in Theorem 7. One next applies a suitable Krull-Schmidt theorem in the category $\underline{H}$ (in this case an analogue of the Baer-Kaplansky-Kulikov-Fuchs-Charles-Kolettis argument used in the theory of torsion-free groups of rank one) to conclude that any balanced projective has a nice direct sum decomposition in the category $\underline{H}$, which will mean that it has the global analogue of a K-basis. We next take two balanced projectives M and N with the same invariants and note (easily) that we can find bases X and Y such that there is a bijection $X \to Y$ which ought, by rights, to extend to an isomorphism of M onto N. It is enough (using Lemma 1) to prove this locally, which we do by an application of the strong local theorem (see the dicussion of Theorem 5). (It was at this stage that the author was forced to prove a better result in the local case than had been needed to prove the classification theorem in the local case.) Having proved the classification theorem, one then proves an existence theorem, which, in particular, says just what restrictions there are on the invariants that can occur. We will not go into this, but will just note that in proving the corresponding existence theorem, one notices that one has constructed all balanced projectives (up to isomorphism) and that they are actually simpler than one had thought, and one gets, as a biproduct, the fact that all balanced projectives have the decomposition stated in Theorem 7.

6. Simply Presented Groups and their Summands. The totally projective p-groups were first defined by Nunke using complicated homological machinery. When Paul Hill proved that these groups could be classified by their Ulm invariants, he did it by giving a new description of these groups using what we have called "Hill's condition". Neither description suggests how one should find a class of mixed groups to which some of these results might extend. For example, all countable groups satisfy Hill's condition, but torsion-free countable groups will clearly continue to defy classification. The work of Crawley and Hales suggested what the right class of mixed groups might be. Crawley and Hales started from the observation that Ulm's theorem gives us a "p-free" description of countable torsion groups. That is, for a given countable 2-group, there is a corresponding 3-group -- the one with the same Ulm invariants. They then looked for other ways of describing groups in such a way that there would be for each p-group so described, a corresponding q-group for every other prime q. The description of a group by generators and relations does not carry over from one prime to another very well, but they noticed that there was more hope if one restricted oneself to generators and relations in which all of the relations were of a special type: $px = 0$ or $px = y$. We use Fuch's terminology, calling such presentations simple presentations, and p-groups with such presentations simply presented p-groups. (Crawley and Hales called these groups "T-groups".) Crawley and Hales proved that these groups could, indeed,

be classified by their Ulm invariants, and they also proved an existence theorem which implies that all countable p-groups are simply presented, and, (using Hill's theorem), that the simply presented groups coincide with the totally projective groups.

From the author's point of view, the interesting thing about this description of totally projective p-groups was that it immediately suggested a class of mixed groups to study. We call a group simply presented if it can be defined by generators and relations in such a way that all of the relations are of the form $nx = y$ or $nx = 0$. Similarly, for modules over a discrete valuation ring with prime p, we define a module to be simply presented if it can be given by generators and relations in such a way that all of the relations are of the form $px = y$ or $px = 0$. It is not hard to see that both are equivalent to the requirement that a presentation exists in which each relation involves at most two generators, a point of view which has recently been exploited further by Hales, in work reported at this conference.

If one takes a simply presented group with a fixed simple presentation and puts on the generators the smallest equivalence relation with the property that two generators involved in a relation are equivalent, then one obtains a direct sum decomposition of the group in which each summand is of rank at most one. This suggests that one should have an easy classification job at hand, since we know what invariants one should have to add to the Ulm invariants to classify groups of rank one. Unfortunately, things don't work out so smoothly.

In the local case, the theory works out reasonably well, the basic results being in two forthcoming papers by the author [51] and Hunter, Richman, and Walker [21]. The author first observed that in the category $\underline{H}$, a module of rank one has local endomorphism ring. Applying a suitable isomorphic refinement theorem [40] for additive categories, one discovers that every summand of a simply presented module is isomorphic in $\underline{H}$ to a direct sum of modules of rank one (though examples going back to Rotman and Yen [36] show that $\underline{H}$-isomorphism cannot be replaced by honest isomorphism). Since we know how to classify rank one objects in $\underline{H}$ (by their Ulm sequences), we can proceed as follows: If M is a module which is $\underline{H}$-isomorphic to a direct sum of modules of rank one, say $M \cong_{\underline{H}} \oplus_{i \in I} A_i$, then for every equivalence class e of Ulm sequences, let g(e,M) be the number of summands A_i such that $U(A_i) = e$. The additive category version [40] of Azumaya's theorem says that this number is independent of the decomposition chosen and thus is an invariant of the module. (These invariants are obtained from counting, and are not functorial or defined for arbitrary groups. This peculiarity was removed by R. O. Stanton, [37], who defined a family of functorial invariants for all modules, which agree with these invariants for summands of simply presented modules.) The classification theorem one obtains, [45,51], is what one should expect:

Theorem 9. [45,51]. Two modules M and N which are summands of simply presented modules are isomorphic if and only if they have the same Ulm invariants and for every equivalence class e of Ulm sequences, g(e,M) = g(e,N).

These modules can be given a homological description similar to the description of balanced projectives, as follows. A short exact sequence $0 \to A \to B \to C \to 0$ is sequentially pure if for all Ulm sequences u, the sequence $0 \to uA \to uB \to uC \to 0$ is exact. A module M is sequentially-pure-projective if it is projective with respect to sequentially pure sequences, i.e., for every such sequence $0 \to A \to B \to C \to 0$, the natural map $\mathrm{Hom}(M,B) \to \mathrm{Hom}(M,C)$ is surjective.

Theorem 10. [45,51]. A module is sequentially pure projective if and only if it is a summand of a simply presented module.

A third description of these modules is frequently more useful technically, and is used in [21]. If M is a module and X a subset,

X is a basis if the submodule [X] generated by X is a free module with X as a set of independent generators and if M/[X] is torsion. X is a decomposition basis if, in addition, whenever $\{x_1,\ldots,x_n\}$ is a set of distinct elements of X and $\{r_1,\ldots,r_n\}$ are elements of the ring,

$$h_p(r_1x_1+\cdots+r_nx_n) = \min_{i=1}^{n} (h_p(r_ix_i)).$$

Theorem 11. [45,51]. A module M is a summand of a simply presented module if and only if it has a decomposition basis [X] such that [X] is a nice submodule and M/[X] is totally projective.

For countably generated modules (in particular, for countable p-local groups) one can say more.

Theorem 12. [45,51]. If M is a countably generated module, then the following properties are equivalent:

(i) M is a summand of a simply presented module.

(ii) M has a decomposition basis.

(iii) There is a countably generated torsion module T such that $M \oplus T$ is simply presented.

(iv) There is a countably generated torsion module T such that $M \oplus T$ is a direct sum of modules of rank one.

I do not know if this result can be substantially improved. In particular, I do not know if a module with a decomposition basis and which satisfies Hill's condition is necessarily a summand of a simply presented module. (Possibly if the rank is countable?)

How badly can a summand of a simply presented module fail to be a direct sum of modules of rank one? Hunter, Richman and Walker [21] give an example of a countable p-local group M which is a summand of a simply presented module, has countably infinite rank, and such that in any direct sum decomposition $M = A \oplus B$, one of the summands is finite. However, they also show that this is the worst that can happen, in the sense that any summand of a simply presented module is the direct sum of a simply presented module and a module of at most countable rank. This remarkable result comes out of their existence theorem, which completes the structure theory in the local case.

The global theory of simply presented groups and their summands is still unfinished, but order is beginning to emerge. It is again easy to see that a simply presented group is a direct sum of groups of rank one. It is also easy to see that a group of rank one is a summand of a simply presented group if and only if its localizations satisfy Hill's condition, so the rank one groups which appear are precisely those considered in Theorem 1. One would next like to consider a direct sum $M = \oplus_{i \in I} M_i$, where each M_i has rank one, and for every equivalence class e of Ulm matrices, to let $g(e,M)$ be the number of indices i such that $U(M_i) = e$. Unfortunately, the numbers so obtained are not invariants of the group M, as was first shown in [47]. If one decides to ignore this problem temporarily, one can at least prove the following.

Theorem 13. Let M and N be groups with the same Ulm invariants which are both summands of simply presented groups. Suppose there is a family $\{A_i, i \in I\}$ of groups of rank one such that both M and N are Walker isomorphic to $\oplus_{i \in I} A_i$. Then $M \cong N$.

This is a strengthening of the result stated and proved in [47], but is proved by the same technique.

This leaves us with two steps needed to complete the classification theory. First one must show that a summand of a simply presented group is Walker isomorphic to a direct sum of groups of rank one. Next one must find suitable invariants to describe direct sums of rank one groups in the category Walk. In effect, the second step of this program has been carried out by R. O. Stanton, and we refer to his article in these proceedings for details. The first part of the program was one which the author conjectured could not be carried out, because he believed that there would be counterexamples for countable groups of finite rank. However, Arnold, Hunter and Walker have carried out large parts of this step, including the finite rank case, so it now seems likely that the suggested theorem is true, though a complete proof is not yet available.

7. Twenty-seven Open Problems

Problem 1. In this paper it was shown that under certain circumstances, Walker isomorphism plus equal Ulm invariants implies isomorphism. How far can this be extended? (It trivially extends to groups G of finite rank such that for each prime p, $G^*_{(p)}$ satisfies Hill's condition.)

Problem 2. In [38], Stratton shows that mixed modules of finite rank over $Z_{(p)}$ which are $\underline{H}$-isomorphic are not necessarily Walker isomorphic. (Equivalently, the modules we get by tensoring with $Z^*_{(p)}$ are no longer $\underline{H}$-isomorphic.) Is there an additional condition which we can put on the $Z_{(p)}$-modules to make this true? For example, does everything work if we restrict ourselves to modules M such that $M \otimes Z^*_{(p)}$ is reduced? (i.e., the quasi-maximal torsion-free subgroups, as in [48] are free.)

Problem 3. In [48], a theory is developed of quasi-maximal torsion-free subgroups in p-local groups ($Z_{(p)}$-modules) of finite rank. This gives a functor from mixed p-local groups of finite rank to quasi-isomorphism classes of finite rank torsion-free modules, and gives an invariant somewhat weaker than $\underline{H}$-isomorphism class. Can the corresponding theory be developed globally (for mixed groups of finite rank)?

Problem 4. Consider p-local groups of finite rank, such that their quasi-maximal torsion-free subgroups (as in [48]) are pure subgroups of the p-adic integers, (i.e., if A is one of these quasi-maximal torsion-free subgroups, then A/pA is cyclic). Can these groups be classified by an Ulm sequence together with the isomorphism type of the quasi-maximal torsion-free subgroups?

Problem 5. Theorems 3 and 4 of this paper are, in effect, old theorems about lattices over orders over Dedekind rings, which were first extended to torsion-free Abelian groups of finite rank in [52], and now to mixed groups. In [1], D. Arnold extends other old results on lattices to torsion-free groups of finite rank. How many of these results extend further to give us interesting results on mixed groups?

Problem 6. Show that if A is a countable group of finite torsion-free rank such that the Ulm invariants of A are finite and such that $A = pA$ for all but a finite number of primes p, and if B and C are arbitrary Abelian groups with $A \oplus B \cong A \oplus C$, then $B \cong C$. In the torsion case, this was proved by Crawley in [10], who shows that in that case A actually has the substitution property studied in [14] and [52]. This suggests that one should show in general that A has the substitution property, i.e., that one is in the stable range for End(A), (see [52]).

Problem 7. Show that if A is a countable group of finite torsion-free rank such that the Ulm invariants of A are finite, and if $A \oplus B \cong A \oplus C$, (without restrictions on the groups B and C) then for some positive integer n, $B^n \cong C^n$. I suspect that the power substitution property studied in [16] would hold for these groups.

Problem 8. What rings occur as endomorphism rings in $\underline{H}$ or in Walk for groups of finite torsion-free rank? (Rather surprisingly, there is a reduced group A of rank one such that $\underline{H}(A,A) \cong Q$.)

Problem 9. What algebras over $Z^*_{(p)}$ arise as the endomorphism rings in the category $\underline{H}$ or Walk of mixed $Z^*_{(p)}$-modules of finite rank? Which ones arise under the additional restriction that the group satisfy Hill's condition? This might shed some light on the nature of $\underline{H}$-indecomposable and Walk-indecomposable modules.

<u>Problem 10</u>. In [9], Corner shows that given a p-group H and a countable subring R of End(A), there is an imbedding $A \to G$, which identifies H with $p^{\omega}G$, such that R is exactly the set of endomorphisms of A induced from endomorphisms of G. Is there a corresponding result for mixed groups, which, for example, would start with a subring of End(A) for some torsion-free group and imbed A into a group G with G/A torsion and with similar properties? This might make more sense if we took A to be a valuated group and took a subring of $\underline{H}(A,A)$, and then looked, for example, at the map $\underline{H}(G,G) \to \underline{H}(A,A)$ or $\underline{Walk}(G,G) \to \underline{H}(A,A)$.

<u>Problem 11.</u> Arnold, Hunter, and Walker have shown that a summand of a finite direct sum of groups of rank one is $\underline{H}$-isomorphic to a direct sum of groups of rank one. Is the summand actually Walker isomorphic to such a direct sum?

<u>Problem 12.</u> The result of Arnold, Hunter and Walker mentioned in the previous problem is really a theorem about projective modules over the ring $\underline{H}(A,A)$, where A is a direct sum of a finite number of groups of rank one. What general theorem on projective modules lies behind this phenomenon?

<u>Problem 13.</u> If G is Walker isomorphic to a direct sum of groups of rank one, is this also true for summands of G? By the additive category version of a theorem of Kaplansky's [40], this is true in general if it is true in the countable rank case. Similarly, if G is $\underline{H}$-isomorphic to a direct sum of groups of rank one, is this also true for summands of G?

<u>Problem 14.</u> In [43], the author showed that if A is a rank one torsion-free group of type t and $\underline{C}$ is the category of all torsion-free groups all of whose elements have heights whose type is greater than or equal to t, and if R is the endomorphism ring of A

(a subring of Q) then $\underline{C}$ is equivalent to the category of torsion-free R-modules. (In particular, this gives a short proof of Baer's theorem [4] on summands of direct sums of isomorphic groups of rank one, since summands of direct sums of copies of A correspond to free R-modules, the "weak projectives" in the category of torsion-free modules -- those whose appearance on the right of a short exact sequence in the category causes the sequence to split. One also can show in this way that if B is a subgroup of a direct sum of copies of A and $B \in \underline{C}$ then B is a direct sum of copies of A.) Can a similar theory be developed in $\underline{H}$ or Walk, starting with a group of rank one (or possibly only certain sorts of groups of rank one?)

Problem 15. If a p-local group is $\underline{H}$-isomorphic to a direct sum of groups of rank one and satisfies Hill's condition, is it a summand of a simply presented group? The same question is open for balanced projective modules, except over a complete discrete valuation ring, where the answer is yes [50]. There are corresponding global questions.

Problem 16. Is a summand of a simply presented group the direct sum of a group of countable rank and a simply presented group? (In the local case, this has been proved by Hunter, Richman and Walker in [21].)

Problem 17. Prove a suitable existence theorem for summands of simply presented groups.

Problem 18. What p-groups are the torsion subgroups of summands of simply presented $Z_{(p)}$-modules? (The torsion subgroups of balanced projectives are precisely (by definition) the S-groups of [49].)

Problem 19. Is a summand of an S-group (defined in the previous problem) an S-group? Roger Hunter points out that the answer to problem 18 is precisely the S-groups if it is true that $A \oplus B$ an S-group and B totally projective implies that A is an S-group.

Problem 20. The proof of 4.4 in [50] suggests that balanced projective modules over a discrete valuation ring should have properties analogous to the transitive and fully transitive properties defined by Kaplansky in [23]. Can one characterize the characteristic and fully invariant submodules? If so, does the result globalize? If so, is there a suitable generalization to simply presented groups and their summands?

Problem 21. If G is a p-local group and α an ordinal, then G is a balanced projective if and only if $p^{\alpha}G$ and $G/p^{\alpha}G$ are, ([49], a much shorter argument is suggested in [50]). The corresponding result fails in the global case (see [53] or the remarks at the end of [50]). In the global case, if h is a height and G is a balanced projective, what kind of group can hG be?

Problem 22. Fuchs and Walker showed ([15, p.191]) that if G is a totally projective p-group and H is a fully invariant subgroup, then H and G/H are totally projective. This will not generalize to balanced projectives, even in the local case, since the torsion subgroup is not necessarily totally projective. What sort of groups arise as fully invariant subgroups or factors of balanced projective modules? The fact that the S-groups keep turning up in such discussions suggests that the proper domain for this discussion might be the class of groups introduced by Wick in [53], and discussed in the next problem.

Problem 23. In [53], Brian Wick considered short exact sequences $0 \to A \to B \to C \to 0$ of groups such that the induced sequence $0 \to cA \to cB \to cC \to 0$ of cotorsion hulls is a balanced exact sequence. The projectives with respect to these include the balanced projectives and also the S-groups of [49]. Is every projective for these sequences the direct sum of a balanced projective and an S-group?

Problem 24. Let λ be a limit ordinal and A a $Z_{(p)}$-module such that $p^{\lambda}A = 0$ and λ is the length of the torsion subgroup of A.

We say that A has the λ-Zippin property if for every pair of modules G and H such that $G/p^{\lambda}G \cong H/p^{\lambda}H \cong A$, and every isomorphism $f\colon p^{\lambda}G \to p^{\lambda}H$, f extends to an isomorphism of G onto H. This property is studied in [31] and [48]. In [48], it is shown that if A is countable and of finite torsion-free rank, and λ is as above, then A has the λ-Zippin property if and only if every torsion-free submodule is free, or, equivalently, if $A \otimes Z^*_{(p)}$ is reduced. Is this true if the finite rank restriction is removed? If not, what is the correct theorem?

Problem 25. In [31], Nunke shows that p-groups with the λ-Zippin property can be quite bizarre, partly because the λ-Zippin property only controls what happens near the "end" of the group. A smoother class of groups might be obtained by following an example set in earlier work of Nunke's [30] and considering groups such that for every limit ordinal λ, $G/p^{\lambda}G$ has the λ-Zippin property. A reasonable conjecture is that the "totally Zippin" p-groups are precisely the S-groups of [49], and that the totally Zippin p-local groups are precisely the class of groups studied by Wick in [53] (see Problem 23).

Problem 26. The λ-Zippin property can be globalized by considering "limit heights". We let $h = \Pi p^{v(p)}$, in which each $v(p)$ is a limit ordinal or zero. We can consider the h-Zippin property for groups A such that $hA = 0$. A beginning, but only that, is made in [48], but the result there is incomplete even in the case in which each $v(p)$ is ω, so that hG is the set of elements of infinite height. (For this h, the result is complete for groups A which are direct sums of torsion groups and torsion-free groups.)

Problem 27. This problem was suggested by David Harrison. In analogy with ideas of Krull in the theory of valuations, say that an extension $0 \to G \to H$ of a p-local group G is an "immediate" extension if G is isotype in H and for every ordinal α, the induced map

$p^{\alpha}G/p^{\alpha+1}G \to p^{\alpha}H/p^{\alpha+1}H$ is an isomorphism. Define a maximal immediate extension in the obvious way, and say that a group is "maximal" if it has no proper immediate extensions. Maximal immediate extensions clearly exist. Under what circumstances are they unique up to isomorphism? What are the maximal groups? At least for groups of countable length, I conjecture that the maximal groups are precisely the cotorsion groups, partly because of the analogy between Harrison's characterization of cotorsion groups of countable length in terms of completeness [19] and Kaplansky's characterization of maximal valuation rings [22]. However, the cotorsion hull of a module is not necessarily an immediate extension.

REFERENCES

1. D. Arnold, Genera and decompositions of torsion-free modules, to appear in these proceedings.

2. D. Arnold and E. Lady, Endomorphism rings and direct sums of torsion-free Abelian groups, Trans. Amer. Math. Soc. 211 (1975), 225-237.

3. R. Baer, The subgroup of the elements of finite order in an Abelian group, Ann. Math. 37 (1936), 766-781.

4. R. Baer, Abelian groups without elements of finite order, Duke Math. J. 3 (1937), 69-122.

5. C. Mo Bang, Countably generated modules over complete discrete valuation rings, J. Alg. 14 (1970), 552-560.

6. C. Mo Bang, Direct sums of countably generated modules over complete discrete valuation rings, Proc. Amer. Math. Soc. 28 (1971), 381-388.

7. H. Bass, K-theory and stable algebra, Pub. Math. I.H.E.S. no. 22 (1964), 5-60.

8. B. Charles, Sous-groupes fonctoriels et topologies, Studies on Abelian Groups (Sympos., Montpellier, 1967) Springer-Verlag, Berlin and Dunod, Paris, 1968, pp. 75-92.

9. A.L.S. Corner, On endomorphism rings of primary Abelian p-groups, II, Quart. J. Math. Ox. 27 (1976), 5-13.

10. P. Crawley, The cancellation of torsion abelian groups in direct sums, J. Alg. 2 (1965), 432-442.

11. P. Crawley and A. W. Hales, The structure of Abelian p-groups given by certain presentations, J. Alg. 12 (1969), 10-23.

12. E. G. Evans, Jr., Krull-Schmidt and cancellation over local rings, Pacific J. Math., 46 (1973), 115-121.

13. L. Fuchs, Notes on Abelian groups, II, Acta Math. Acad. Sci. Hungar. 11.(1960), 117-125.

14. L. Fuchs, On a substitution property for modules, Monatshefte fur Math., 75 (1971), 198-204.

15. L. Fuchs, Infinite Abelian Groups. vol. II, Academic Press, New York, 1973.

16. K. R. Goodearl, Power cancellation of groups and modules, Pacific J. Math. 64 (1976), 387-411.

17. P. Griffith, A solution to the splitting mixed group problem of Baer, Trans. Amer. Math. Soc., 139 (1969), 261-269.

18. D. K. Harrison, Infinite Abelian groups and homological methods, Ann. Math. 69 (1959), 366-391; Correction, ibid. 71 (1960), 197.

19. D. K. Harrison, On the structure of Ext, Topics in Abelian Groups, (Proc. Sympos., New Mexico State Univ., 1962), Scott, Foresman, Chicago, 1963, pp. 195-210.

20. R. Hunter, Balanced subgroups of abelian groups, Trans. Amer. Math. Soc. 215 (1976), 81-98.

21. R. Hunter, F. Richman, and E. A. Walker, Existence theorems for Warfield groups, Trans. Amer. Math. Soc. (to appear).

22. I. Kaplansky, Maximal fields with valuation, Duke Math. J. 9 (1942), 303-321.

23. I. Kaplansky, Infinite Abelian Groups, U. Mich. Press, Ann Arbor, 1954; rev. ed. 1969.

24. I. Kaplansky and G. W. Mackey, A generalization of Ulm's theorem, Summa Brasil. Math. 2 (1951), 195-202.

25. G. Kolettis, Homogeneously decomposable modules, Studies on Abelian Groups (Sympos. Montpellier, 1967) Springer-Verlag, Berlin, and Dunod, Paris, 1968, pp. 223-238.

26. E. Lady, Nearly isomorphic torsion-free Abelian groups, J. Alg. 35 (1975), 235-238.

27. C. Megibben, On mixed groups of torsion-free rank one, Ill, J. Math. 11 (1967), 134-144.

28. C. Megibben, Modules over an incomplete discrete valuation ring, Proc. Amer. Math. Soc. 19 (1968), 450-452.

29. R. J. Nunke, Modules of extensions over Dedekind rings, Ill. J. Math. 3 (1959), 222-241.

30. R. J. Nunke, Homology and direct sums of countable Abelian groups, Math. Zeits. 191 (1967), 182-212.

31. R. J. Nunke, Uniquely elongating modules, Proc. Conf. on Abelian Groups, (November, 1972), Symposia Mathematica 13, Academic Press, London, 1974, pp. 315-330.

32. F. Richman, The constructive theory of KT-modules, Pacific J. Math. 61 (1975), 263-274.

33. R. Richman and E. A. Walker, Valuated Groups, to appear.

34. J. Rotman, Mixed modules over valuation rings, Pacific J. Math. 10 (1960), 607-623.

35. J. Rotman, Torsion-free and mixed Abelian groups, Ill. J. Math. 5 (1961), 131-143.

36. J. Rotman and Ti Yen, Modules over a complete discrete valuation ring, Trans. Amer. Math. Soc. 98 (1961), 242-254.

37. R. O. Stanton, An invariant for modules over a discrete valuation ring, Proc. Amer. Math. Soc. 49 (1975), 51-54.

38. A. E. Stratton, Mixed modules over an incomplete discrete valuation ring, Proc. London Math. Soc. 21 (1970), 201-218.

39. A. E. Stratton, A splitting theorem for mixed Abelian groups, Proc. Conf. on Abelian Groups (November, 1972), Symposia Mathematica 13, Acadmemic Press, London, 1974, pp. 109-125.

40. C. Walker and R. B. Warfield, Jr., Unique decomposition and isomorphic refinement theorems in an additive category, J. Pure and Appl. Alg. 7 (1976), 347-359.

41. E. A. Walker, Ulm's theorem for totally projective groups, Proc. Amer. Math. Soc. 37 (1973), 387-392.

42. K. Wallace, On mixed groups of torsion-free rank one with totally projective primary components, J. Alg. 17 (1971), 482-488.

43. R. B. Warfield, Jr., Homomorphisms and duality for torsion-free groups, Math. Zeits. 107 (1968), 189-200.

44. R. B. Warfield, Jr., An isomorphic refinement theorem for Abelian groups, Pacific J. Math., 34 (1970), 237-255.

45. R. B. Warfield, Jr., Invariants and a classification theorem for modules over a discrete valuation ring, University of Washington notes, 1971.

46. R. B. Warfield, Jr., Classification theorems for p-groups and modules over a discrete valuation ring, Bull. Amer. Math. Soc. 78 (1972), 88-92.

47. R. B. Warfield, Jr., Simply presented groups, Proc. Sem. Abelian group theory, Univ. of Arizona lecture notes, 1972.

48. R. B. Warfield, Jr., The uniqueness of elongations of Abelian groups, Pacific J. Math. 52 (1974), 289-304.

49. R. B. Warfield, Jr., A classification theorem for Abelian p-groups, Trans. Amer. Math. Soc. 210 (1975), 149-168.

50. R. B. Warfield, Jr., Classification theory of Abelian groups I: Balanced projectives, Trans. Amer. Math. Soc. 222 (1976), 33-63.

51. R. B. Warfield, Jr., Classification theory of Abelian groups II: Local theory, to appear.

52. R. B. Warfield, Jr., Cancellation for modules and the stable range of endomorphism rings, to appear.

53. B. Wick, Classification theorems for infinite Abelian groups, Dissertation, Univ. of Washington, 1972.

DECOMPOSITION BASES AND ULM'S THEOREM

Robert O. Stanton

1. Introduction. In 1933, Ulm [19] classified countable reduced primary groups via cardinal invariants. Kaplansky and Mackey [8] gave a new proof of Ulm's theorem, which served as a basis for future generalizations. Kolettis [9] generalized this result to direct sums of countable p-groups. Nunke [13] introduced the class of totally projective p-groups and Parker and Walker [14] extended Ulm's theorem for a subclass of the totally projective groups. Hill [5] then proved Ulm's theorem for the entire class of totally projective groups, and that the class of totally projective groups was closed under direct summands. (Also see Griffith [4], chapter 5.) Crawley and Hales [2] defined a class of groups now known as simply presented groups, and showed that a reduced p-group is simply presented if and only if it is totally projective. E. Walker [20] gave a new proof of Hill's theorem, generalizing it in a way significant to the study of mixed groups.

In 1937, Baer [1] used a cardinal invariant called type to classify completely decomposable torsion-free groups. Baer also showed that direct summands of completely decomposable groups of finite rank are completely decomposable. This result was extended by Kulikov [10] to groups of countable rank and by Kaplansky [7] to all completely decomposable groups. It is natural to ask if the theory of p-groups initiated by Ulm, and the theory of completely decomposable groups of Baer, can be unified in the setting of mixed groups. We will see that a positive, although not yet completely satisfactory answer, can be given to this question.

The local case (i.e. modules over a discrete valuation ring) does have a satisfactory classification theorem. The first steps were taken by Kaplansky and Mackey [8], and their work was extended by Rotman [15] and Megibben [12]. Warfield [24] extended the Kaplansky-Mackey invariant to the class of summands of simply presented modules, and used this invariant along with the Ulm invariants to classify these modules. Rotman [16] introduced the basic concepts necessary to study the global case, and his structure theorem was generalized by Megibben [11], Wallace [21] and Warfield [22]. All of these results were restricted to groups of rank one. A limited class of groups of arbitrary rank was classified by Warfield [23].

Our goal is to determine the structure of a class of mixed groups which contains the simply presented groups. The first task is to define a new set of invariants which can be used to characterize direct sums of groups of torsion-free rank one. These invariants are then related to the concept of decomposition basis defined by Rotman [15]. Finally, the new invariants and the Ulm invariants are used to arrive at a new classification theorem.

2. Invariants. In this section, a new set of invariants is defined. These invariants will be needed in the main classification theorem.

A height matrix $M = [\alpha_{pi}]$ is an $\omega \times \omega$ matrix, where the rows are indexed by the set of all primes, and the columns are indexed by the non-negative integers. The entries are either ordinals or the symbol ∞, and satisfy $\alpha_{pi} < \alpha_{p,i+1}$. We say $\alpha < \infty$ if α is an ordinal, and $\infty < \infty$. If q is a prime, qM is obtained from M by shifting the q-row one position to the left. If n is a positive integer, nM is obtained inductively. If a is an element of a group A, then the height matrix H(a) of a is formed by letting α_{pi} be the (generalized) p-height of $p^i a$. (See Fuchs [3], Section 37.) The p-row M_p of a height matrix M is called its p-indicator. Denote the p-indicator of an element a by $H_p(a)$, and the p-height of a by $h_p(a)$.

The height matrices M and N are said to be equivalent if there are positive integers m and n such that mM = nN. If A is a group of torsion-free rank one, then all elements of infinite order in A have equivalent height matrices. (Henceforth torsion-free rank will be abbreviated rank.) Therefore equivalence yields an invariant for groups of rank one. Warfield [22] has demonstrated groups

A, B, C, D, all of rank one, such that $A \oplus B \simeq C \oplus D$, but the height matrices representing each group are nonequivalent. Therefore equivalence must be replaced by a new relation.

<u>Definition</u>. Two height matrices M and N are <u>compatible</u> if there are positive integers m and n such that $mM \geq N$ and $nN \geq M$.

Compatibility is an equivalence relation. The resulting equivalence classes are called compatibility classes. If M and N are compatible height matrices, then the p-indicators of M and N are equal for all but finitely many primes p. If A is a group, its torsion part will be denoted A_t.

<u>Definition</u>. Let M be a height matrix and A be a group. Then

$$M(A) = \{a \in A: \ H(a) \geq M\}, \ \text{ and}$$

$$M^*(A) = [\{a \in M(A): \ H(a) \text{ is not compatible with } M \text{ or } a \in A_t\}].$$

If M is a height matrix, let P_M be the multiplicative closed set containing those primes p for which M_p contains an ∞. The principal ideal domain Z_M is defined to be

$$Z_M = \{a/b \in Q: \ a \in Z, \ b \in P_M\}.$$

For any height matrix M, $M(A)/M^*(A)$ is a torsion-free Z_M-module. If A has rank one, $a \in A$ has infinite order and $H(a) = M$, then $M(A)/M^*(A) \simeq Z_M$. If N is any height matrix not compatible with M, then $N(A)/N^*(A) = 0$.

Let c and c' be compatibility classes. We say $c \leq c'$ if there are height matrices $M \in c$ and $M' \in c'$ such that $M \leq M'$. Because $\leq$ is a partial order on compatibility classes, the proof of the following lemma is straightforward.

<u>Lemma 1.</u> Let $A = \oplus_{i \in I} A_i$ be a group, and let M be a height matrix. Then

$$M(A)/M^*(A) \simeq \oplus_{i \in I}(M(A_i)/M^*(A_i)).$$

Hence if A is a direct sum of rank one groups, $M(A)/M^*(A)$ is a free Z_M-module. We now define a new invariant, which generalizes the concept of type for completely decomposable torsion-free groups.

Definition. Let A be a direct sum of rank one groups and c be a compatibility class. Then T(c,A) is the supremum of the ranks of the free modules $M(A)/M^*(A)$, where $M \in c$.

To be useful in the classification of mixed groups, the invariant T needs to be localized. This is done by combining it with the invariant S defined in [18]. If $\mu = \{\alpha_0, \alpha_1, \ldots\}$ is a p-indicator, and A is a group, then

$$\mu A = \{a \in A\colon\ H_p(a) \geq \mu\}, \quad \text{and}$$

$$\mu^* A = [\{a \in \mu A\colon\ h_p(p^i a) > \alpha_i \text{ for infinitely many } i \text{ or } a \in A_t\}].$$

If μ is the p-indicator of the height matrix M, let

$$(M;p)^* A = \mu^* A \cap M(A) + M^*(A).$$

Then $M(A)/(M;p)^*(A)$ is a free module over the integers localized at p if μ contains ∞. Otherwise $M(A)/(M;p)^*(A)$ is a vector space over the finite field of order p. In either case, we may speak of the rank of $M(A)/(M;p)^*(A)$.

Definition. Let A be a group, c a compatibility class, and e an equivalence class of p-indicators. Then ST(c,p,e,A) is the supremum of the ranks of $M(A)/(M;p)^*(A)$, where $M \in c$ and $M_p \in e$.

The invariant ST is especially useful in groups which have a decomposition basis.

Definition. A subset $X = \{x_i\}_{i \in I}$ of a group A is a decomposition basis (Rotman [15]) if [X] is the free subgroup generated by X, A/[X] is torsion and $H(\Sigma r_i x_i) = \min\{H(r_i x_i)\}$, whenever the r_i are integers. If, for non-zero integers s_i, $i \in I$, $y_i = s_i x_i$, the set $Y = \{y_i\}_{i \in I}$ is a subordinate decomposition basis to X. If $A = \oplus_{i \in I} A_i$, where $x_i \in A_i$ for all i, then X is called a splitting decomposition basis.

Lemma 2. Let $X = \{x_i\}_{i \in I}$ be a decomposition basis of a group A. If the height matrix of x_1 is M, then $x_1 \notin (M;p)^*(A)$.

Proof. Suppose $x_1 \in (M;p)^*(A)$. Let $\mu = M_p$. Write $x_1 = \Sigma r_i a_i$, where r_i are non-zero integers and $a_i \in (\mu^* A \cap M(A)) \cup M^*(A)$, the generating set of $(M;p)^*(A)$. Since X is a decomposition basis, there is an integer s such that $sr_i a_i$ is in [X] for all i. If we write

$$sr_i a_i = \Sigma_{j\varepsilon I} t_{ij} y_j,$$

then

$$sy_1 = \Sigma_{i\varepsilon I} t_{i1} y_1.$$

Then there is at least one non-zero t_{i1} for which the p-factor of s is equal to the p-factor of t_{i1}. For this i, we have

$$H_p(sy_1) \leq H_p(sr_i a_i) \leq H_p(t_{i1} y_1) = H_p(sy_1),$$

so $a_i \notin \mu^* A \cap M(A)$. But

$$H(sy_1) \leq H(sr_i a_i) \leq H(t_{i1} y_1),$$

implies $H(a_i)$ is compatible with M. Therefore $a_i \notin M^*(A)$, a contradiction.

Corollary 3. If the group A has a decomposition basis X, then

$$ST(c,p,e,A) = |\{x \in X: \ H(x) \in c \text{ and } H_p(x) \in e\}|.$$

As a result of Corollary 3, if A is a direct sum of rank one groups, the invariant ST can be used to classify the rank one summands independently of the particular decomposition.

3. Equivalence of Decomposition Bases. This section is devoted to the proof of a theorem which provides the fundamental relation between decomposition bases and the invariant ST.

In a collection of height matrices, the same matrix may be included more than once. If S is a collection of height matrices, A is a group with decomposition basis X, and there is a bijection α: $X \to S$ such that H(x) is equivalent to αx for all $x \in X$, we write $H(X) \sim S$. If the height matrix M is in the compatibility class c and M_p is in the equivalence class e, we write $M \in [c,p,e]$. The following definition, as well as Lemma 4, are due to Hunter [6].

Definition. Let S and S' be two collections of height matrices. Suppose A is a group having decomposition bases X and Y such that $H(X) \sim S$ and $H(Y) \sim S'$. Then S and S' are called basis-equivalent, denoted $S \sim_b S'$.

Lemma 4. Let S and S' be two collections of height matrices, and suppose $S \sim_b S'$, with respect to a group A. If B is any group with a decomposition basis X such that $H(X) \sim S$, then there is a decomposition basis Y of B such that $H(Y) \sim S'$.

Aside from ensuring that the concept of basis-equivalence is independent of a particular group, Lemma 4 also guarantees that basis-equivalence is an equivalence relation. Given any height matrix M, there is a rank one group A with an element $x \in A$ of infinite order such that $H(x) = M$. Consequently, given any collection of height matrices S, there is a group A with splitting decomposition basis X such that $H(X) \sim S$. If S is a collection of height matrices, then $ST(c,p,e,S)$ denotes the cardinality of the set of height matrices in S that are also in $[c,p,e]$. We now state the main result of this section.

Theorem 5. Let S and S' be two collections of height matrices. Then the following three conditions are equivalent.

(a) S is basis-equivalent to S'.

(b) For all c, p and e, $ST(c,p,e,S) = ST(c,p,e,S')$.

(c) (i) There is a bijection ϕ: $S \to S'$ such that M and $\phi(M)$ are compatible for all $M \in S$.

(ii) For each prime p, there is a bijection ψ_p: $S \to S'$ such that $\psi_p(M)$ is compatible with M and $(\psi_p(M))_p$ is equivalent to M_p. Moreover, $\psi_p(M) = \phi(M)$ for all but finitely many primes p.

Proof. Corollary 3 provides the proof of (a) implies (b). The proof of (b) implies (c) is based on the following set theoretical argument. We claim we may write S as a disjoint union $S = T \cup (\bigcup_p T_p)$ subject to the following two properties.

(α) If $M \in T_p$, with $M \in [c,p,e]$, then $ST(c,p,e,S)$ is infinite.

(β) If $ST(c,p,e,S)$ is infinite, then $ST(c,p,e,S) = ST(c,p,e,T_p) = ST(c,p,e,T)$.

A similar decomposition $S' = T' \cup (\bigcup_p T'_p)$ is formed.

We now justify our claim that the decomposition can be accomplished. By transfinite induction we may write S as a disjoint union of countable subsets, $S = \bigcup_\sigma S_\sigma$, where each S_σ has the property that, whenever $ST(c,p,e,S)$ is infinite, then either $ST(c,p,e,S_\sigma) = 0$ or $ST(c,p,e,S_\sigma) = \aleph_0$.

For any given σ, there are at most countably many $[c,p,e]$ with $ST(c,p,e,S_\sigma) = \aleph_0$. Order these $[c,p,e]$ and let K_i be the set of those elements of S_σ belonging to the i-th $[c,p,e]$. We now define subsets $T_{ij}(i \geq 1,\ j \geq i)$ of S_σ as follows.

Let T_{11} be a subset of S_σ such that

(1) $T_{11} \subseteq K_1$, and

(2) $|T_{11}| = \aleph_0$.

Let M_{11} be any element of T_{11}. Suppose that T_{ij} along with elements $M_{ij} \in T_{ij}$ have been defined for all $j < n$. If

$$|K_n \cap (S_\sigma \setminus (\bigcup_{i<n} T_{i,n-1}))| = \aleph_0, \qquad (*)$$

then we can define T_{nn} from elements in $S_\sigma \setminus (\bigcup_{i<n} T_{i,n-1})$ so that properties analogous to (1) and (2) hold. In this case, define $T_{in} = T_{i,n-1}$, $(i < n)$, and the element M_{in} $(i \leq n)$ is chosen in T_{in} to be distinct from any previously selected element.

Suppose (*) fails to hold. Then for some i, $|T_{i,n-1} \cap K_n| = \aleph_0$. Now T_{nn} will be a countable subset of $T_{i,n-1}$ such that

$$T_{nn} \cap \{M_{ii}, M_{i,i+1}, \ldots, M_{i,n-1}\} = \phi,$$

$$|T_{nn}| = \aleph_0, \quad \text{and}$$

$$|T_{i,n-1} \setminus T_{nn}| = \aleph_0.$$

Define $T_{in} = T_{i,n-1} \setminus T_{nn}$, and let $T_{jn} = T_{j,n-1}$ for $j \neq i$, $j < n$. Select arbitrary elements M_{in} ($i \leq n$) distinct from the previously selected elements.

For each i, define $V_i = \bigcap_{j \geq i} T_{ij}$. Since M_{ij} is in V_i for all $j \geq i$, $|V_i| = \aleph_0$. Let V_i be a disjoint union of V_{i1} and V_{i2}, where $|V_{i1}| = |V_{i2}| = \aleph_0$. For a fixed prime p, let $V_{p\sigma}$ be the union of those V_{i1} for which K_i is represented by some [c,p,e]. Let $T_p = \bigcup_\sigma V_{p\sigma}$, and let T be the set of elements not in any T_p. Note that T contains all V_{i2}. Then $S = T \cup (\bigcup_p T_p)$ satisfies the required conditions.

By (b) and the above construction, we may define ϕ: $S \to S'$ satisfying (i), and also requiring that the restriction to T (respectively T_p) maps onto T' (T_p'). It follows from (b) that maps ψ_p satisfying the first sentence of (ii) can be defined. We also claim that the ψ_p can be required to satisfy the condition that if $M \notin T_p$, and $M_p = (\phi(M))_p$, then $\psi_p(M) = \phi(M)$. This condition clearly allows the second sentence of (ii) to be satisfied. Let K be the subset of S consisting of all elements in [c,p,e], and let L be the subset of K such that $M \notin T_p$ and $M_p = (\phi(M))_p$. If K is finite, then ψ_p can be defined as indicated. If K is infinite, define ψ_p on L first. Let K' be the subset of S' consisting of all elements in [c,p,e]. By the above construction, $|K \setminus L| = |K| = |K'| = |K' \setminus \phi(L)|$, so ψ_p can be defined to map $K \setminus L$ onto $K' \setminus \phi(L)$.

We now begin the proof of (c) implies (a). The proof will be completed by Lemmas 6 and 7. Represent S by a decomposition basis X of a group A, where $H(X) \sim S$, and α: $X \to S$ is the corresponding bijection. We will find a decomposition basis Y of A such that $H(Y) \sim S'$. For a fixed prime p, $\alpha^{-1}\psi_p^{-1}\phi\alpha$ is a permutation of X, and so splits X into equivalence classes consisting of countably infinite cycles, finite cycles, and fixed points. (The element x in X is a fixed point exactly if $\psi_p \alpha x = \phi\alpha x$.) Using transfinite induction, X may be written as a disjoint union of countable subsets, $X = \bigcup X_\sigma$, such that if $x \in X_\sigma$, p is any prime and k is any integer, $(\alpha^{-1}\psi_p^{-1}\phi\alpha)^k x$ is also in X_σ. We will replace each X_σ by a

Y_σ to obtain the new decomposition basis Y.

We proceed from X_σ to Y_σ by introducing a sequence of intermediate sets W_i (i = 1, 2, ...), such that $(X \setminus X_\sigma) \cup W_i$ is a decomposition basis for every i. Let S_i be the collection of height matrices of W_i, and let $\alpha^{(i)}$: $W_i \to S_i$ be a canonical bijection. For each i, we will have maps $\phi^{(i)}$: $S_i \to S'$, $\psi_p^{(i)}$: $S_i \to S'$ with properties analogous to (i) and (ii). (We will drop the superscripts when there is no danger of confusion.) An element $x \in W_i$ is called permanent if $\phi^{(i)}\alpha^{(i)}x = \psi_p^{(i)}\alpha^{(i)}x$ for all primes p. We will require that if $x \in W_i$ is permanent, then $x \in W_j$ and that $\phi^{(i)}\alpha^{(i)}x = \phi^{(j)}\alpha^{(j)}x$ for all $j \geq i$. A prime p is said to be repaired (with respect to W_i) if $\phi^{(i)} = \psi_p^{(i)}$. We require that if p is repaired with respect to W_i, then p is repaired with respect to W_j for $j \geq i$. Essentially, W_{i+1} is obtained from W_i by selecting and repairing a finite set of primes. We will indicate how this finite set is chosen after the proof of the following lemma.

Lemma 6. Let F be a finite subset of W_i. Then we may choose W_{i+1} so that each element of F is a linear combination of permanent elements of W_{i+1}.

Proof. First we note that any prime p can be repaired. As indicated previously, $\alpha^{-1}\psi_p^{-1}\phi\alpha$ divides W_i into equivalence classes consisting of countably infinite cycles, finite cycles, and fixed points. Each finite cycle can be written as a product of two-cycles. An infinite cycle may be represented by the permutation ξ: $Z \to Z$ defined by $\xi(i) = i+1$. We may write $\xi = \xi_2\xi_1$, where $\xi_1(i) = -i$ and $\xi_2(i) = 1-i$, each of which is an infinite product of disjoint transpositions. (This representation of ξ was pointed out by R. Hunter, and replaces a longer proof by the author.) By repeated application of Lemma 7, p may be repaired, and the other primes are left unchanged. The repair scheme can, and must be specified so that it is independent of any particular basis. That is, if U and V are two decomposition bases for which p has not yet been repaired, and $x_1, x_2 \in U$, $y_1, y_2 \in V$ with $\phi\alpha x_i = \phi\alpha y_i$, then x_1 and x_2 are in the same transposition exactly if y_1 and y_2 are. If x_1 and y_1 are in an infinite cycle, then they are indexed by Z in exactly the same way.

Call a prime good if $\psi_p\alpha x = \phi\alpha x$ for all $x \in F$. The remaining primes are bad and form a finite set $\{q_1, q_2, \ldots, q_n\}$. Suppose q_1 were to be repaired immediately in the manner described above. Then there would be a new decomposition basis T, each of the elements of F would be linear combinations of elements of T which in turn would be a linear combination of a finite set G_1 of elements in W_i. Starting with W_i, repair all the good primes p for which there is an element x in G_1 such that $\phi\alpha x \neq \psi_p\alpha x$. A new decomposition basis U_1 results. Since only good primes were repaired, the elements of F remain in U_1. An element x of G_1 is replaced by a new element y such that

$$\phi\alpha x = \phi^*\alpha^* y,$$

$$\psi_q\alpha x = \psi_q{}^*\alpha^* y, \text{ whenever } q \text{ is a bad prime, and}$$

$$\psi_p{}^*\alpha^* y = \phi^*\alpha^* y, \text{ whenever } p \text{ is a good prime.}$$

(ϕ^*, ψ_p^* , ψ_q^* have the obvious meaning.)

Let G_1^* be the set of elements replacing G_1. At this point we actually repair q_1, in the same manner as we did when obtaining T. The set G_1^* is replaced by a new set F_1. If $x \in G_1^*$, the corresponding y in F_1 has the properties

$$\phi^*\alpha^* x = \phi^{**}\alpha^{**} y,$$

$$\psi_{q_1}^{**}\alpha^{**} y = \phi^{**}\alpha^{**} y, \text{ and}$$

$$\psi_p^*\alpha^* x = \psi_p^{**}\alpha^{**} y,\ p \neq q_1.$$

By the construction of G_1, the elements of F are a linear combination of elements of F_1.

The same routine is continued, with F_1 replacing F, and the decomposition basis obtained after the repair of q_1 replacing W_i. The set of bad primes is now $\{q_2, \ldots, q_n\}$. At the last step, we obtain a set F_n, consisting of permanent elements, such that F is a linear combination of the elements of F_n. The decomposition basis at this step will be W_{i+1}.

We now complete the proof of Theorem 5. Write the elements of the countable set $X_\sigma = \{x_0, x_1, \ldots\}$. Let $W_0 = X_\sigma$. To obtain W_1, let $F = \{x_0\}$ and use the lemma. Given W_i, write x_i as a linear combination of elements of W_i. Let the set of these elements be F, and again use Lemma 6 to obtain W_{i+1}. Let Y_σ be the set of permanent elements of the sequence $\{W_i\}$.

Now $Y = \bigcup Y_\sigma$ is easily seen to be a decomposition basis of A, because of the above construction. The map χ: $Y \to S'$ defined by the appropriate $\phi^{(i)}\alpha^{(i)}$ for each $y \in Y$ manifests the fact that $H(Y) \sim S'$.

Theorem 5 will be completed with the statement and proof of Lemma 7.

Lemma 7. Let X be a decomposition basis of a group A, let x_1 and x_2 be elements of X with compatible height matrices, and let p be a prime. Then there are elements y_1 and y_2 in [X] such that $H_p(y_1) = H_p(x_2)$, $H_p(y_2) = H_p(x_1)$, $H_q(y_1) = H_q(x_1)$, $H_q(y_2) = H_q(x_2)$ for all primes $q \neq p$. Moreover,

$$Y = (X \setminus \{x_1,x_2\}) \cup \{y_1,y_2\}$$

is a decomposition basis and [X] = [Y].

Proof. There are non-negative integers i(q), j(q) for every prime q such that

$$H_q(q^{i(q)}x_1) \geq H_q(x_2), \text{ and}$$

$$H_q(q^{j(q)}x_2) \geq H_q(x_1).$$

For all but finitely many primes q, we may choose i(q) = j(q) = 0. Also we require that $j(p) \geq 1$. Let $d_1 = p^{i(p)}$, $d_2 = \Pi_{q\neq p}q^{j(q)}$, $b_1 = \Pi_{q\neq p}q^{i(q)}$, and $b_2 = p^{j(p)}$. Choose integers s and t such that $sd_2b_1 - td_1b_2 = 1$ and let $a_1 = td_1$ and $a_2 = sd_2$. Choose $y_1 = a_1x_1+a_2x_2$ and $y_2 = b_1x_1+b_2x_2$. It is routine to show that the indicators of y_1 and y_2 have the desired properties and that if $Y = (X\setminus\{x_1,x_2\}) \cup \{y_1,y_2\}$, then [Y] = [X]. Only the height property is needed for Y to be a decomposition basis.

Clearly we need only consider elements of the form $z = m_1y_1 + m_2y_2$, where m_1 and m_2 are integers. We first consider p-indicators. Since X is a decomposition basis,

$$H_p(z) = \min \{H_p((m_1a_1 + m_2b_1)x_1), H_p((m_1a_2 + m_2b_2)x_2)\}.$$

Let m_1', m_2', a_1' be the p-factors of m_1, m_2, a_1 respectively. Consider the following two conditions:

(i) $m_2' < m_1'a_1'$;

(ii) $m_1' < m_2'b_2$.

Condition (i) implies

$$H_p(m_2y_2) = H_p((m_1a_1 + m_2b_1)x_1).$$

Condition (ii) implies

$$H_p(m_1y_1) = H_p((m_1a_2 + m_2b_2)x_2).$$

If both (i) and (ii) hold, then

$$H_p(z) = \min \{H_p(m_1y_1), H_p(m_2y_2)\},$$

as desired. Moreover, at least one of the conditions must hold, since $b_2 \neq 1$. Suppose (i) does not hold and (ii) does. Then it can be shown that $H_p(m_2y_2) \geq H_p(m_1y_1)$ and $H_p(z) = H_p(m_1y_1)$. So the desired result holds for p-indicators.

For primes $q \neq p$, the proof is similar, except that it may happen that the analogues to both (i) and (ii) fail to hold. In this case $H_q(y_1) = H_q(y_2) = H_q(x_1) = H_q(x_2)$. Also the q-factor of m_1 must equal the q-factor of m_2. We can thus assume $(m_1,q) = (m_2,q) = (m_1,m_2) = 1$. We wish to have $H_q(z) = H_q(y_1)$. If not, then q must divide both $(m_1a_1 + m_2b_1)$ and $(m_1a_2 + m_2b_2)$. This implies q divides $a_2b_1 - a_1b_2 = 1$, a contradiction.

4. The Classification Theorem. We will prove a structure theorem for a class of mixed groups containing the totally projective p-groups and the completely decomposable torsion-free groups. This will simultaneously generalize the classification theorems of Hill [5] and Baer [1].

A group is called simply presented if it can be described in terms of generators and relations in such a way that all of the relations involve at most two generators. We now define a class of groups containing the simply presented groups.

Definition. A group A is affable if it has a splitting decomposition basis X such that A/[X] is a simply presented torsion group.

A mixed group is affable if and only if it is a direct sum of affable groups of rank one. A rank one group is affable exactly if it is a summand of a simply presented group. A torsion group is affable if and only if it is simply presented. The torsion-free affable groups are precisely the completely decomposable groups. A group A is called almost-affable if there is a torsion group T such that $A \oplus T$ is an affable group.

If α is an ordinal, the α-th Ulm invariant $f(\alpha,p,A)$ of the group A with respect to the prime p is the rank of the vector space $(p^{\alpha}A)[p]/(p^{\alpha+1}A)[p]$ over the field of order p. Let $p^{\infty}A = \bigcap p^{\alpha}A$, where α runs over all ordinals. We say $f(\infty,p,A)$ is the rank of $(p^{\infty}A)[p]$. If G is a subgroup of A, the α-th relative Ulm invariant $f(\alpha,p,A,G)$ of A with respect to G and p is the rank of

$$(p^{\alpha}A)[p]/((p^{\alpha+1}A+G)\cap (p^{\alpha}A)[p]).$$

The invariant $f(\infty,p,A,G)$ is the rank of

$$(p^{\infty}A)[p]/(G\cap (p^{\infty}A)[p]).$$

Let $A(p,G)$ denote

$$\{a \in A: \; p^{k}a \in G \text{ for some } k \geq 0\}.$$

The following version of Ulm's theorem for simply presented groups is a direct consequence of Walker [20], Theorem 2.8 and Warfield [23], Lemma 1.16.

Theorem 8. Let A and B be groups containing subgroups G and H respectively such that G is p-nice in $A(p,G)$ and H is p-nice in $B(p,H)$, and such that $f(\alpha,p,A,G) = f(\alpha,p,B,H)$ for all primes p, and for α an ordinal or ∞. If A/G and B/H are simply presented, and ϕ: $G \to H$ is a height preserving isomorphism, then ϕ extends to an isomorphism ψ: $A \to B$.

If A is a group with a splitting decomposition basis X, then X is p-nice in $A(p,[X])$ for any prime p.

In attempting to generalize Ulm's theorem, care must be taken to preserve relative Ulm invariants. This necessitates the following definition, due to Warfield [24].

Definition. A decomposition basis X of a group A is a lower decomposition basis if, whenever the relative Ulm invariant $f(\alpha,p,A,[X])$ is infinite (for α an ordinal or ∞), then $f(\alpha,p,A) = f(\alpha,p,A,[X])$.

Lemma 9. If A has a decomposition basis X, then there is a decomposition basis Y subordinate to X which is a lower decomposition basis.

Proof. The local version of this lemma was proved by Warfield [24], Theorem 5.1. Warfield's proof can easily be adapted for use here.

We now state the theorem which generalizes the theorems of Hill and Baer.

Theorem 10. Let A and B be almost-affable groups. Then $A \simeq B$ if and only if $f(\alpha,p,A) = f(\alpha,p,B)$ and $ST(c,p,e,A) = ST(c,p,e,B)$, for all ordinals α and for $\alpha = \infty$, for all primes p, compatibility classes c and equivalence classes e.

Proof. There are torsion groups T and T' such that $A \oplus T$ and $B \oplus T'$ are affable, and hence direct sums of rank one groups. Then we may find splitting decomposition bases X and Y for $A \oplus T$ and $B \oplus T'$ by taking an element of infinite order from each rank one summand. X and Y have subordinate decomposition bases X_1 and Y_1 which are lower decomposition bases for A and B, by Lemma 9. By Theorem 5, there is a decomposition basis X_2 of A such that $[X_2] = [X_1]$ and $H(X_2) \sim H(Y_1)$. Choose decomposition bases X_3 and Y_2 subordinate to X_2 and Y_1 so that there is a bijection $\gamma\colon X_3 \to Y_2$ such that $H(x) = H(\gamma x)$ for all $x \in X_3$. Since $[X_3]$ and $[Y_2]$ are free on X_3 and Y_2, γ extends to a height preserving isomorphism $\delta\colon [X_3] \to [Y_2]$. Since X_3 and Y_2 are lower decomposition bases, $f(\alpha,p,A,[X_3]) = f(\alpha,p,B,[Y_2])$ for all p and α. The hypotheses of Theorem 8 are satisfied, so $A \simeq B$. The converse is trivial.

We list some standard consequences in the next corollary.

Corollary 11. Let A, B, and C be almost-affable groups. Then

(a) If $A \oplus A \simeq B \oplus B$, then $A \simeq B$.

(b) If the Ulm invariants and the ST-invariants of C are all finite, and $A \oplus C \cong B \oplus C$, then $A \cong B$.

(c) If A is isomorphic to a summand of B and B is isomorphic to a summand of A, then $A \cong B$.

Corollary 12. Let A and B be almost affable groups such that $ST(c,p,e,A) = ST(c,p,e,B)$ for all c,p and e. Then there are simply presented torsion groups T and T' such that $A \oplus T \cong B \oplus T'$.

Proof. T and T' may be selected so that the Ulm invariants of $A \oplus T$ and $B \oplus T'$ are equal. Since the ST invariants remain the same, Theorem 10 applies.

Theorem 13. Suppose A is an affable group and X is a decomposition basis of A. Then A has a splitting decomposition basis X' which is subordinate to X.

Proof. Let Y be a splitting decomposition basis for A. We may take subordinate decomposition bases X' and Y' of X and Y respectively such that $[X'] = [Y']$. Write $X' = \{x_i\}_{i \in I}$, $Y' = \{y_i\}_{i \in I}$, $A = \oplus_{i \in I} A_i$, with $y_i \in A_i$ for all i. For each prime p, let $\psi_p\colon X' \to Y'$ be a bijection such that $H_p(x_i) = H_p(\psi_p x_i)$. This is possible by the proof of Theorem 5. ψ_p induces a permutation ζ_p of I. We now define affable groups C_i of rank one. Each C_i contains an element c_i of infinite order such that $H(c_i) = H(x_i)$, and if $j = \zeta_p(i)$, $f(p,\alpha,C_i) = f(p,\alpha,A_j)$ for all α. The existence of such a C_i is guaranteed by Hunter [6], Proposition 5.2. Then if $C = \oplus_{i \in I} C_i$, the map $\gamma\colon x_i \to c_i$ extends to an isomorphism $\delta\colon A \to C$, by Theorem 10. Since $\{c_i\}_{i \in I}$ is a splitting decomposition basis for C, X' is a splitting decomposition basis for A.

Four classes of groups are related according to the following diagram, each class being contained in the class below it.

Simply presented groups

|

Affable groups

|

Almost-affable groups

|

Summands of affable groups

An example of an affable group that is not simply presented is given by Warfield ([24], 2.7). Rotman and Yen ([17], p. 251) have demonstrated an almost-affable group that is not affable. We may conjecture that the class of almost-affable groups coincides with the summands of affable groups. This is true in the local case, a corollary of Warfield [24], Theorem 5.2. If the two classes were to coincide, then the class $\mathcal{A}$ of summands of affable groups would be the largest class of groups containing $\mathcal{A}$ and closed under summands that could be classified via Ulm and ST-invariants. For if $\mathcal{C}$ is a larger class and $G \in \mathcal{C}$, then there is an affable group H such that $ST(c,p,e,H) = ST(c,p,e,G \oplus H)$ and $f(\alpha,p,H) = f(\alpha,p,G \oplus H)$ for all c, p, e and α. By the classification theorem, we would then have $G \oplus H \approx H$, implying that $G \in \mathcal{A}$.

REFERENCES

1. R. Baer, Abelian groups without elements of finite order, Duke Math. J. 3(1937), 68-122.

2. P. Crawley and A.W. Hales, The structure of Abelian p-groups given by certain presentations, J. Alg. 12(1969), 10-23.

3. L. Fuchs, Infinite Abelian Groups, Two volumes, Academic Press, New York, 1970, 1973.

4. P. Griffith, Infinite Abelian Group Theory, University of Chicago Press, Chicago, 1970.

5. P. Hill, On the classification of Abelian groups, preprint.

6. R.H. Hunter, Balanced Subgroups of Abelian Groups, Doctoral dissertation, Australian National University, Canberra, 1975.

7. I. Kaplansky, Projective modules, Ann. Math. 68(1958), 372-377.

8. I. Kaplansky and G.W. Mackey, A generalization of Ulm's theorem, Summa Brazil. Math. 2(1951), 195-202.

9. G. Kolettis, Direct sums of countable groups, Duke Math. J. 27(1960), 111-125.

10. L. Ya. Kulikov, On direct decompositions of groups, (Russian), Ukrain. Mat. Z. 4(1952), 230-275, 347-372.

11. C. Megibben, On mixed groups of torsion free rank one, Ill. J. Math. 11(1967), 134-144.

12. C. Megibben, Modules over an incomplete discrete valuation ring, Proc. Amer. Math. Soc. 19(1968), 450-452.

13. R. Nunke, Homology and direct sums of countable Abelian groups, Math. Zeit. 101(1967), 182-212.

14. L.D. Parker and E.A. Walker, An extension of the Ulm-Kolettis theorems, Studies on Abelian Groups, Paris, 1968, 309-325.

15. J. Rotman, Mixed modules over valuation rings, Pac. J. Math. 10(1960), 607-623.

16. J, Rotman, Torsion free and mixed Abelian groups, Ill. J. Math. 5(1961), 131-143.

17. J. Rotman and T. Yen, Modules over a complete discrete valuation ring, Trans. Amer. Math. Soc. 98(1961), 242-254.

18. R.O. Stanton, An invariant for modules over a discrete valuation ring, Proc. Amer. Math. Soc. 49(1975), 51-54.

19. H. Ulm, Zur Theorie der abzählbar-unendlichen abelschen Gruppen, Math. Ann., 107(1933), 774-803.

20. E.A. Walker, Ulm's theorem for totally projective groups, Proc. Amer. Math. Soc. 37(1973), 387-392.

21. K.D. Wallace, On mixed groups of torsion-free rank one with totally projective primary components, J. Alg. 17(1971), 482-488.

22. R.B. Warfield, Jr., Simply presented groups, Proceedings of the Special Semester on Abelian Groups, Spring 1972, University of Arizona, Tucson.

23. R.B. Warfield, Jr., Classification theory of Abelian groups I, Balanced projectives, to appear.

24. R.B. Warfield, Jr., Classification theory of Abelian groups II, Local theory, to appear.

THE STRUCTURE OF p-TREES:

ALGEBRAIC SYSTEMS RELATED TO ABELIAN GROUPS

Laurel Rogers

1. Introduction. Given a prime p and an abelian group G, a natural multiplication by p on G is defined by setting $px = x + x + \ldots + x$, where p is the number of terms in the sum. If G is a p-group, when we forget about the addition on G but keep the multiplication by p, we have the algebraic structure called a p-basic tree, which was studied in [3]. This notion led to an alternate proof of Ulm's Theorem for simply presented (i.e. totally projective) p-groups. In addition, it was shown that two reduced p-basic trees generate the same p-group iff there is a mapping between them which is composed of two pairs of functions, each pair being a stripping function followed by the inverse of a stripping function. The principal object of study in this paper is the p-tree, which differs from a p-basic tree in that it is not required to be torsion. The algebraic structure of a p-tree can be very easily described. A p-tree X generates a mixed group whose structure can be described in terms of that of X. If the group is reduced, it is a member of the class known as Warfield groups, which are studied in [1] and [4]. Such groups are completely characterized by their Ulm and Warfield invariants. We give an alternate proof of this result for those groups which are generated by p-trees, and we show furthermore that, as in the reduced torsion case, two p-trees generate the same group iff there is a mapping between them which is composed of two pairs of functions, each pair being a stripping function followed by the inverse of a stripping function.

2. Preliminaries. Throughout the discussion, the letter p shall denote a fixed but arbitrary prime, and the word "group" shall mean "abelian group". A p-tree (or simply tree) is a set X, with a distinguished element 0, upon which a multiplication by p is defined such that $p^n x = x$ iff $n = 0$ or $x = 0$. We reserve the letter X for a tree.

Examples of trees include: any group whose torsion subgroup is p-primary, any basis of a vector space over the field Z_p, and typical sets of generators of $Z(p^\infty)$, the Prüfer group, and simply presented groups.

For an element x of a tree, if there is a nonnegative integer n for which $p^n x = 0$, the least such is called the exponent of x. If no such integer exists, we say x has infinite exponent. A tree in which every element has finite exponent is a torsion tree; such trees were studied in [3] under the name of p-basic trees. A subtree of X is a subset S containing 0, and for which $s \in S$ implies $ps \in S$. The collection of all elements of X with finite exponent is a subtree called the torsion subtree of X: it is denoted X_t. The cyclic subtree generated by an element x of X is $\{0, x, px, p^2x, \ldots\}$ and is denoted (x). The direct sum of a set of trees is their disjoint union with the zero elements identified. A morphism $\phi: X \to Y$ of trees is a function such that $p\phi(x) = \phi(px)$ for each x in X. It is not difficult to show that ϕ is a monomorphism iff it is one-to-one; an epimorphism iff it is onto; and an isomorphism iff it is a bijection.

3. Natural decomposition. An order relation on a tree arises naturally by setting $y > x$ if $p^n y = x$ for some positive integer n. Clearly, $\geq$ is a partial order on the tree. The graph of this partial order provides an excellent visual model for the theory, and accounts for the name "tree". The partial order enables us to define an equivalence relation $\equiv$ on the nonzero elements. Say $x \equiv y$ if there is a nonzero element z in X for which both $x \geq z$ and $y \geq z$. As usual, an equivalence class is called a component.

Observe that if $x \equiv y$, then x has finite exponent iff y does. A component whose elements have finite exponent is a torsion component, and one whose elements have infinite exponent is a torsion-free component. Our first theorem, whose proof is an easy exercise, shows that the algebraic structure of a tree can be completely described in terms of its components.

Theorem 1. A subtree T is an indecomposable summand of X iff $T \setminus \{0\}$ is a component of X. Furthermore, X is the direct sum of all its indecomposable summands.

The decomposition of X as the direct sum of all its indecomposable summands is the natural decomposition of X. Henceforth, if C is a component of X, the summand $C \cup \{0\}$ will also be referred to as a component; the context will ensure that no confusion arises.

4. Generated groups. As in [1], [2], and [3], we may associate a group $[X]$ with a given tree X. It is defined as follows. Let $F_X = \bigoplus \underline{Z} \langle x \rangle$ be the free group on the nonzero elements of X. Let R_X be the subgroup of F_X generated by elements of the form:

$$p \langle x \rangle \qquad \text{if} \quad px = 0 ,$$

$$p \langle x \rangle - \langle px \rangle \text{ if} \quad px \neq 0 \quad \text{in} \quad X.$$

Set $[X] = F_X / R_X$; this is the group generated by X. It is clear that $[X]$ is a simply presented group. The image of X in $[X]$, consisting of 0 and the elements $\langle x \rangle$, is called a spanning tree of $[X]$. If $\phi\colon X \to Y$ is a morphism of trees, it induces a group homomorphism from F_X into F_Y which takes $\sum n_i \langle x_i \rangle$ into $\sum n_i \langle \phi(x_i) \rangle$. Clearly, the image of R_X under this map is contained in R_Y; hence ϕ induces a group homomorphism $[\phi]\colon [X] \to [Y]$. Thus $[\ ]$ is a functor from the category of trees to the category of groups. It is easy to see that $[\ ]$ is a monofunctor which commutes with direct sums. In particular, the natural decomposition of X provides a decomposition of [X], which is called the decomposition induced by X.

Obviously, if X and Y are isomorphic trees, then [X] and [Y] are isomorphic groups, but the converse is false. If X and Y are trees for which $[X] \cong [Y]$, say that X and Y are equivalent. Necessary and sufficient conditions for two reduced torsion trees to be equivalent were given in [3]; one of the purposes of this paper is to generalize these results to arbitrary trees.

Every element g of [X] clearly has a representation in the form $g = \sum_{i=1}^{k} n_i\langle x_i\rangle$ with the x_i distinct nonzero elements of X and each coefficient n_i an integer such that $|n_i| < p$. A representation in this form is a normal representation of g. It is not unique (for example, $\langle x\rangle = \langle px\rangle + (1-p)\langle x\rangle$), but we do have the following useful result.

Lemma 2. If $g = \sum_{i=1}^{k} n_i\langle x_i\rangle$ is a normal representation of g, and $g = 0$, then each $n_i = 0$.

Proof. Since $\sum n_i\langle x_i\rangle = 0$ in [X], the element $n_i\langle x_i\rangle$ of F_X must lie in R_X. Thus there are integers u_i, t_i and elements y_i, z_i of X such that in F_X, $\sum_{i=1}^{k} n_i\langle x_i\rangle = \sum_{i=1}^{r} u_i p\langle y_i\rangle + \sum_{i=1}^{s} t_i(p\langle z_i\rangle - \langle pz_i\rangle)$, where $py_i = 0$ and $pz_i \neq 0$ in X. Because no n_i is divisible by p unless it is 0, all the u_i and t_i must be 0, so each $n_i = 0$.////

Let $\{F_i : i \in I\}$ be the collection of distinct torsion-free components of X, and let $\{T_j : j \in J\}$ be the collection of distinct torsion components. Then the torsion-free rank of X is $|I|$, and the residual rank of X is $|J|$. The former term is justified by the next theorem.

Theorem 3. If X has torsion-free rank r_0, then the group $[X]$ has torsion-free rank r_0.

Proof. Let $\{x_i : i \in I\}$ be a set of representatives from the distinct torsion-free components of X; then $r_0 = |I|$. Suppose $\sum_{j=1}^{k} m_j' \langle x_j \rangle = 0$ in $[X]$, where each m_j' is an integer and the x_j are distinct representatives. Express each m_j' as $m_{j0} + m_{j1}p + \ldots + m_{j,u(j)}p^{u(j)}$, where $|m_{ji}| < p$ and $u(j)$ is a nonnegative integer. Then $\sum_{j=1}^{k} \sum_{i=0}^{u(j)} m_{ji}\langle p^i x_j \rangle = 0$. If $p^i x_j = p^n x_q$, then $x_j \equiv x_q$, which is impossible since the x_i come from distinct components. So the above is a normal representation, hence by the lemma each m_j' must be 0. Thus $\{\langle x_i \rangle : i \in I\}$ is an independent set of elements in $[X]$. Suppose g is an element of $[X]$ with infinite order, and that $g = \sum_{j=1}^{k} n_j \langle y_j \rangle + \sum_{j=1}^{m} q_j \langle z_j \rangle$ is a normal representation of g, where each y_j has infinite exponent and each z_j has finite exponent. It is easy to see that there is an integer e for which each $p^e y_j$ is less than one of the representatives x_i, and each $p^e z_j = 0$. Thus $p^e g$ depends on $\{\langle x_i \rangle : i \in I\}$, showing that this set is a maximal independent system of elements in $[X]$ of infinite order. So the torsion-free rank of $[X]$ is r_0.////

The notion of height plays an important role in what follows. For an ordinal α, the subset $p^\alpha X$ of X is defined inductively by setting $p^0 X = X$, and $p^\alpha X = \bigcap_{\beta<\alpha} p(p^\beta X)$ when $\alpha > 0$. Then an element x of $p^\alpha X \setminus p^{\alpha+1} X$ has height α; we write $hx = \alpha$. If $x \in p^\alpha X$ for all ordinals α, then we set $hx = \infty$. The symbol ∞ satisfies $\infty < \infty$ and $\alpha < \infty$ for all ordinals α. In order to show the relationship between heights of elements in X and heights of elements in $[X]$, we first need a lemma.

Lemma 4: If G is a nonzero element of $[X]$, then there are distinct nonzero elements $x_1, \ldots, x_k$ of X and integers $r_1, \ldots, r_k$ and $t_1, \ldots, t_k$ for which we have:

(1) $g = \sum_{i=1}^{k} r_i \langle x_i \rangle + p \sum_{i=1}^{k} t_i \langle x_i \rangle$,

(2) no two of the x_i are comparable,

(3) for each i, $0 < r_i < p$. The elements $x_1, \ldots, x_k$ and the integers $r_1, \ldots, r_k$ are unique.

Proof. Let $g = n_1 \langle x_1 \rangle + \ldots + n_k \langle x_k \rangle$ be a normal representation of g with each $n_i \neq 0$. If some $n_i < 0$, replace $n_i \langle x_i \rangle$ with $(p + n_i) \langle x_i \rangle - p \langle x_i \rangle$. If x_i and x_j are comparable, say $p^n x_i = x_j$ with $n > 0$, replace $\langle x_j \rangle$ with $p \cdot p^{n-1} \langle x_i \rangle$. Thus we can get an expression of the desired type.

Suppose g has two expressions of this type. Temporarily allowing coefficients to be 0 if necessary, we may assume $\sum_{i=1}^{k} r_i \langle x_i \rangle + p \sum_{i=1}^{k} t_i \langle x_i \rangle = \sum_{i=1}^{k} r_i' \langle x_i \rangle + p \sum_{i=1}^{k} t_i' \langle x_i \rangle$. Consequently, $\sum_{i=1}^{k} (r_i - r_i') \langle x_i \rangle + p \sum_{i=1}^{k} (t_i - t_i') \langle x_i \rangle = 0$. The second summation may be rewritten in a normal representation, and no term will contain an x_i. Thus by Lemma 2, each $r_i - r_i' = 0$, giving the result.////

If g is a nonzero element written in the form given in the above lemma, then $\sum_{i=1}^{k} r_i \langle x_i \rangle$ is its maximal part. The following theorem, showing the relationship between heights in X and in $[X]$, is analogous to Proposition 1 in [3]. The proof is very similar, and involves using the maximal part of an element in place of the unique representation of the element as a linear combination of elements in the image of X, which exists in the torsion case.

Theorem 5. Let g be a nonzero element of $[X]$, with a normal representation $g = \sum n_i \langle x_i \rangle$ where no $n_i = 0$. Then $g \in p^{\alpha}[X]$ iff each $x_i \in p^{\alpha} X$. Consequently, the height of g in $[X]$ is the minimum of the heights of the x_i in X.

5. Stripping functions. The concept of a stripping function for torsion trees was introduced in [3]. For arbitrary trees X and Y, a stripping function $\sigma: X \to Y$ is a height-preserving bijection which satisfies:

(1) $p\sigma(x) \neq \sigma(px)$ implies $p\sigma(x) = 0$, and

(2) for each x in X, there is an element $z < x$ such that $p^n\sigma(z) = \sigma(p^n z)$ for all positive integers n.

The action of a stripping function on a tree is easily visualized by examining the graphs of the tree and of its image. Note that the composition of two stripping functions is a stripping function. The significance of stripping functions is that the equivalence of two trees may be characterized in terms of them. A function ξ is a strip-graft if $\xi = \gamma\sigma$, where γ^{-1} and σ are stripping functions. It was found in [3] that two reduced torsion trees are equivalent iff there is a bijection between them which is the composition of two strip-grafts; we shall show that this result can be extended to arbitrary trees. The next theorem describes the action of a stripping function in terms of its effect on the torsion-free and residual ranks.

Theorem 6. Suppose $\sigma: X \to Y$ is a stripping function. Then the torsion-free ranks of X and Y are equal. Let r be the residual rank of X, and let c be the cardinality of the set $\{x \in X: p\sigma x \neq \sigma px\}$. Then the residual rank of Y is $r + c$.

Proof. If $x \in X$ and $p^n\sigma x \neq 0$, then $p^n\sigma x = \sigma p^n x$. Thus if $\sigma x \equiv \sigma y$ in Y, then $x \equiv y$ in X. On the other hand, if $x \equiv y$ in X but $\sigma x \not\equiv \sigma y$ in Y, then σx or σy has finite exponent. Thus if $x \equiv y$ in X and both σx and σy have infinite exponent, then $\sigma x \equiv \sigma y$ in Y. It follows that the torsion-free ranks of X and Y are equal.

The residual rank of a tree is the cardinality of the set of elements of exponent 1. If $\sigma x \in Y$ and has exponent 1, then either x has exponent 1 in X, or x is such that $p\sigma x \neq \sigma px$. ////

The most important feature of a stripping function is that it preserves equivalence of trees. This fact was proved for torsion reduced trees in [3], Theorem 2. The ideas here are similar, but the lack of finite exponents and the lack of unique representation for each group element complicates the proof. Before we proceed, we need another lemma.

Lemma 7. Let $\sigma: X \to Y$ be a stripping function. Then X has a maximum subtree B for which $\sigma|B$ is an isomorphism. This subtree has the following properties:

(1) if $px \in B$ and $p\sigma x = \sigma px$ then $x \in B$, and

(2) if $x \notin B$, there is a positive integer e for which $p^e x \in B$.

Proof. The class of subtrees on which σ is an isomorphism, ordered by set inclusion, is inductive, hence contains a maximal element B. If A is any subtree in this class, then $A \cup B$ is in the class; hence B is maximum. If $px \in B$ and $p\sigma x = \sigma px$, then B must contain the cyclic subtree (x), so (1) holds. If $x \notin B$, by the properties of σ there is e such that $p^n\sigma(p^e x) = \sigma(p^{n+e}x)$ for all $n > 0$, hence $p^e x \in B$ and (2) holds.////

Call this subtree B the subtree fixed by σ.

Theorem 8. If $\sigma: X \to Y$ is a stripping function, then the groups $[X]$ and $[Y]$ are isomorphic; consequently, X and Y are equivalent.

Proof. Denote by $\langle X \rangle$ the image of X in the group $[X]$. For x in X, let $e(x)$ be the least nonnegative integer such that $p^{e(x)}x$ is in the fixed subtree B of σ. Define $d(x)$ inductively by setting $d(x) = 0$ if $e(x) = 0$, and for $e(x) = k + 1$, set

$$d(x) = \begin{cases} d(px) & \text{if } p\sigma x = \sigma px, \\ d(px) + 1 & \text{if } p\sigma x \neq \sigma px. \end{cases}$$

We shall use both $e(x)$ and $d(x)$ to define a function $\pi: X \to X$ which will have these properties for each x in X: (i) $p\pi x = \pi px$ iff $p\sigma x = \sigma px$; (ii) $h(\pi x) \geq hx$; (iii) there is a nonnegative integer n such that $\pi^n x = 0$; (iv) $p\pi x = \pi px$ or $p\pi x = px$; and (v) $\pi x \neq 0$ implies $d(\pi x) < d(x)$. The construction of π is nearly the same as in Theorem 2 of [3]; to begin, set $\pi x = 0$ for each x such that $d(x) = 0$. Assuming π has been defined for all y such that $d(y) < k$, and for all y with $d(y) = k$ and $p^{m-1}y$ in B, suppose x is an element for which $e(x) = m$ and $d(x) = k$. Proceed now as in Theorem 2 of [3].

We now define a function $\bar{\sigma}: X \to [X]$. Set $\bar{\sigma}(0) = 0$, and set $\bar{\sigma}(\langle x \rangle) = \langle x \rangle - \langle \pi x \rangle$. It is routine to check that the properties of π insure that $\bar{\sigma}$ is one-to-one. Furthermore, property (iv) of π allows us to conclude that $X' = \bar{\sigma}(\langle X \rangle)$ is itself a p-tree under the multiplication by p in the group [X]. It is a laborious but fairly routine exercise to show that [X] is isomorphic to the group $[X']$ under the correspondence $\langle x \rangle \leftrightarrow (\langle x \rangle - \langle \pi x \rangle) + (\langle \pi x \rangle - \langle \pi^2 x \rangle) + \ldots + (\langle \pi^{m-1} x \rangle - \langle \pi^m x \rangle)$, where m is large enough so that $\pi^m x = 0$. Consequently, we may regard X' as a spanning tree of [X].

Finally, define $\eta: Y \to X'$ by $\eta(0) = 0$ and $\eta(\sigma x) = \bar{\sigma}(\langle x \rangle)$ for nonzero x. It is clear that η is a bijection, and it is easy to check that η preserves multiplication by p; thus η is a tree isomorphism. Hence $[Y] \cong [X']$, giving the desired result.////

As a corollary, if $\sigma: X \to Y$ is a stripping function, the decomposition of [X] induced by Y is a refinement of the decomposition of [X] induced by X.

6. Divisibility. An element x in X is divisible by p if $hx > 0$. The tree X is divisible if every element of X is divisible by p. If $x \in X$ and X is divisible, given a positive integer m, it is clear that there is y in X such that $x = p^m y$. Thus when X is divisible, $h^{-1}(\infty) = X$. Recall that a group G is p-divisible if $p^m G = G$ for every positive integer m. The next result is easily verified.

Theorem 9. If X is a divisible p-tree, then [X] is a p-divisible group.

If we are given the additional information that for a prime $q \neq p$, multiplication by q is an automorphism, then X is a p-local tree. In this case,

$[X]$ is a divisible group when X is a divisible tree.

A tree X is <u>reduced</u> if $h^{-1}(\infty) = 0$. It is easy to show, using Theorem 5, that if X is reduced, then $[X]$ is a reduced group. The usual decomposition of the group $[X]$ into its divisible and reduced parts may be accomplished by the application of a stripping function to X. More precisely, we have the following.

<u>Theorem 10</u>. Let X be a p-local tree. Then there is a tree Y and a stripping function $\sigma: X \to Y$ such that $Y = D \oplus R$, where D is a divisible subtree and R a reduced subtree of Y.

<u>Proof</u>. In X, set $D' = \{x \in X: hx = \infty\}$; clearly, D' is a subtree of X which has this property: if $x \in D'$, then there is an element z of D' such that $pz = x$. For x in X, denote by $C(x)$ the set of elements which are comparable to x. Set $R' = \{x \in X: C(x) \cap D' \subseteq \{0\}\}$. Now define the tree Y to be the elements of X with multiplication $*$ by p defined as follows:

$$p * y = \begin{cases} py & \text{if } y \in D' \cup R', \\ py & \text{if } y \notin D' \cup R' \text{ and } py \notin D', \text{ and} \\ 0 & \text{if } y \notin D' \cup R' \text{ and } py \in D'. \end{cases}$$

Define $\sigma(x) = x$. By induction on α, it is easy to show that $x \in p^{\alpha}X$ iff $\sigma(x) \in p^{\alpha}Y$, for any ordinal α. Thus σ preserves heights. It is easy to check that σ is a stripping function. In Y, set $D = \sigma(D')$ and $R = (Y \sim D) \cup \{0\}$. Obviously, D is a divisible subtree of Y. Furthermore, R is a subtree of Y, which is clearly reduced.////

A tree X is <u>fully stripped</u> if, whenever there are distinct elements x and y with $px = py$, then either $px = 0$ or $h(px)$ is a limit ordinal. If X is divisible and fully stripped, then $x \neq y$ and $px = py$ imply $px = 0$.

Divisible trees may be characterized in much the same way as divisible groups; in outline form, the procedure is this. If X is divisible, a stripping function can be defined whose image Y is the direct sum of a torsion-free, fully stripped subtree F, and a torsion fully stripped subtree T. Each nonzero element x of Y has the property that $p^{-1}x$ contains exactly one element, so any stripping function defined on Y is an isomorphism. Thus the torsion-free rank and the residual rank of Y completely characterize it; if X is p-local, these cardinal numbers correspond to the torsion-free rank and the p-rank of the divisible part of $[X]$.

7. Reduced trees. In view of the preceding remarks and Theorem 10, we assume unless otherwise stated that all trees are reduced.

A principal result of [3] (Theorem 4) is the following, which we refer to as "the W-Theorem": Let X and Y be torsion reduced trees. Then X and Y are equivalent iff there is a function $\theta: X \to Y$ which is the composition of two strip-grafts. In this section, we shall generalize this result to arbitrary trees.

Let X be a tree. For each ordinal α set $U(\alpha,X) = \{x \in X: hx = \alpha$ and $h(px) > \alpha + 1\}$. For each x in X of height $\alpha + 1$, choose an element z_x in $p^{-1}x$ of height α. Define $D(\alpha,X) = \{y \in X: hy = \alpha, h(py) = \alpha + 1$, and $y \neq z_{py}\}$. We now define the α-th Ulm invariant of X to be the cardinal number $f(\alpha,X) = |U(\alpha,X)| + |D(\alpha,X)|$. It is shown in [1] and [3] that the Ulm invariants of X are the same as those of the group $[X]$, i.e. we have for each ordinal α, $f(\alpha,X) = f(\alpha,[X])$.

Say that X is sharp if it has the following property for each ordinal α: whenever an element x in X has height $\alpha + 1$, the set $p^{-1}x$ has a unique element y of height α. Notice that if X is sharp, each of the sets $D(\alpha,X)$ is empty; consequently, $f(\alpha,X) = |U(\alpha,X)|$.

Lemma 11. Suppose X is torsion-free and $x \in X$. Then X has a sharp subtree S which contains x, and which has the property that for each ordinal α, $p^{\alpha}S = S \cap p^{\alpha}X$ (i.e. S is isotype).

Proof. We first show that for any element y in X, there is a set $A(y)$ with these properties: $A(y)$ is sharp, $A(y) \cap p^{\alpha}X = p^{\alpha}A(y)$ for each α, if $z \in A(y)$ then $z \geq y$, and if $x \geq y$ then $x \in A(y)$. Assume that this is true for elements of height $\beta < \gamma$, and that $hy = \gamma$. If γ is a limit ordinal, take $A(y) = \bigcup \{A(z): pz = y\} \cup \{y\}$. If $\gamma = \delta + 1$ and $y = p^m x$ for some $m > 0$ with

$\delta = h(p^{m-1}x)$, take $A(y) = A(p^{m-1}x) \cup \{y\}$. If $\gamma = \delta + 1$ and $y = p^m x$ for some $m > 0$ with $\delta > h(p^{m-1}x)$, choose z_y of height δ and take $A(y) = A(z_y) \cup A(p^{m-1}x) \cup \{y\}$. If $\gamma = \delta + 1$, and $y > x$, choose z_y of height δ and take $A(y) = A(z_y) \cup \{y\}$. Clearly, $A(y)$ has the stated properties. Set $S = \bigcup_{n=0}^{\infty} A(p^n x)$. It is easy to see that S gives the desired results.////

Theorem 12. If X is a tree, then there is a tree Y and a stripping function $\sigma: X \to Y$ such that Y is sharp.

Proof. Let $\{x_i\}$ be a set of representatives of the torsion-free components of X. Let S_i be the sharp subtree of the component to which x_i belongs, which exists by lemma 11, and contains x_i. Define the tree Z to be the elements of X with multiplication $*$ by p defined by:

$$p * y = \begin{cases} 0 & \text{if } y \notin \bigcup S_i \text{ but } py \in \bigcup S_i, \\ py & \text{otherwise.} \end{cases}$$

Set $\psi: X \to Z$ to be the identity. Since $p^\alpha X \cap S_i = p^\alpha S_i$ for each α and each S_i, we see that ψ preserves heights. Property (2) of a stripping function is satisfied by ψ because the restriction of ψ to $\bigcup S_i$ is an isomorphism; obviously then, ψ is a stripping function.

Now the set of torsion elements Z_t of Z is a torsion tree, hence there is a fully stripped tree V and a stripping function $\tau: Z_t \to V$, by Proposition 4 of [3]. Letting $F = (Z \backslash Z_t) \cup \{0\}$, define $Y = F \cup V$ and define $\phi: Z \to Y$ be setting $\phi(y) = y$ if $y \in F$, and $\phi(y) = \tau(y)$ if $y \in Z_t$. It is clear that Y is sharp, and by setting $\sigma = \phi\psi$, we have the result.////

It is necessary at this point to introduce the notion of the Warfield invariants for trees. Our definition is similar to the definition of these invariants for valuated groups given in [1]. The height sequence of an element x is the sequence $\{hx, hpx, hp^2x, \ldots\}$. Any sequence $\mu = \{\mu_0, \mu_1, \ldots\}$ of ordinals and symbols ∞ such that $\mu_i < \mu_{i+1}$ for $i = 0, 1, \ldots$ will be called a value sequence, and the sequence $\{\mu_1, \mu_2, \ldots\}$ will be denoted by $p\mu$. Value sequences

μ and ν are <u>equivalent</u> if there are nonnegative integers m and n such that $p^m\mu = p^n\nu$.

Define a relation H on X as follows: $x \mathrel{H} y$ iff the height sequence of x is equivalent to the height sequence of y. Clearly, H is an equivalence relation. It is easy to verify the following.

<u>Lemma 13.</u> The relation $\equiv$ is a subset of the relation H.

Suppose R is an H-equivalence class of X, and let μ denote the height sequence of one of its elements. Then $\equiv$ partitions R into components, say $R = \bigcup_{j\in J} C_j$ where each C_j is a $\equiv$-equivalence class. Define $w(\mu,X) = |J|$; this is the <u>Warfield invariant</u> of X at μ. It can be shown (see [1], remarks following lemma 4) that $w(\mu,X)$ is equal to the (dimension of the) Warfield invariant of the group $[X]$ at μ.

Suppose now that X and Y are trees and that $\theta\colon X \to Y$ is a height-preserving bijection which has the property that, if $\theta(px) \neq p\theta(x)$ then both $h(px) > hx + 1$ and $h(p\theta x) > h\theta x + 1$. Then θ will be called a <u>$\bar{T}$-function</u>. Observe that the composition of two $\bar{T}$-functions is a $\bar{T}$-function; the inverse of a $\bar{T}$-function is a $\bar{T}$-function; and if both X and Y are sharp, and $\sigma\colon X \to Y$ is a stripping function, then both σ and σ^{-1} are $\bar{T}$-functions. The notion of a T-function, defined in [3], is related to this; every T-function is obviously a $\bar{T}$-function.

<u>Theorem 14.</u> If X and Y are sharp, and if they have the same Ulm invariants and the same Warfield invariants, then there is a $\bar{T}$-function $\theta\colon X \to Y$. Furthermore, X has a subtree B which contains a complete set of representatives of the torsion-free components of X, and for which $\theta|B$ is an isomorphism.

<u>Proof.</u> Let $\{x_i\}$ be a complete set of representatives from the torsion-free components of X. The relation H partitions this set into subsets R_j, and if μ is the height sequence of an element in R_j, then $w(\mu,X) = |R_j|$. Since

$w(\mu,X) = w(\mu,Y)$, there is a bijection ζ from R_j onto the set of Ξ-components of Y, the elements of each of which have height sequences equivalent to μ. For each x_i in R_j, pick an element y_i in the component $\zeta(x_i)$. Now y_i and x_i have equivalent height sequences, so without loss of generality we shall assume their height sequences are equal (either x_i or y_i could be replaced by a smaller element if necessary, since they both have infinite exponent). Define θ on $\bigcup(x_i)$ by $\theta(p^n x_i) = p^n y_i$, for $n = 0, 1, \ldots$.

If $x \notin \bigcup(x_i)$, then there is a least positive integer $n = e(x)$ such that $p^n x = 0$ or $p^n x \in \bigcup(x_i)$, since $\{x_i\}$ was a complete set of representatives of the torsion-free components of X. Let A denote the set $\{x \in X \setminus \bigcup(x_i)$: $h(p^k x) + 1 = h(p^{k+1} x)$ for $k = 0, 1, \ldots, n-1\}$. Note A contains no torsion elements. We can extend the definition of θ to A as follows. Let $x \in A$, suppose that θ has been defined for elements y with $e(y) < n$, and that $n = e(x)$. Then $\theta(px)$ is already defined; and $hx + 1 = h(px)$ implies that x is the unique element of height hx in $p^{-1}(px)$, since X is sharp. Thus there is an element z in $p^{-1}\theta(px)$ of height hx, which is unique since Y is sharp. So define $\theta x = z$; thus θ is defined on $\bigcup(x_i) \cup A$. Set $B = \bigcup(x_i) \cup A$ and note that B is a subtree and that $\theta|B$ is an isomorphism.

If $x \notin B$ but $px \in B$, this means $h(px) > hx + 1$. Consequently, $x \in U(hx, X)$. The fact that X and Y have the same Ulm invariants allows us to extend the definition of θ to all of X by the method which was used in Proposition 5 of [3] to define a T-function from one tree to another. Thus θ can be defined from all of X onto Y, and it is easy to see that θ is a $\bar{T}$-function. ////

We are now able to show that if we have a $\bar{T}$-function $\theta: X \to Y$ for which a subtree B of the type described in Theorem 14 exists, then θ is the composition of two strip-grafts. Under these conditions, set $o(x) = n$ if n is the least nonnegative integer for which $p^n x \in B$. The decomposition of θ into strip-grafts is accomplished in nearly the same way that a T-function is decomposed into strip-grafts in the torsion case (see [3], Propositions 5 - 8). The differences are as follows; the reader is invited to consult [3].

The definition of a suitably partitioned subset S of X must be modified as follows: S must contain the subtree B, and the function e must have the property that $e(x) = 0$ if $x \in B$. In the proof of lemma 2, the integer m must be taken to be the least positive integer such that $p^m x \in B$ or $p^m x \notin E$. In the statements of Propositions 6 and 7, and in their proofs, each phrase "$px \neq 0$" must be replaced by "$h(px)$ is a limit ordinal" and each phrase "$p\theta x \neq 0$" must be replaced by "$h(p\theta x)$ is a limit ordinal". The proof of Proposition 8 requires the use of lemma 1, whose analog is clearly true in this more general context. Also, the function $e(x)$ will no longer be the exponent of the element (x,y) in the tree Z, but will be the function $o(x)$ which was defined above. We obtain the following result.

Lemma 15: If X and Y are sharp, and if they have the same Ulm invariants and the same Warfield invariants, then there is a function $\theta: X \to Y$ which is the composition of two strip-grafts.

We now drop the hypothesis that all trees are reduced.

Theorem 16 (The W-Theorem). Let X and Y be arbitrary trees. Then X and Y are equivalent iff there is a function $\theta: X \to Y$ which is the composition of two strip-grafts.

Proof. By Theorem 10, there are trees Z_1 and Z_2 and stripping functions $\sigma_1: X \to Z_1$ and $\sigma_2: Y \to Z_2$ such that $Z_i = D_i \oplus R_i$ $(i = 1, 2)$, with D_i divisible and R_i reduced. By the discussion following Theorem 10, there are trees Z_3 and Z_4 and stripping functions $\sigma_3: Z_1 \to Z_3$ and $\sigma_4: Z_2 \to Z_4$ which are isomorphisms on R_1 and R_2 respectively, and such that the images of D_1 and D_2 are fully stripped. The reduced parts of Z_3 and Z_4 are equivalent, so by Theorem 12 there are trees Z_5 and Z_6 and stripping functions $\sigma_5: Z_3 \to Z_5$ and $\sigma_6: Z_4 \to Z_6$ whose reduced parts are sharp. By Lemma 15, there is a function θ between the reduced parts of Z_5 and Z_6, which can be extended to all of Z_5 and Z_6 in the obvious way, and which is the composition of two strip-grafts, say $\Theta = \sigma_7^{-1}\sigma_8\sigma_9^{-1}\sigma_{10}$. The maps $\sigma_7\sigma_6\sigma_4\sigma_2 = \phi_1$ and $\sigma_{10}\sigma_5\sigma_3\sigma_1 = \phi_2$ are stripping functions, so the function $\phi_1^{-1}\sigma_8\sigma_9^{-1}\phi_2$ is the desired mapping.////

Our final result is a corollary to the proof of Lemma 15.

Theorem 17 (Ulm's Theorem for Trees). Two reduced trees are equivalent iff they have the same Ulm invariants and the same Warfield invariants. Consequently, two Warfield, groups which are simply presented are isomorphic iff they have the same Ulm invariants and the same Warfield invariants.

REFERENCES

1. R. Hunter, F. Richman, and E. Walker, Existence Theorems for Warfield Groups, Trans. Amer. Math. Soc., to appear.

2. R. Hunter, F. Richman, and E. Walker, Simply Presented Valuated Abelian p-Groups, J. Alg., to appear.

3. L. Rogers, Ulm's Theorem for Partially Ordered Structures Related to Simply Presented Abelian p-Groups, Trans. Amer. Math. Soc., to appear.

4. R. Warfield, Jr., Classification Theory of Abelian Groups, II: Local Theory, to appear.

A GUIDE TO VALUATED GROUPS

Fred Richman[1]

0. Introduction. This paper outlines the development of the concept of a valuated group and describes the present state of the art. Valuated groups have appeared implicitly in much of abelian group theory but their explicit use grew out of two ideas. The first of these was the idea of studying an abelian p-group by looking at its socle. The second was the constructive approach to mathematics, in the sense of Bishop [1], applied to structure theorems associated with the names Ulm, Zippin, Hill, and Warfield. These two developments are discussed in Sections 1 and 2.

Once the concept of a valuated group is isolated, the category of valuated groups becomes an object of interest. Section 3 deals with this category and its relation to questions in abelian group theory. If we restrict ourselves to finite valuated groups we have a new combinatorial gadget, or at least a new way of looking at an old one. Section 4 contains some of the basic facts about such things. Much of the work described in this paper was done jointly with Elbert Walker whose influence on the subject has been enormous. I would like to express my pleasure in having been able to work with Roger Hunter who sparked the most intensely productive six months of mathematics at New Mexico State University that I can remember.

If you stare at heights long enough you will begin to see valuations. To get in a state of "valuation readiness" we shall consider an easy theorem about heights, motivated by constructive considerations. For the time being we restrict ourselves to a fixed prime p. If you are going to prove Ulm's theorem in a finitistic setting you have to be able to compute heights. Generally speaking there is no way to tell whether a given element is divisible by p, let alone what its exact height is, so you must restrict yourself to groups where you *can* compute heights, of which there are plenty. Thus height is added data -- a group comes equipped with a height function h.

[1] This research was partially supported by NSF Grant MCS76-23082.

This function h must satisfy:

H1) hx is an ordinal or ∞;

H2) $hx < hpx$ (Note: $\alpha < \infty$ and $\infty < \infty$);

H3) If $\alpha < hx$, then $x = py$ for some y such that $hy \geq \alpha$.

From a constructive point of view H3 entails a finite procedure for computing y given x and α. Now if height is added data, there is a uniqueness problem. This is settled in [17] by proving that height, so defined, is an isomorphism invariant. Stripped of constructive subtleties, the argument is as follows.

Theorem. If h_1 and h_2 satisfy H1, H2, and H3, then $h_1 = h_2$.
Proof. It suffices by symmetry to show that $h_1 x \leq h_2 x$ for all x. If $h_2 x = \infty$ we are done. Otherwise proceed by induction on $h_2 x$. If $\alpha < h_1 x$, then $x = py$ with $h_1 y \geq \alpha$ by H3 and $h_2 y < h_2 x$ by H2. Thus $h_1 y \leq h_2 y$ by induction, so $\alpha \leq h_1 y \leq h_2 y < h_2 x$. Since $\alpha < h_1 x$ implies $\alpha < h_2 x$, we have $h_1 x \leq h_2 x$.

Reference to this theorem can spare many an induction on height. The strategy is, whenever you want to prove that heights behave in a certain way, define a function that behaves that way and show that it satisfies H1, H2, and H3. An example of this will also prepare us for cokernels of valuated groups.

Corollary. The following two conditions on a subgroup A of a group B are equivalent:

i) Every coset of A has an element of maximal p-height.

ii) The map $p^\alpha B \to p^\alpha(B/A)$ is onto for each α.

Proof. To go from ii) to i) we simply take $x \in b + A$ such that $hx = h(b + A)$. The hard part is going from i) to ii). Define h_0 on B/A by $h_0(b + A) = \max\{hx : x \in b + A\}$. Clearly h_0 satisfies H1, H2, and H3, so $h = h_0$ and thus ii) holds.

The point of all this in the present context is that we are considering the height function, at least a priori, as an independent entity. If we insist on H1, H2, and H3, then we get nothing new; but if we relax our demands we arrive at the notion of a valuated group. The prototype situation is a group A contained in a group B. If we define vx for each $x \in A$ to be the p-height of x in B, then v satisfies the following:

V1) vx is an ordinal or ∞;

V2) $vx < vpx$;

V3) $v(x - y) \geq \min(vx, vy)$.

A function v on a group A satisfying V1, V2, and V3 is called a p-valuation. A valuated group is a group A together with a p-valuation v_p on A for each prime p. The prototype situation is exhaustive in the sense that any valuated group A can be embedded in a group B so that the restriction to A of the p-height functions on B are the p-valuations on A [21; Theorem 23]. This generalizes the "crude existence theorem" of Rotman and Yen [22; Theorem 2] which treats the case where A is torsion-free cyclic and only one prime is involved.

We shall be mostly concerned with the local case and will find the following definitions useful. The valuated groups that are modules over the integers localized at p form the category V_p of valuated Z_p-modules. Since $v_q x = \infty$ if $q \neq p$ we need only consider the p-valuation. The morphisms in V_p are homomorphisms f such that $vf(x) \geq vx$. The direct sum of a family of valuated Z_p-modules is their group direct sum, the value of an element being the minimum of the values of its coordinates. This is the coproduct in V_p. An infinite cyclic Z_p-module A is said to be rank-one free in V_p if $vpx = vx + 1$ for each x in A. A free valuated Z_p-module is a direct sum of a family of rank-one free valuated Z_p-modules. A submodule of a valuated Z_p-module is said to be nice if every coset of it contains an element of maximal value.

The submodules $A(\alpha) = \{x \in A : vx \geq \alpha\}$ of a valuated Z_p-module A completely determine the valuation v. These submodules satisfy:

1) if $\alpha < \beta$, then $A(\alpha) \supseteq A(\beta)$;

2) if β is a limit ordinal, then $A(\beta) = \bigcap_{\alpha<\beta} A(\alpha)$;

3) $p(A(\alpha)) \subseteq A(\alpha + 1)$.

Conversely any family of submodules, indexed by the ordinals, that satisfies these three conditions gives rise to a valuation on A. Thus we may think of a valuated Z_p-moule as a group with a filtration.

1. The Structure of Socles. The socle of a p-group G comes with a natural filtration given by the subgroups $p^{\alpha}G[p]$. The associated graded group is composed of the Ulm invariants $p^{\alpha}G[p]/p^{\alpha+1}G[p]$. Perhaps the first theorem dealing with the socle as a filtered group rather than a graded group was Kulikov's criterion for a p-group to be a direct sum of cyclic groups [7; Theorem 17.1], namely that the socle be a union of an increasing sequence of subgroups whose nonzero elements are of bounded height. Charles [3] showed that the proof of this theorem may be broken into two parts through the notion of a decomposable S-structure. An S-structure, in our terminology, is a p-bounded valuated group. An S-structure is decomposable if it is the direct sum of subgroups Q_{α} where $vx = \alpha$ for all nonzero x in Q_{α}. Thus the socle of a group is a decomposable S-structure if and only if it is summable in the sense of Hill and Megibben [12]. In the language of Fuchs [9], a decomposable S-structure is a free valued vector space with ordinal values. Charles appears to be the first to treat these objects in a systematic way as independent entities.

Charles' proof that a p-group satisfying Kulikov's criterion is a direct sum of cyclics proceeds as follows. First show that an S-structure satisfying Kulikov's criterion is decomposable [3; Théorème 1]. Then show that any p-group with a decomposable socle and no elements of infinite height is a direct sum of cyclics [3; Théorème 8]. The first part is a theorem about valuated groups (generalized by Fuchs [9; Theorem 1]). The second relates valuated group structure to group structure. This strategy is as efficient as any other in establishing Kulikov's criterion, and it provides additional insight. Prüfer's theorem [7; Theorem 11.3] that a countable p-group with no elements of infinite height is a direct sum of cyclics follows from the valuated groups theorem that every countable S-structure is decomposable, which is proven in [3; Théorème 2] for the finite values case and in [4; Théorème 3] for the general case. An even more general theorem, first proven by Brown [2], follows from Fuchs' generalized Kulikov criterion [9; Theorem 1].

One other theorem of Charles worth mentioning is that every S-structure of type $\leq \omega$ (that is, with no elements of infinite value) is the socle of some p-group [3; Théorème 4]. This may be thought of as a sort of "fine existence theorem" in the spirit of Rotman and Yen [22] in that it gives an efficient embedding (no superfluous socle) of a valuated group in a group. The question then arises whether p-groups with no elements of infinite height are characterized by their socles (viewed as valuated groups). The answer is no, and the first example was given by Hill [11; Example 2] who constructed two nonisomorphic pure subgroups A and B of a torsion complete group such that $A[p] = B[p]$.

Charles' notion of an S-structure of type $\leq \omega$ was exploited in [16], under the name ω-filtered vector space, to classify p-groups G such that $pG^1 = 0$ and G/G^1 is torsion complete. Such a group is determined by the valuated group structure of its socle [16; Corollary 1]. If G^1 is also finite, then any two p-groups with the same socle structure as G are isomorphic [16; Corollary 4], the point being that if H has this socle structure, then H/H^1 must be torsion complete. The idea of using socle structure as an invariant for groups was also employed by Fuchs and Irwin who showed that a $p^{\omega+1}$-projective p-group is determined by the valuated group structure of its socle [10; Theorem 3] and characterized those valued vector spaces (p-bounded valuated groups) that are the socles of $p^{\omega+1}$-projective p-groups [10; Theorem 4]. In a subsequent paper Fuchs showed that a $p^{\omega+n}$-projective p-group is determined by the valuated group structure of its p^n-socle

[8; Theorem 2].

Valued vector spaces are a generalization of S-structures in that the ground field is not restricted to the p-element field and the values need not be ordinals. The reader interested in going in this direction should consult Fuchs' paper [9], and the references therein, which also links the subject with the theory of nonarchimedean Banach spaces. In this paper Fuchs establishes various properties of projectives (they are free) and injectives (they are the s-complete spaces) which, of course, specialize to theorems about p-bounded valuated groups. The reader should be aware that Fuchs' definition of injective, unlike his definition of projective, is a priori stronger than the one suggested by general consideration of pre-abelian categories which we shall discuss in Section 3. Also [9; Theorem 7] appears to be in error, as does its corollary [9; Theorem 3] which states that s-closed (nice) subspaces of free spaces are free (see [21; Theorem 19]).

2. The Influence of Constructive Mathematics.

Valuated groups play a central role in the finitistic approach to constructing countable p-groups with prescribed Ulm invariants (Zippin's theorem). In this approach we can examine only a finite number of Ulm invariants at a time, so at no point in the construction do we have a complete picture of what the group G looks like. Yet we must start enumerating the elements of G, together with their heights, on the basis of the information we have. The result is a chain of finite valuated groups whose union is G, the basic construction being a "fine existence theorem" [17; Theorem 12] which embeds a given finite valuated p-group in a p-group with specified Ulm invariants. For the statement of this theorem, and other similar theorems, it is convenient to define the αth Ulm invariant $f_G(\alpha)$ of a valuated group G to be (the dimension of) the following vector space:

$$\frac{\{x : vx \geq \alpha \text{ and } vpx > \alpha + 1\}}{\{x : vx \geq \alpha + 1\}}.$$

This definition, which essentially appears in Kaplansky's proof of Ulm's theorem [4; Theorem 14], was introduced in [17; Section 4] where it was stated for groups with a subheight function, that is, a p-valuation.

If $f(\alpha)$ is the αth Ulm invariant of a fixed countable reduced p-group, then

i) $f(\alpha) \leq \aleph_0$,

ii) $f(\alpha) = 0$ for all α beyond some countable ordinal,

iii) if $f(\alpha + n) = 0$ for all $n < \omega$, then $f(\beta) = 0$ for all $\beta \geq \alpha$.

A function f satisfying i), ii), and iii) is said to be admissible. The existence theorem says that if G is a finite valuated p-group with countable values, and f is an admissible function such that $f_G(\alpha) \leq f(\alpha)$ for all α, then G can be embedded in a countable reduced p-group with Ulm invariants $f(\alpha)$.

The constructive theory of countable p-groups may be extended to include Warfield's KT-modules [24]. Valuated modules play an even more conspicuous role in this theory. A countable KT-module is a countable Z_p-module which contains a nice free valuated Z_p-submodule with torsion quotient. The fact that finitely generated submodules are not finite complicates the decision procedures and focuses attention on those valuated modules that are presented as the cokernel of a map between finite rank free valuated modules [18; Section 2]. The fine existence theorem here [18; Theorem 3] tells when such a valuated module can be embedded in a countable KT-module with specified invariants, and is analogous to the countable p-group case.

The techniques developed in [17] and [18] of growing countable groups out of finitely generated valuated groups, to meet the demands of a constructive theory, found application in an existence theorem [13; Theorem 12] which directly generalizes the fine existence theorem of Rotman and Yen [22; Theorem 3] and is the basic construction for Warfield groups (T*-modules [24] or summands of simply presented Z_p-modules [25]). A Warfield group is a Z_p-module which contains a nice direct sum of cyclic valuated Z_p-modules with a totally projective quotient. The fundamental existence theorem for countable Warfield groups tells under what circumstances a countable direct sum H of cyclic valuated Z_p-modules can be embedded as a nice valuated subgroup of a countable reduced group G, with specified Ulm invariants relative to H, such that G/H is torsion. Whereas KT-modules are all direct sums of modules of torsion-free rank one, there are Warfield modules of countably infinite torsion-free rank such that any summand is either finite or of finite index [13; Example after Theorem 12]. It is this richer structure, coupled with the fact that the countable case is at the heart of the problem, that makes the constructive, valuated groups approach so suitable. This existence theorem is formulated in terms of invariants of valuated groups which are the right derived functors of the Ulm invariants [13; Section 3].

A fine existence theorem for (global) valuated groups that can be embedded in finitely generated groups is proven in [21; Theorem 32] which says that the group can be chosen to have the same Ulm invariants and torsion-free rank as the embedded valuated group. The proof generalizes and uses the corresponding theorem for finite p-groups which is a special case of the constructive Zippin's theorem [17; Theorem 12].

We end with the following, somewhat isolated, example of the influence of the constructive program on the development of valuated groups. The problem of computing heights in Tor(A, B), if you can compute them in A and in B, suggests looking at Tor(S, T) where S and T are finite subgroups of A and B valuated by the height functions on A and B. You then want to recover the valuation on Tor(S, T), induced by the height function on Tor(A, B), purely in terms of the valuations on S and T. This leads to the definition of a valuation on Tor(S, T) by $\mathrm{Tor}(S, T)(\alpha) = \mathrm{Tor}(S(\alpha), T(\alpha))$. A refinement of the usual identity relating Tor and intersection shows that this is the desired valuation [19].

3. <u>Categorical Properties</u>. The category V_p of valuated Z_p-modules is pre-abelian in the sense that it is additive and that every map has a kernel

(the usual one with the induced valuation) and a cokernel (the usual one with the coinduced valuation). The general theory of Ext in pre-abelian categories [20] was developed to determine what Ext(C, A) should be in V_p. One certainly wants to look at sequences

$$A \xrightarrow{f} B \xrightarrow{g} C$$

which are exact in the sense that $f = \ker g$ and $g = \operatorname{coker} f$. However such sequences do not remain exact under the pushout and pullback operations that make Ext a functor and a group. In any pre-abelian category there is a largest class of exact sequences that do remain exact, and these constitute the elements of Ext(C, A). For V_p these are the exact sequences such that every element of C comes from an element of B of the same value [21; Theorem 6]. An equivalent condition is that f(A) be nice in B. With this definition of Ext one can do homological algebra in a pre-abelian category in much the same way as in an abelian one.

Free valuated Z_p-modules may be thought of as images of the adjoint of the forgetful functor from V_p to "valuated sets." Clearly they are projective in V_p since you need only test them on nice sequences, and there are enough of them since they can be mapped nicely onto anything. Moreover they are closed under summands, by general nonsense [23; Theorem 4], since rank-one frees have local endomorphism rings. Thus they are precisely the projectives in V_p. The injectives in V_p are the algebraically compact groups (valuated by their height functions) [21; Theorem 9].

An exact sequence $A \subseteq B \to B/A$ of p-groups is in Ext(B/A, A) in V_p if and only if it is balanced, that is, A is isotype and nice in B. Thus the restriction of Ext from V_p to the subcategory of p-groups yields the relative homological algebra of total purity. We would therefore expect to be able to characterize totally projective groups as objects in V_p. This is most conveniently done using the fact that a p-group is totally projective if and only if it has a nice composition series [7; Theorem 82.3], a notion that generalizes immediately to V_p. A valuated p-group A has a <u>nice</u>

<u>composition series</u> if it admits a well-ordered ascending chain of nice subgroups N_λ such that

1) $N_0 = 0$,
2) $\bigcup N_\lambda = A$,
3) $|N_{\lambda+1} : N_\lambda| = p$,
4) $N_\lambda = \bigcup_{\alpha<\lambda} N_\alpha$ for λ a limit ordinal.

These are precisely the valuated p-groups of projective dimension one [21; Theorem 13].

Every $A \in V_p$ can be embedded nicely in a group B so that B/A is totally projective [21; Theorem 1], so if $\dim A \geq 1$, then $\dim A = \dim B$. Moreover, if B is a p-group, then $\dim B$ in V_p will be one more than the dimension of B in the category of p-groups relative to total purity (balanced exact sequences). Call the latter relative theory B_p and let tV_p be the category of valuated p-groups. Then the global dimension of tV_p is one more than the global dimension of B_p. This allows us to settle the question whether $\dim B_p \leq 1$, that is, whether isotype nice subgroups of totally projective groups are totally projective, by showing that $\dim tV_p > 2$. In fact there exists a p-bounded valuated group of dimension greater than two [21; Theorem 19]. Such an example must have cardinality at least $\aleph_2$ by [21; Theorem 16] which says that every $A \in tV_p$ of cardinality not exceeding $\aleph_n$ has dimension not exceeding $n + 1$.

We end this section with a brief look at the global theory and its relation to the stacked bases theorem of Cohen and Gluck [6]. Suppose F is a free group and K is a subgroup of F such that F/K is torsion. Consider the following three possibilities:

1) F/K is a direct sum of cyclic groups.
2) F has a basis $\{x_i\}$ so that $\{n_i x_i\}$ is a basis for K.
3) K is a free valuated group (under the valuation induced by F).

By a <u>free</u> (global) <u>valuated group</u> K we mean a direct sum of infinite cyclic valuated groups such that $v_p(px) = v_p(x) + 1$ for each $x \in K$ and prime p. Clearly 2) implies 3) and 1). It is easy to show that 3) implies 2) by simply dividing each element of the basis of K by as large an integer as possible, and verifying that the result is a basis for F. That 1) implies 2) is the stacked bases theorem, the only difficult implication.

We may use the stacked bases theorem to show that summands of type zero free valuated groups are free [21; Theorem 26]. By <u>type zero</u> we mean that if $x \neq 0$, then $v_p x < \omega$ for all primes p and $v_p x = 0$ for almost all primes p. Conversely, a routine application of Schanuel's trick derives the stacked bases theorem from the fact that summands of type zero frees are free. Thus the essence of the stacked bases theorem is that summands of type zero free valuated groups are free. An abstract nonsense theorem like this should have an abstract nonsense proof. So far we still rely on Cohen and Gluck.

The general problem of whether summands of frees are free may be reduced to the case where all values are finite and all cyclic summands are of the same type, that is, can be embedded in the same rank-one torsion-free group. The reduction is effected through a slight generalization of Charles' notion of a semi-rigid system [5; Theorem 2.13] which was developed to treat completely decomposable torsion-free groups. This class of groups is clearly intimately related to free valuated groups. There should be some generalization of the stacked bases theorem, with free groups replaced by completely decomposable groups, which would imply that summands of free valuated groups are free. Half the battle here will be finding the right theorem.

4. <u>Finite Valuated Groups</u>. Consider the problem of describing how a subgroup A sits inside a group B. Part of the information is contained in the valuations v_p on A that are induced by the p-height functions on B. If B is finitely generated, then the valuated group A is a complete description provided you agree that A sits in $B \oplus K$ in the same way that A sits in B alone.

To make this precise let C be the category of pairs of finitely generated groups $A_0 \subseteq A_1$ with a morphism from $A_0 \subseteq A_1$ to $B_0 \subseteq B_1$ being a homomorphism from A_0 to B_0 that can be extended to a homomorphism from A_1 to B_1. Then C is equivalent to the full exact subcategory F of valuated groups that can be embedded in finitely generated groups. To go from a valuated group A_0 to a pair $A_0 \subseteq A_1$ we let A_1 be an injective envelope of A_0. The injectives in F are precisely the groups, the injective envelope being a group with the same Ulm invariants and torsion-free rank (the fine existence theorem) [21; Theorem 32]. The projectives of F are the free valuated groups in F, and there are enough [21; Theorem 31].

In particular, to study the category F_p of finite valuated p-groups is to study subgroups of finite p-groups. The change in point of view allows us to bring much machinery and many insights to the latter subject. For example, the endomorphism ring of an object in F_p is finite and so, since a finite ring with no nontrivial idempotents is local, the Krull-Schmidt theorem holds [23; Theorem 4] and every object in F_p is uniquely a sum of indecomposable objects.

The simplest indecomposable objects in F_p are the cyclics. Functorial invariants $f(\mu, A)$ can be defined for $A \in F_p$ that measure the number of cyclics of isomorphism type μ in A if A is a direct sum of cyclics [14; Lemma 1]. Moreover $\sum_{\mu} f(\mu, A) \geq \text{rank } A$ with equality holding if and only if A is a direct sum of cyclics [14; Lemma 3 and Theorem 3]. If $A \in F_p$ is indecomposable and $p^2A = 0$, then A is cyclic [14; Theorem 4]. However it is easy to construct indecomposable $A \in F_p$ such that $p^3A = 0$ and A is not cyclic [14; after Lemma 2].

In analogy with the theory of abelian p-groups, the natural generalization of a direct sum of cyclics in F_p is a simply presented valuated p-group. These are constructed from valuated trees, that is, valuated sets X that admit multiplication by p and have an element 0 satisfying:

1) $p0 = 0$;

2) for each x in X there is n such that $p^n x = 0$;

3) $vpx > vx$ for all x in X.

Given a valuated tree X we get a valuated group $S(X)$ by using X as a set of generators and relations, and setting

$$v(\sum n_x x) = \min\{vx : n_x \neq 0\}$$

where $0 \leq n_x < p$. This gives a functor S from the category of trees (with the obvious morphisms) to the category of simply presented valuated p-groups.

Any nontrivial retraction of a valuated tree X induces a decomposition of $S(X)$ into simply presented valuated p-groups [15; Lemma 1]. If X does not admit a nontrivial retraction, then $S(X)$ is indecomposable [15; Theorem 7]. Thus every finite simply presented valuated p-group is a direct sum of indecomposable ones. Moreover, to each indecomposable finite simply presented valuated p-group G, there is a unique valuated tree X such that $G \cong S(X)$ [15; Theorem 6]. Infinite trees do not add to the supply of indecomposables since every infinite (height reduced) valuated tree has a nontrivial retraction [15; Theorem 11].

REFERENCES

1. Bishop, E., Foundations of constructive analysis, McGraw-Hill, 1967.

2. Brown, R., Valued vector spaces of countable dimension, Publ. Math. Debrecen 18(1971) 149-151.

3. Charles, B., Etude des groups abéliens primaires de type ≤ ω, Ann. Univ. Saraviensis, IV, 3(1955) 184-199.

4. ________, Sous-groupes de base des groupes abéliens primaires, Séminaire Dubreil-Pisot, 13e année, 1959/60, n° 17.

5. ________, Sous-groupes fonctoriels et topologies, Studies on abelian groups, Dunod, Paris 1968, 75-92.

6. Cohen, Joel M., and Herman Gluck, Stacked bases for modules over principal ideal domains, J. Algebra 14(1970) 493-505.

7. Fuchs, L., Infinite abelian groups, Volumes I & II, Academic Press 1970 and 1973.

8. _________, On $p^{\omega+n}$-projective p-groups, Publ. Math. Debrecen (to appear)

9. _________, Vector spaces with valuations, J. Algebra 35(1975) 23-38.

10. Fuchs, L., and J. M. Irwin, On $p^{\omega+1}$-projective p-groups, Proc. London Math. Soc., 30(1975) 459-470.

11. Hill, Paul, Certain pure subgroups of primary groups, Topics in abelian groups, Irwin & Walker eds. Scott, Foresman 1963, 311-314.

12. Hill, P., and C. Megibben, On direct sums of countable groups and generalizations, Studies on abelian groups, Dunod, Paris 1968, 183-206.

13. Hunter, R., F. Richman, and E. A. Walker, Existence theorems for Warfield groups, Trans. Amer. Math. Soc. (to appear).

14. _________, Finite direct sums of cyclic valuated p-groups, Pac. J. Math. 69 (1977) 97-104.

15. _________, Simply presented valuated p-groups, J. of Algebra (to appear).

16. Richman, F., Extensions of p-bounded groups, Arch. der Math. 21(1970) 449-454.

17. _________, The constructive theory of countable abelian p-groups, Pac. J. Math. 45(1973) 621-637.

18. _________, The constructive theory of KT-modules, Pac. J. Math. 61(1975) 621-637.

19. _________, Computing heights in Tor, Houston J. Math. (to appear).

20. Richman, F., and E. A. Walker, Ext in pre-abelian categories, Pac. J. Math. (to appear).

21. _________, Valuated groups, Trans. Amer. Math. Soc. (to appear).

22. Rotman, J., and Ti Yen, Modules over a complete discrete valuation ring, Trans. Amer. Math. Soc. 98(1961) 242-254.

23. Walker, C. L., and R. B. Warfield, Jr., Unique decomposition and isomorphic refinement theorems in additive categories, J. Pure and Appl. Algebra 7(1976) 347-359.

24. Warfield, R. B. Jr., Classification theorems for p-groups and modules over a discrete valuation ring, Bull. Amer. Math. Soc. 78(1972) 88-92.

25. _________, Classification theory of abelian groups, II: Local theory, (to appear).

WARFIELD MODULES

Roger Hunter, Fred Richman[1] and Elbert Walker[1]

1. Introduction. Fix a prime p and denote the ring of integers localized at p by Z_p. Throughout, all modules will be Z_p-modules. Of course, our use of Z_p-modules is just a convenient device for dealing with p-local abelian groups, that is, abelian groups for which multiplication by each prime $q \neq p$ is an automorphism.

In [15], Warfield studied the class of modules which arise as summands of simply presented modules. This paper is a survey of the existing theory of such modules (which we call Warfield modules) together with a number of new results. The classification of these modules in terms of numerical invariants represents the most recently completed stage in a natural progression which began with the work of Ulm and Zippin on countable p-groups and went through successive generalizations to direct sums of countable p-groups and simply presented (also called totally projective) p-groups.

The central theme throughout is the notion of height - both the Ulm and Warfield invariants are defined in terms of height, and height concepts are at the heart of nearly all the proofs. For this reason, we have taken the valuated viewpoint, treating height as an entity apart from a module, thereby separating and emphasizing its role. We have also taken a different approach from that of Warfield in a number of other respects and because of this, provide new proofs of many known theorems. Perhaps the most essential difference is our starting point - Warfield modules are defined as extensions, in the category of valuated modules, of direct sums of cyclics by simply presented torsion modules. This definition suggests that the theory of such modules will include both the theory of completely decomposable torsion free modules (these correspond to special direct sums of cyclic valuated modules) and the rather extensive theory of simply presented torsion modules. This is indeed the case. There is an isomorphism theorem which tells us, in terms of numerical invariants, when two Warfield modules are

[1]These authors were supported by NSF-MPS 71-02773-A04.

ismomorphic, and an existence theorem giving necessary and sufficient conditions for the existence of a Warfield module with prescribed numerical invariants. These two results are the main feature, and in combination enable us to prove many decomposability properties.

In many other respects, the theory of Warfield modules parallels the theory of simply presented torsion modules. Aside from the reasons already given, this is to be expected in view of the fact that Warfield modules arose by examining the various characterizations of simply presented torsion modules and trying to generalize them in some way. The most obvious approach, and indeed, the one used by Warfield, is to drop the torsion requirement from the definition of a simply presented torsion module. The problem is that, in general, summands of simply presented modules with elements of infinite order are not simply presented, and closure under taking summands is responsible for many of the nice properties of the class of simply presented torsion modules. Since Warfield included summands of simply presented modules in his study, this closure was assured and the resulting class of modules turned out to be the correct generalization.

The dominant 'torsion free' feature of Warfield modules is their possession of a nice decomposition basis, and a result of this is that the pathology of torsion free groups is avoided. The need for a nice decomposition basis rather than just a decomposition basis allows the extension of maps from the basis to the containing module (of course, other conditions must also be satisfied).

With this in mind, we have included a fairly close analysis of the conditions which force a decomposition basis to be nice, or have a nice subordinate. An important technique for dealing with modules having a decomposition basis is the construction of categories whose objects are modules and whose morphisms, in effect, ignore torsion. A Krull-Schmidt theorem is then proved in such a category and translated into a theorem about modules. Two categories of this type are the category $\mathcal{C}$, discussed in detail in this paper, and the category $\mathcal{H}$ which appears in [15]. The reason for introducing the category $\mathcal{H}$ in [15] is to prove that a summand of a simply presented module has a nice decomposition basis. Until this is established, very little progress can be made, and, in particular, the isomorphism theorem cannot be proved. On the other hand, our definition of a Warfield module gives us the nice decomposition basis for free and the isomorphism theorem follows directly. Of course, the difficulty is then to show closure under summands, and in particular that summands of simply presented modules are exactly our valuated extensions. The category C is introduced for just that purpose.

A brief outline of the paper follows. The word 'module' alone as opposed to 'valuated module' will always mean a valuated module with the height valuation (see Section 2). This convention will be strictly observed and should be kept in mind at all times.

Section 2 outlines the basic concepts of valuated trees and modules, all of which have been discussed previously in [4], [5] and [7]. In Section 3, Ulm and derived Ulm invariants are defined and the relationship between them explored in detail. As the name suggests, the derived Ulm invariants arise from the right derived functor associated with the Ulm invariants. Although derived Ulm invariants vanish for modules, they are required in the proofs of the existence theorem [4] and the various decomposition theorems which follow from it.

Section 4 records some results concerning valuated modules with composition series; these are just the torsion valuated modules with homological dimension one in the category of valuated modules. The modules with composition series are the simply presented torsion modules and we state the well-known isomorphism and existence theorems, due collectively to Hill [3], and Crawley and Hales [1] , which will be extended to Warfield modules in Sections 7 and 10 respectively.

In Section 5 we extend to valuated modules an invariant defined by Warfield [15] for modules. The treatment here is a generalization of that given by Stanton [14]. This invariant (which we call the Warfield invariant) gives the essential information we require about the torsion free structure of Warfield modules and it is shown that if a valuated module has a decomposition basis, then that basis determines this invariant.

Section 6 contains the proof of a set theoretic lemma which is used to 'juggle' relative Ulm invariants in the proof of the isomorphism theorem. A special case of this lemma has appeared in the literature in various forms and references are given.

In Section 7, Warfield modules are defined as valuated extensions and the isomorphism theorem is proved. Also included is a generalization of this theorem due to Stanton [12]. Sections 8 and 9 are in the main directed toward showing that our Warfield modules are indeed the summands of simply presented modules. Section 8 introduces the category C which was discovered by Elbert Walker. A Krull-Schmidt theorem in C is shown to follow from the very general results of C. Walker and Warfield [13]. Some additional properties of this category are also given. In particular, it is shown that two valuated modules A and B are C-isomorphic if and only if there are torsion valuated modules S and T so that $A \oplus S$ and $B \oplus T$ are isomorphic as valuated modules. Thus C provides the correct categorical setting for the notion of almost - isomorphism introduced by Rotman and Yen [6]. For further discussion and applications of C to the theory of mixed groups, see the papers of Warfield and C. Walker in these proceedings.

In Section 9 it is shown that a summand of a Warfield module is Warfield and that, given a Warfield module A, there is a simply presented torsion module B so that $A \oplus B$ is simply presented. It follows from these two results that the Warfield modules as defined in Section 7 are indeed the summands of direct sums of simply presented modules studied by Warfield in [15].

Section 10 deals with the existence and decomposition theorems for Warfield modules. Necessary and sufficient conditions for the existence of a Warfield module with prescribed Ulm and Warfield invariants were given in [4], and this and other results and examples from that paper are stated. The question of when a Warfield module is a direct sum of rank one modules is examined and a fairly satisfactory answer in terms of invariants is found. This is used to generalize a result of Rotman [9].

Section 11 is a study of decomposition bases and in particular, answers the question of whether a given decomposition basis necessarily has a nice subordinate. For example, it is shown that every countable decomposition basis has a nice subordinate, and this allows us to obtain the results of Warfield [15] concerning countable modules with decomposition bases in a more direct fashion. An example of a module with a decomposition basis which has no nice subordinate is provided, and a result of Warfield [15,Lemma 6] is applied to show that such a module cannot have a nice decomposition basis.

Projective characterizations are discussed in Section 12, and the class of Warfield modules known as balanced projectives or KT-modules, is scrutinized. An important result here is that any decomposition basis X of a balanced projective with the property that every element x of X has no gaps in its value sequence (such a basis is called a K-basis in [19]) generates a submodule $\langle X \rangle$ of A such that $A/\langle X \rangle$ is simply presented. This is proved by first showing that a balanced projective has homological dimension at most one in the category of valuated modules.

In Section 13 we examine the torsion submodule of a Warfield module. It is shown that a Warfield module A can be decomposed $A = B \oplus C$ so that B is balanced projective and the torsion submodule of C is simply presented. In particular, the torsion submodule of A is an S-module in the sense of [18], and that part of the torsion submodule of A which is not simply presented lies in a summand (B above) which is simply presented and has a particularly simple structure. Thus there is a trade-off between the complexity of the torsion part of a Warfield module and the complexity of its torsion free structure.

We remark that [20] is a revised version of [15], and also contains a discussion of extensions of the theory of Warfield modules to a global setting. Another more recent discussion is also given in the paper of Warfield which appears in these proceedings.

2. <u>Valuated trees and modules</u>. A <u>tree</u> is a set X, with a distinguished element 0, that admits a multiplication by p satisfying:

1) $p0 = 0$;

2) $p^n x = x$ only if $n = 0$ or $x = 0$.

For a tree X and ordinal α, the subset $p^\alpha X$ is defined inductively by setting $p^0 X = X$, $pX = \{px : x \in X\}$ and

$$p^{\alpha}X = \bigcap_{\beta<\alpha} p(p^{\beta}X)$$

when $\alpha > 0$. The height hx of an element x in X is α if $x \in p^{\alpha}X \backslash p^{\alpha+1}X$. If $x \in p^{\alpha}X$ for all ordinals α, then we set $hx = \infty$. The symbol ∞ satisfies $\infty > \infty$ and $\infty > \alpha$ for all ordinals α.

By a valuated tree we mean a tree X together with a function v on X (called a valuation) such that:

1) vx is an ordinal or ∞;

2) $vpx > vx$.

A valuated tree X is reduced if $vx = \infty$ implies $x = 0$. The height function h is clearly a valuation satisfying $hx \leq vx$ for all x in X. Any tree is naturally valuated by setting $vx = hx$ for all x. Conversely, if $vx = hx$ for all x, we say that the valuated tree X is a tree. A map $f:X \to Y$ of valuated trees is a function such that:

1) $f(px) = pf(x)$;

2) $vf(x) \geq vx$.

The resulting category is called the category of valuated trees. An embedding of valuated trees is a one-to-one map f such that $vf(x) = vx$ for all x. If the inclusion $X \subset Y$ is an embedding, we say that X is a valuated subtree of Y. If X is a valuated tree, and α is an ordinal, we set

$$X(\alpha) = \{x \in X: vx \geq \alpha\}.$$

A valuated module is a module that is a valuated tree and satisfies

$$v(x + y) \geq \min(vx,vy).$$

It follows that $vnx = vx$ if p does not divide n. A map of valuated modules is a module homomorphism that is a map of valuated trees. A valuated module which is a tree is called a module. The valuated modules form a category V_p. An embedding $f : A \to B$ of valuated modules is a map that is an embedding of valuated trees. If the inclusion map $A \to B$ is an embedding, we say that A is a valuated submodule of B.

The category V_p is pre-abelian (additive with kernels and cokernels), and the theory of [8] provides a natural definition of Ext. In [7] it is shown that a sequence

$$0 \to A \to B \to C \to 0$$

of valuated modules is in $\mathrm{Ext}(C, A)$ if and only if

$$0 \to A(\alpha) \to B(\alpha) \to C(\alpha) \to 0$$

is an exact sequence of modules for each α. If $A \subset B$ is an inclusion of valuated modules, then an element b of B is A-proper if b has maximal value among the elements in the coset $b + A$. In case each coset of A contains an element of maximal value, A is said to be nice. It is not difficult to see that $0 \to A \to B \to C \to 0$ is in $\mathrm{Ext}(C,A)$ if and only if the inclusion $A \subset B$ is a nice embedding.

We associate with each valuated tree X a valuated module $S(X)$ in the following way. Let $F_X = \oplus Z_p(x)$ be the free Z_p-module on the nonzero elements of X. Let R_X be the submodule of F_X generated by the elements of the form

$$p(x) \qquad \text{where } px = 0, \text{ and}$$

$$p(x) - (px) \qquad \text{where } px \neq 0,$$

and set $S(X) = F_X/R_X$. Each element s of $S(X)$ can be written in the form $s = \sum u_i x_i$ where the u_i's are units in Z_p. Setting $vs = \min\{v_i x_i\}$ makes $S(X)$ into a valuated module. This valuated module is called the simply presented valuated module on X. If a valuated module A is isomorphic to $S(X)$, then the image of X in A is said to be a spanning tree for A. The notion of a spanning tree coincides with that of a T-basis or a standard presentation (see [1],[2],[15]). The usual definition of a simply presented module is a module that can be defined in terms of generators and relations so that the only relations are of the form $px = 0$ or $px = y$. However, it is straightforward to prove (see, for example [20,Lemma 2.1]) that such a module has a spanning tree, so there is no conflict. The rank of $A \in V_p$ is the dimension of the vector space $Q \otimes A$ over the field of rational numbers Q.

Lemma 1. A simply presented module is a direct sum of modules of rank at most one.

Proof. Let X be a spanning tree of A. Define $x,y \in X$ to be equivalent if there are positive integers m and n such that $p^m x = p^n y$, and let $\{X_i\}$ be the set of equivalence classes so obtained. Then it is easy to see that $A = \oplus S(X_i)$.

Observe that the functor S is the adjoint of the forgetful functor from V_p to the category of valuated trees. If $X \subseteq Y$ is an embedding of valuated trees, and Y/X is the valuated tree gotten from Y by identifying X with zero, then

$$0 \to S(X) \to S(Y) \to S(Y/X) \to 0$$

is exact in V_p.

Let $\{A_i, i \in I\}$ be a family of valuated modules such that A_i has valuation v_i. Then the direct sum (coproduct) of the A_i's in V_p is readily seen to be the module direct sum with valuation $va = \min\{v_i a_i\}$. If X is a subset of a valuated module A, then $\langle X \rangle$ denotes the valuated submodule of A generated by X. A valuated module is <u>cyclic</u> if it is cyclic as a module. A <u>basis</u> of a direct sum of valuated cyclics A is a Z_p-independent subset X of A such that A is the direct sum of $\{\langle x \rangle : x \in X\}$. If X and Y are bases for A, we say that X is <u>subordinate</u> to Y if every element of X is a multiple of an element of Y. A valuated module A is <u>free</u> if A is a direct sum of cyclics of infinite order with basis X such that $vp^n x = vx + n$ for $n = 1, 2, \ldots$.

The <u>value sequence of an element</u> a of a valuated module is defined by

$$V(a) = va, vpa, vp^2 a, \ldots .$$

A <u>value sequence</u> is a sequence $\alpha_0, \alpha_1, \ldots$ of ordinals and symbols ∞ satisfying $\alpha_i < \alpha_{i+1}$ for $n = 0, 1, \ldots$. Clearly the value sequence of an element is a value sequence. If $\underline{\alpha} = \alpha_0, \alpha_1, \ldots$ is a value sequence, we write $p\underline{\alpha} = \alpha_1, \alpha_2, \ldots$ and if $\underline{\beta} = \beta_0, \beta_1, \ldots$ we write $\underline{\alpha} \geq \underline{\beta}$ in case $\alpha_n \geq \beta_n$ for $n = 0, 1, \ldots$ Value sequences $\underline{\alpha}$ and $\underline{\beta}$ are said to be <u>equivalent</u> if there are positive integers m and n such that $p^m \underline{\alpha} = p^n \underline{\beta}$. Observe that a cyclic valuated module A is determined up to isomorphism by the value sequence of any generator, and the value sequences of the elements of A all lie in the same equivalence class.

If A has rank one then the value sequences of the elements of infinite order in A all lie in the same equivalence class.

Each submodule A of a module B is a valuated module with valuation given by restricting the height function on B. This makes A a valuated submodule of B. Conversely, each $A \in V_p$ can be obtained this way:

__Theorem__ 2. There is a functor $T : V_p \to V_p$ such that, for each $A \in V_p$,

1) TA is a module containing A,

2) $A \subseteq TA$ is a nice embedding,

3) TA/A is a simply presented torsion module.

The proof appears in [7].

3. __Ulm invariants.__ Let A be a valuated module and α an ordinal. Multiplication by p induces a natural map

$$A(\alpha)/A(\alpha + 1) \to A(\alpha + 1)/A(\alpha + 2).$$

We denote the kernel and cokernel of this map by $F_A(\alpha)$ and $G_A(\alpha + 1)$ respectively. Both $F_A(\alpha)$ and $G_A(\alpha + 1)$ are vector spaces over the p-element field. Clearly

$$F_A(\alpha) = \frac{\{a \in A(\alpha) : pa \in A(\alpha + 2)\}}{A(\alpha + 1)}$$

and

$$G_A(\alpha + 1) = \frac{A(\alpha + 1)}{A(\alpha + 2) + pA(\alpha)} .$$

The appropriate definition of $G_A(\alpha)$ in general seems to be

$$G_A(\alpha) = \frac{A(\alpha)}{\bigcap_{\beta<\alpha} (A(\alpha + 1) + A(\alpha) \cap pA(\beta))} .$$

This definition agrees with the other for non-limit ordinals. The dimensions of $F_A(\alpha)$ and $G_A(\alpha)$ will be denoted $f_A(\alpha)$ and $g_A(\alpha)$, respectively. We call $f_A(\alpha)$ the α- __th__ __Ulm__ __invariant__ of A and $g_A(\alpha)$ the α- __th__ __derived__ __Ulm__ invariant of A. If A is a module, $f_A(\alpha)$ agrees with the usual definition of Ulm invariant while $g_A(\alpha)$ vanishes for all α. On the other hand, there do exist valuated modules A which do not have the height valuation, and for which $g_A(\alpha) = 0$ for all ordinals α - for example, the valuated Z_p submodules of the p-adic integers generated by 1 and an irrational.

We now examine the connection between these two invariants. Let $0 \to A \to B \to C \to 0$ be exact in V_p. Then for each α,

$$0 \to \frac{A(\alpha)}{A(\alpha + 1)} \to \frac{B(\alpha)}{B(\alpha + 1)} \to \frac{C(\alpha)}{C(\alpha + 1)} \to 0$$

is an exact sequence of vector spaces. We arrive at the following commutative diagram:

$$\begin{array}{ccccccccc}
0 & \to & F_A(\alpha) & \to & F_B(\alpha) & \to & F_C(\alpha) & & \\
 & & \downarrow & & \downarrow & & \downarrow & & \\
0 & \to & \frac{A(\alpha)}{A(\alpha + 1)} & \to & \frac{B(\alpha)}{B(\alpha + 1)} & \to & \frac{C(\alpha)}{C(\alpha + 1)} & \to & 0 \\
 & & \downarrow & & \downarrow & & \downarrow & & \\
0 & \to & \frac{A(\alpha + 1)}{A(\alpha + 2)} & \to & \frac{B(\alpha + 1)}{B(\alpha + 2)} & \to & \frac{C(\alpha + 1)}{C(\alpha + 2)} & \to & 0 \\
 & & \downarrow & & \downarrow & & \downarrow & & \\
 & & G_A(\alpha + 1) & \to & G_B(\alpha + 1) & \to & G_C(\alpha + 1) & \to & 0.
\end{array}$$

The snake lemma applied to this diagram gives the exact sequence

$$0 \to F_A(\alpha) \to F_B(\alpha) \to F_C(\alpha) \to G_A(\alpha + 1) \to G_B(\alpha + 1) \to G_C(\alpha + 1) \to 0$$

This is the fundamental sequence relating the F's and the G's. Now F_A is zero when A is projective (=free) in V_p, and G is zero on modules. Furthermore, there are enough projectives in V_p (see [7]), and by Theorem 2 each $A \in V_p$ has a nice embedding in a module. Using this it is easy to show that $F_A(\alpha)$ is the left derived functor of $G_A(\alpha + 1)$, and that $G_A(\alpha + 1)$ is the right derived functor of $F_A(\alpha)$.

For an arbitrary embedding $A \subset B$ (that is, we do not insist that A be nice in B) the embedding $\phi : F_A(\alpha) \to F_B(\alpha)$ is still defined. The cokernel of ϕ is denoted $F_{B,A}(\alpha)$ and its dimension $f_{B,A}(\alpha)$ is called the α- th Ulm invariant _of_ B _relative_ _to_ A. It is straightforward to check that if B is a module, then this definition agrees with the usual one. Notice that

$$F_B(\alpha) \simeq F_A(\alpha) \oplus F_{B,A}(\alpha).$$

There are three further relations of interest. Let A be a nice valuated submodule of the module B. Then $G_B(\alpha + 1) = 0$, so the exact sequence

$$0 \to F_{B,A}(\alpha) \to F_{B/A}(\alpha) \to G_A(\alpha + 1) \to 0$$

gives

$$F_{B/A}(\alpha) \simeq F_{B,A}(\alpha) \oplus G_A(\alpha + 1).$$

Taking the direct sum with $F_A(\alpha)$, and using the previous isomorphism, we have

$$F_{B/A}(\alpha) \oplus F_A(\alpha) \simeq F_B(\alpha) \oplus G_A(\alpha + 1).$$

Finally, if $A \oplus B$ is a valuated submodule of C, then composing the isomorphisms $F_A(\alpha) \oplus F_{C,A}(\alpha) \simeq F_C(\alpha)$ and $F_C(\alpha) \simeq F_{A \oplus B}(\alpha) \oplus F_{C,A \oplus B}(\alpha) \simeq F_A(\alpha) \oplus F_B(\alpha) \oplus F_{C,A \oplus B}(\alpha)$ provides an isomorphism $F_A(\alpha) \oplus F_{C,A}(\alpha) \simeq F_A(\alpha) \oplus F_B(\alpha) \oplus F_{C,A \oplus B}(\alpha)$. It is readily checked that the latter isomorphism takes $F_A(\alpha)$ to $F_A(\alpha)$, whence

$$F_{C,A}(\alpha) \simeq F_B(\alpha) \oplus F_{C,A \oplus B}(\alpha).$$

Our exclusion of ∞ to this point has been a device to simplify discussion, and is not rooted in any fundamental difference. Indeed, if we define $F_A(\infty)$ and $G_A(\infty)$ to be the kernel and cokernel, respectively, of the natural map $A(\infty) \overset{p}{\to} A(\infty)$, then with the obvious modifications to the proofs, all the results of this section can be seen to hold when $\alpha = \infty$. Of course, the convention $\infty + 1 = \infty$ must be observed. In particular, the fundamental sequence becomes

$$0 \to F_A(\infty) \to F_B(\infty) \to F_C(\infty) \to G_A(\infty) \to G_B(\infty) \to G_C(\infty) \to 0.$$

To avoid numerous trivial special cases involving these and other invariants, we assume from this point on that all valuated modules are _reduced_.

4. _Modules with composition series._ A valuated module has a (nice) _composition series_ if it admits a well-ordered ascending chain of nice submodules N_α such that:

1) $N_0 = 0$;
2) $\bigcup N_\alpha = A$;

3) $| N_{\alpha+1} : N_\alpha | = p$;

4) $N_\alpha = \bigcup_{\beta<\alpha} N$ if α is a limit.

If A has a composition series then it is clear that A is torsion. As every countable valuated torsion module has a composition series, we can replace condition 3) by $| N_{\alpha+1} : N_\alpha | \le \aleph_0$ and the requirement that A be torsion. Modules with composition series can be characterized in terms of their homological dimension in V_p:

<u>Theorem</u> 3. (Richman and Walker [7]). A nonzero valuated torsion module has homological dimension one in V_p if and only if it has a composition series.

A module has a composition series if and only if it is simply presented and torsion [2, p99]; we shall be using these two characterizations interchangeably.

Perhaps the most important result concerning modules with composition series was proved by Hill in an unpublished paper [3], and in the following generalized form by Walker [14].

<u>Theorem</u> 4. Let A, B be modules and G, H nice valuated submodules of A and B, respectively, such that A/G and B/H have composition series. Then any (V_p) isomorphism $G \to H$ extends to an isomorphism $A \to B$ if (and only if) $f_{A,G} = f_{B,H}$.

Note that A and B are not required to be torsion. If $G = H = 0$ then A and B are modules with composition series and the theorem says that A is isomorphic to B if and only if $f_A = f_B$. The latter result was also proved independently by Crawley and Hales [1]. The argument of Fuchs [2, p84] yields:

<u>Corollary</u> 5. Let $A \subset B$ be a nice valuated embedding with B a module, and $f : A \to C$ a homomorphism. If B/A has a composition series then f can be extended to a homomorphism $f^* : B \to C$.

Since a module A with a composition series is determined (up to isomorphism) by f_A, it is natural to seek a complement to Theorem 4 which describes those functions which can arise as f_A. This was accomplished by Crawley and Hales. Before stating their theorem, we need some definitions. Let f and g be two functions from ordinals to cardinals which vanish beyond some ordinal. We say that f <u>dominates</u> g if

$$\sum_{\alpha \le \beta < \alpha + \omega} f(\beta) \ge \sum_{\beta \ge \alpha} g(\beta)$$

for all ordinals α. A function which dominates itself is said to be <u>admissible</u>.

<u>Theorem</u> 6. (Crawley and Hales [1]). If A has a composition series, then f_A is admissible. Conversely, if f is admissible then there is a module A with a composition series such that $f_A = f$.

We refer to the version of Theorem 4 in which $G = H = 0$ as the isomorphism theorem for modules with composition series, and to Theorem 6 as the existence theorem for such modules. In what follows, both of these theorems will be extended to Warfield modules.

5. <u>More invariants</u>. It is clear from the definition that Ulm invariants are a property of the torsion submodule of a module. In this section we define a complementary invariant which is dependent on the elements of infinite order. This invariant was first introduced by Warfield [15] for a restricted class of modules, and later generalized by Stanton [11] to include all modules. We make a further extension to valuated modules.

Let A be a valuated module, $\underline{\alpha} = \alpha_0, \alpha_1, \ldots$ a value sequence. Let $A(\underline{\alpha})$ be the valuated submodule of A defined by

$$A(\underline{\alpha}) = \{a \in A : V(a) \ge \underline{\alpha}\},$$

and let $A(\underline{\alpha})^*$ be the submodule of $A(\underline{\alpha})$ generated by those elements a of $A(\underline{\alpha})$ such that $vp^k a \ne \alpha_k$ for infinitely many k. Multiplication by p induces natural maps

$$\frac{A(\underline{\alpha})}{A(\underline{\alpha})^*} \to \frac{A(p\underline{\alpha})}{A(p\underline{\alpha})^*} \to \frac{A(p^2\underline{\alpha})}{A(p^2\underline{\alpha})^*} \quad \cdots,$$

forming a direct system whose limit we denote $W_A(\underline{\alpha})$. It is not difficult to see [11, Lemma 1] that the maps of this system are all one-to-one. When A is a module, these maps are also isomorphisms - this follows from the fact that, if $a \in A$ and $ha = \alpha > 0$, then for each $\beta < \alpha$ there is an element b in A such that $hb \geq \beta$ and $pb = a$. Note that $W_A(\underline{\alpha})$ is a vector space over the p-element field. We call the dimension of $W_A(\underline{\alpha})$ the <u>Warfield invariant</u> of A at $\underline{\alpha}$, and denote it $w_A(\underline{\alpha})$. If A and B are valuated modules whose Warfield invariants are the same for all value sequences, we write $w_A = w_B$. That these invariants provide a measure of the torsion free structure is evident from:

<u>Lemma</u> 7. Let A be a valuated submodule of B. If B/A is torsion then $W_A \cong W_B$. In particular, $W_A = 0$ if A is torsion.

<u>Proof.</u> If $b \in B(p^n\underline{\alpha})$ then $p^m b \in A$ for some m, so $p^m b \in A(p^{m+n}\underline{\alpha})$ represents an element of $W_A(\underline{\alpha})$ if and only if b represents an element of $W_B(\underline{\alpha})$. The lemma now follows easily.

We list three easily verified properties of Warfield invariants.

(a) If $A = \oplus A_i$ is a valuated direct sum, then $w_A(\underline{\alpha}) = \sum w_{A_i}(\underline{\alpha})$.

(b) If A is torsion free cyclic then $w_A(\underline{\alpha}) = 1$ or 0. Further, $w_A(\underline{\alpha}) = 1$ if and only if $\underline{\alpha}$ is equivalent to V(a), where a is a generator of A.

(c) If A is a (valuated) direct sum of torsion free cyclics with basis X, then $w_A(\underline{\alpha})$ is the number of elements in X with value sequence equivalent to $\underline{\alpha}$.

Let A be a valuated submodule of B with B/A torsion. If A is a (valuated) direct sum of torsion free cyclics and X is a basis for A , then X is called a <u>decomposition basis</u> of B. Modules with decomposition bases play a vital role in the sequel, and one of their most important aspects is their behavior with respect to Warfield invariants.

Lemma 8. Let X and Y be decomposition bases for the valuated modules A and B respectively. Then $w_A = w_B$ if and only if there are subordinates X' of X and Y' of Y such that $\langle X' \rangle \cong \langle Y' \rangle$ (as valuated modules).

Proof. Assume that $w_A = w_B$. Using Lemma 7 we have $w_{\langle X \rangle} = w_A = w_B = w_{\langle Y \rangle}$ so there is a bijection $\phi : X \to Y$ such that $V(x)$ is equivalent to $V(\phi(x))$ for each $x \in X$. Now let $X' = \{m_x x : x \in X\}$ and $Y' = \{m_y y : y \in Y\}$, where m_x, m_y are powers of p chosen so that if $\phi(x) = y$ then $V(m_x x) = V(m_y y)$.

The converse follows directly from Lemma 7.

6. A set theoretic lemma. Before starting the description of Warfield modules, we prove a set theoretic lemma which will be needed for 'juggling' invariants. Proofs of this lemma for the case $m = \aleph_0$ can be extracted from the proofs of [15,Lemma 21], and [16, Lemma 11]. The case $m = \aleph_0$ is all that is needed in the proof of the isomorphism theorem for Warfield modules.

Lemma 9. Let $\underline{X}$ be a set, m an infinite cardinal, and $\mathcal{F}$ a family of subsets of $\underline{X}$ such that card $F \leq m$ for each F in $\mathcal{F}$. Then there exists a function $f : \mathcal{F} \to \underline{X}$ such that $f(F) \in F$ for all $F \in \mathcal{F}$ and such that, for each $x \in \underline{X}$,

$$\text{card}\,\{F : f(F) = x\} = \text{card}\{F : x \in F\}$$

whenever the latter is $\geq m$.

Proof. First, each element $x \in \underline{X}$ can be assumed to lie in exactly m members of $\mathcal{F}$. Indeed, if x is in fewer than m members of $\mathcal{F}$, enlarge $\mathcal{F}$ with m sets whose only element is x. If x is in more than m members of $\mathcal{F}$, partition the family of sets F that contain x into subfamilies S of size m and in each $F \in S$, replace x by x_S. Of course, this process changes $\underline{X}$ and $\mathcal{F}$, but after building our function f, we revert to the original $\underline{X}$ and $\mathcal{F}$ in the obvious way and still have a function satisfying the conditions of the lemma.

Define F, $F' \in \mathcal{F}$ to be equivalent if there is a finite chain $F = F_1, F_2, \ldots, F_n = F'$, $F_i \in \mathcal{F}$, such that F_i and F_{i+1} have an element in common for $i = 1, \ldots, n-1$. The equivalence classes produced in this way have size m. We can construct f separately on each equivalence class, and therefore assume that $\mathcal{F}$ itself is a single equivalence class. Let α be the first ordinal such that $\operatorname{card}\alpha = m$ and initially well order $\underline{X}$ with the ordinals $\beta < \alpha$. Using transfinite induction, it is an easy matter to partition $\mathcal{F}$ into subfamilies $\{F_{\beta 0}, F_{\beta 1}, \ldots F_{\beta\beta}\}$, $\beta < \alpha$, so that $x_\beta \in F_{\beta\gamma}$, $\gamma \le \beta$. This can always be done, because at each stage in the induction we have used fewer than m members of $\mathcal{F}$. Now let $f(F_{\beta\gamma}) = x_\beta$.

7. Warfield modules and the isomorphism theorem. A valuated module A is *Warfield* if there is an exact sequence

$$0 \to C \to A \to B \to 0 \qquad (*)$$

in $\mathcal{V}_p$ with C a direct sum of cyclics and B a valuated module with a composition series. We call a sequence of this kind a *representing sequence* for A. Thus a Warfield module is an extension (in $\mathcal{V}_p$) of a direct sum of cyclics by a simply presented torsion valuated module. If A is Warfield and has representing sequence (*), then we can obtain another representing sequence $0 \to C' \to A \to A/C' \to 0$ for A by letting C' be the submodule of C generated by any subordinate of a basis for C. To see this, note that

$$0 \to C/C' \to A/C' \to B \to 0$$

is exact in $\mathcal{V}_p$ and C/C' and B have composition series. Hence A/C' has a composition series as required. In particular, we may as well assume C is torsion free and so a basis for C is a nice decomposition basis for A.

We turn now to the isomorphism theorem for Warfield modules.

Theorem 10. (Warfield [15]) Let A and B be Warfield modules. Then A and B are isomorphic if and only if $f_A = f_B$ and $w_A = w_B$.

Proof. Only sufficiency needs proof. Let X and Y be nice decomposition bases for A and B, respectively, such that $A/<X>$, $B/<Y>$ have composition series. By Theorem 4 it is enough to choose X and Y so that $<X> \simeq <Y>$ and $f_{A,<X>} = f_{B,<Y>}$. Perhaps surprisingly, we can do this by replacing any first choice of X and Y with suitable subordinates. Examining the equation

$$f_A(\alpha) = f_{<X>}(\alpha) + f_{A,<X>}(\alpha)$$

from Section 3, we see that if $f_A(\alpha)$ is infinite, then suitably multiplying the elements of X by powers of p we can replace X by a subordinate such that $f_A(\alpha) = f_{A,<X>}(\alpha)$. The trick is to do this for all α at once, and this is where our set theoretic lemma comes in. Let $\underline{X}$ be the set of α such that $f_A(\alpha) \neq 0$, let $m = \aleph_0$ and let $F = \{F_x : x \in X\}$, where F_x consists of those α for which $f_{<x>}(\alpha) \neq 0$. If f is the function from F to $\underline{X}$ given by Lemma 9 and if $f(F_x) = \alpha$, then we multiply x by p^n so that $f_{<p^n x>}(\alpha) = 0$. Doing the same for Y and B, we have

$$0 \to <X> \to A \to A/<X> \to 0$$

and

$$0 \to <Y> \to B \to B/<Y> \to 0$$

exact in V_p with $f_A(\alpha) = f_{A,<X>}(\alpha) = f_B(\alpha) = f_{B,<Y>}(\alpha)$ whenever $f_A(\alpha)$ is infinite. Note that taking further subordinates of X and Y will not alter these equations. Therefore Lemma 8 allows us to also arrange that $<X> \simeq <Y>$. But this ensures that $f_{A,<X>}(\alpha) = f_{B,<Y>}(\alpha)$ for those α for which $f_A(\alpha)$ is finite. Hence $f_{A,<X>}(\alpha) = f_{B,<Y>}(\alpha)$ for all α. Finally, the remarks preceding this theorem show that our new X and Y are such that $A/<X>$ and $B/<Y>$ have composition series.

Two definitions are needed before we can give the fullest extension (due to Stanton [12]) of Theorem 4 to Warfield modules.

Let A be a valuated submodule of B and $\underline{\alpha}$ a value sequence. If ϕ is the natural map $W_A(\underline{\alpha}) \to W_B(\underline{\alpha})$, then Coker ϕ is denoted $W_{B,A}(\underline{\alpha})$ and its rank is called the Warfield invariant of B relative to A at the value sequence $\underline{\alpha}$. Now let A be a nice valuated submodule of B . Then A is said to be quasi- sequentially nice in B if, for each coset $b + A$, there is an integer n and an element a in A such that $V(p^n b + A) = V(p^n b + a)$.

Theorem 11 . (Stanton[12]) Let $A, B \in V_p$ and let G, H be sequentially nice valuated submodules of A and B , respectively, such that A/G and B/H are Warfield modules. Then any isomorphism $G \to H$ extends to an isomorphism $A \to B$ if (and only if) $f_{A,G} = f_{B,H}$ and $w_{A,G} = w_{B,H}$.

Proof. We use the proof of Theorem 10 as a guide, detailing only those arguments that require the additional hypotheses. Again, only sufficiency need be proved. Let $\{x + G : x \in X\}$ be a nice decomposition basis for A/G . Using the fact that G is quasi-sequentially nice in A and taking subordinates if necessary, we may assume $V(x) = V(x + G)$ for all x in X. Then $v(\sum r_x x + G) = \min\{v(r_x x + G)\} = \min\{vr_x x\}$, and so $v(\sum r_x x + g) = \min\{v\sum r_x x, vg\}$ for all $g \in G$. Hence $\langle X\rangle = \oplus \langle x\rangle$ and $\langle X\rangle \oplus G$ are direct in V_p. Set $K = \langle X\rangle$. As $A/(K \oplus G)$ is torsion, Lemma 7 shows that the vector spaces $W_A(\underline{\alpha})$ and $W_{K \oplus G}(\underline{\alpha})$ are the same, so $W_{A,G}(\underline{\alpha}) = \text{Coker}\left[W_G(\underline{\alpha}) \to W_{K \oplus G}(\underline{\alpha})\right] = W_K(\underline{\alpha})$. It was observed in Section 3 that

$$f_{A,G}(\alpha) = f_K(\alpha) + f_{A,K \oplus G}(\alpha).$$

Arguing with this equation and Lemma 9 as we did in the proof of Theorem 10, we can ensure that $f_{A,G}(\alpha) = f_{A,G \oplus K}(\alpha)$ whenever $f_{A,G}(\alpha)$ is infinite. There is also a corresponding direct sum of cyclics $L \subset B$. We may further arrange that $K \cong L$ and hence that $f_{A,G \oplus K}(\alpha) = f_{B,H \oplus L}(\alpha)$ for all α. Now $G \oplus K \cong H \oplus L$ and we are once again in the situation of Theorem 4.

Of course, setting $G = H = 0$ in the preceding theorem gives Theorem 10.

8. The category C. We interrupt the study of Warfield modules momentarily to introduce a useful category. Let B denote the torsion submodule of the valuated module B. Define C to be the category whose objects are the objects of V_p and

whose morphism sets are

$$\mathrm{Hom}_{C}(A, B) = \mathrm{Hom}_{V_p}(A, B)/\mathrm{Hom}_{V_p}(A, B_t).$$

Note that C is additive with kernels and arbitrary infinite direct sums; the kernel of $f + \mathrm{Hom}_{V_p}(A, B_t) \in \mathrm{Hom}_{C}(A, B)$ is $f^{-1}(B_t)$, while the direct sum in C of the family $\{A_i\}$ is just the direct sum in V_p. The next Theorem characterizes C-isomorphism in terms of the category V_p.

<u>Theorem</u> 12. Let $A, B \in C$. Then A is isomorphic to B in C if and only if there are torsion S and T in V_p such that $A \oplus S$ and $B \oplus T$ are isomorphic as valuated modules.

<u>Proof</u>. Suppose $f : A \oplus S \to B \oplus T$ is an isomorphism in V_p. Let i_A be the injection of A into $A \oplus S$, π_A the projection of $A \oplus S$ onto A; we denote the corresponding injections and projections associated with S, B and T in a similar fashion. We claim that $\pi_B \circ f \circ i_A + \mathrm{Hom}_{V_p}(A, B_t)$ is a C-isomorphism with inverse $\pi_A \circ f^{-1} \circ i_B + \mathrm{Hom}_{V_p}(A, B_t)$. Now

$$\begin{aligned} 1_A &= \pi_A \circ f^{-1} \circ f \circ i_A = \pi_A \circ f^{-1} \circ (i_B\pi_B + i_T\pi_T) \circ f \circ i_A \\ &= \pi_A \circ f^{-1} \circ i_B \circ \pi_B \circ f \circ i_A + \pi_A \circ f^{-1} \circ i_T \circ \pi_T \circ f \circ i_A \\ &= (\pi_A \circ f^{-1} \circ i_B) \circ (\pi_B \circ f \circ i_A) + \pi_A \circ f^{-1} \circ i_T \circ \pi_T \circ f \circ i_A \end{aligned}$$

while $\pi_A \circ f^{-1} \circ i_T \circ \pi_T \circ f \circ i_A \in \mathrm{Hom}_{V_p}(A, A_t)$. Similarly for 1_B.

Conversely, if A is isomorphic to B in C, there are valuated maps $f : A \to B$ and $g : B \to A$ such that

$$g \circ f + \mathrm{Hom}_{V_p}(A, A_t) = 1_A + \mathrm{Hom}_{V_p}(A, A_t)$$

and

$$f \circ g + \mathrm{Hom}_{V_p}(B, B_t) = 1_B + \mathrm{Hom}_{V_p}(B, B_t)$$

Thus there is $\alpha \in \mathrm{Hom}_{V_p}(A, A_t)$ and $\beta \in \mathrm{Hom}_{V_p}(B, B_t)$ with $g \circ f + \alpha = 1_A$ and $f \circ g + \beta = 1_B$. We complete the proof by showing that $A \oplus \mathrm{Im}\beta$ is isomorphic (in V_p) to $B \oplus \mathrm{Im}\alpha$. First observe that composing the last mentioned equations with f yields $f \circ g \circ f + f \circ \alpha = f$ and $f \circ g \circ f + \beta \circ f = f$, respectively, and hence $f \circ \alpha = \beta \circ f$ and $g \circ \beta = \alpha \circ g$. Using these, it is readily checked that the required isomorphism is given by $(a,s) \longmapsto (f(a) + s, \alpha(a) - g(s))$ with inverse given by

$(b,t) \mapsto (g(b) + t, \beta(b) - f(t))$.

We have the following isomorphic refinement theorem in $\mathcal{C}$.

<u>Theorem</u> 13. If A is a summand (in $\mathcal{V}_p$) of a direct sum of rank one modules $\{M_i : i \in I\}$, then A is $\mathcal{C}$-isomorphic to the direct sum of modules $\{M_i : i \in J\}$ for some $J \subset I$.

<u>Proof</u>. Since $\mathcal{C}$ satisfies the conditions of [13,Theorem 5], it suffices to prove that the endomorphism ring in $\mathcal{C}$ of a rank one module is local. Let B have rank one, $b \in B$ an element of infinite order in B, and $f + \mathrm{Hom}_{\mathcal{V}_p}(B,B_t)$ a non-zero $\mathcal{C}$-endomorphism of B. If $pB \neq B$ then there are positive integers m, n such that $p^n f(b) = up^m b$ for some unit u. Obviously $m \geq n$ and up^{m-n} is unique, so the assignment $f \to up^{m-n}$ defines an isomorphism of the $\mathcal{C}$-endomorphism ring of B with Z_p. Similarly, when $pB = B$ it can be shown that the $\mathcal{C}$-endomorphism ring of B is isomorphic to Q.

Our next result is an immediate consequence of the preceding two theorems.

<u>Corollary</u> 14. Let $A \in \mathcal{V}_p$ be a summand of a direct sum of modules of rank one. Then there is a torsion module T such that $A \oplus T$ is a direct sum of modules of rank one.

9. <u>Summands of Warfield modules are Warfield</u>. In order to prove this assertion, we first explore a little of the relationship between Warfield modules and simply presented modules. Let $A = S(Y)$ be simply presented and let $A = \oplus S(Y_i)$ be the decomposition given in Lemma 1. In each Y_i choose an element x_i so that, if possible, x_i has infinite order. If X is the valuated subtree of Y consisting of the x_i's, all their p-multiples, and 0, then

$$0 \to S(X) \to S(Y) \to S(Y/X) \to 0$$

is a representing sequence for $S(Y)$ as a Warfield module. Thus A is Warfield. However, Warfield modules need not be simply presented - the exact relationship between these two classes of modules will be described in Section 10 ; for the present, we show that with the 'addition' of a suitable torsion module, a Warfield module becomes simply presented.

Lemma 15. Let A be a Warfield module. Then there is a simply presented torsion module B such that $A \oplus B$ is simply presented.

Proof. Let $0 \to C \to A \to A/C \to 0$ be a representing sequence for A. Then $T(C)$ is simply presented (recall that the functor T described in Theorem 2 takes a valuated module to a module). By Theorem 6 there is a simply presented torsion module B such that $f_B \geq f_{T(C)} \aleph_0$ and $f_B \geq f_A \aleph_0$. Now $T(C) \oplus B$ and $A \oplus B$ have the same Ulm and Warfield invariants, so Theorem 10 implies $T(C) \oplus B \cong A \oplus B$.

There are in fact a number of different ways to prove Lemma 15. We present one other, based on the following lemma which is also needed for later work.

Lemma 16. Let A and B be modules, $f : A \to B$ and $g : B \to A$ homomorphisms, and C a submodule of A such that $g \circ f$ is the identity on C. Then $(A/C) \oplus B \cong A \oplus (B/f(C))$. The isomorphism is given by

$$(a + C, b) \mapsto (a + g(b - f(a)), b - f(a) + f(C)),$$

with inverse given by

$$(a, b + f(C)) \mapsto (a - g(b) + C, b + f(a - g(b))).$$

The proof is trivial.

Second proof of Lemma 15. Let $0 \to C \to A \to A/C \to 0$ be a representing sequence for A. As we observed in Corollary 5, there are maps $f : A \to T(C)$ and $g : T(C) \to A$ which are the identity on C. Lemma 16 yields $(A/C) \oplus T(C) \cong$

$A \oplus (T(C)/C)$ and we are done.

Theorem 17. A summand of a Warfield module is Warfield.

Proof. Let $M = B \oplus D$ be Warfield and suppose $B \neq 0$. By Lemma 15, we may assume M is simply presented. Thus $M = \oplus_I M_i$ where each M_i is simply presented of rank one. Theorem 13 implies B is C-isomorphic to $A = \oplus M_i$ for some $J \subset I$, so there are maps $f : A \to B$, $g : B \to A$, $\alpha : A \to A_t$, and $\beta : B \to B_t$ with $g \circ f + \alpha = 1_A$ and $f \circ g + \beta = 1_B$. Now choose a representing sequence $0 \to C \to A \to A/C \to 0$ for A. As $\mathrm{Im}\alpha$ is torsion, we can arrange that C is chosen so that $\alpha(C) = 0$ and hence $g \circ f$ is the identity on C. We show that $0 \to f(C) \to B \to B/f(C) \to 0$ is a representing sequence for B. Lemma 16 gives an isomorphism

$$h : (A/C) \oplus B \to A \oplus (B/f(C))$$

and hence an isomorphism

$$h' : (A/C) \oplus B \oplus D \to A \oplus (B/f(C)) \oplus D$$

Using the fact that $\mathrm{Im}\beta$ is torsion and the description of h given in Lemma 16, it is readily checked that $B/f(C)$ is torsion. Thus it is possible to choose $E \subset B \oplus D = M$ so that M/E is simply presented and $h'(E) \subset A \oplus D$. But then

$$\frac{A}{C} \oplus \frac{B \oplus D}{E} \simeq \frac{B}{f(C)} \oplus \frac{A \oplus D}{h'(E)}$$

with the left hand side simply presented and torsion, and therefore $B/f(C)$ is simply presented and torsion. Obviously $f(C)$ is a direct sum of cyclics, and it remains to show that $f(C)$ is nice in B. If $c \in C$ then $h^{-1}(c, 0) = (0, f(c))$. Thus the isomorphism h^{-1} sends $C \subset A$ to $f(C) \subset B$ so $f(C)$ is indeed nice in B.

Corollary 18. A module is Warfield if and only if it is a summand of a simply presented module.

Proof. We have already seen that if A is Warfield then A is a summand of a simply presented module (Lemma 15). Conversely, if A is a summand of a simply presented module, then A is a summand of a Warfield module and Theorem 17 implies A is Warfield.

10. <u>Existence theorems</u>. Necessary and sufficient conditions for the existence of a Warfield module with prescribed Ulm and Warfield invariants were given in [4]. Since [4] is written from the valuated viewpoint and the relevant notation and definitions are those we are presently employing, we report the results of [4] without proofs.

First, the main existence theorem.

<u>Theorem</u> 19. Let C be a direct sum of infinite cyclic valuated modules and f a function from the ordinals to the cardinals that vanishes beyond some ordinal. Then there exists a Warfield module A with $f_A = f$ and $w_A = w_C$ if and only if f dominates $f + g_C$ and $f \geq f_C$.

Note that if C is a direct sum of cyclics of the kind mentioned in the theorem, then $\sum_{\alpha+\omega\leq\beta} g_C(\beta)$ is determined, to within a finite cardinal, by the Warfield invariants of C. Thus Theorem 19 does indeed give conditions for the existence of a Warfield module with prescribed Ulm and Warfield invariants. The proof in [4] makes extensive use of the derived Ulm invariants and in fact constructs a representing sequence

$$0 \to C \to A \to A/C \to 0$$

for A. Of course, Theorem 19 generalizes Theorem 6.

It was stated in Section 9 that a Warfield module need not be simply presented. The reason for this is a lack of relative Ulm invariants in sense made precise by:

<u>Theorem</u> 20. ([4]) Let A be a Warfield module and X a nice decomposition basis of A with $A/\langle X\rangle$ simply presented. Then A is simply presented with spanning tree containing X if and only if $f_{A,\langle X\rangle}$ dominates $f_{A,\langle X\rangle} + g_{\langle X\rangle}$.

The first example of a Warfield module which is not simply presented was given by Rotman and Yen [10]. Their example has rank two. The following rank one example is due to Warfield [20].

Example 21. Let A be the Warfield module with decomposition basis x such that $V(x) = 0, 2, 4, \ldots$ and $f_{A,\langle x\rangle}(\alpha) = 1$ if $\alpha = \omega$ and 0 otherwise. Such a module exists (Theorem 19), while Theorem 20 shows that x cannot be extended to a spanning tree. Since every element of infinite order in A has value sequence equivalent to V(x), it follows that A is not simply presented.

Theorem 19 can be used to construct a Warfield module of countably infinite rank which is not simply presented.

Example 22. ([4]). Let $C = \bigoplus_{n=0}^{\infty} C_n$ where C_n is generated by x such that

$$vp^m x_n = n + 2m.$$

By Theorem 19, C can be embedded nicely in a Warfield module A with the same Ulm invariants as C, such that A/C is torsion and simply presented (actually, we are using the fact that, in the proof of Theorem 19, A is constructed with representing sequence $0 \to C \to A \to A/C \to 0$). We claim that each summand of A is either finite or of finite index. To see this, let $A = B \oplus D$ and let $\underline{\alpha} = 0, 2, 4, \ldots$. Since $w_A(\underline{\alpha}) \neq 0$ we may assume that $w_B(\underline{\alpha}) \neq 0$. Then B must contain an element y such that $vp^n y = 2m + 2n$ for some m and all n. Hence $f_B(2m + 2n) \neq 0$ for all n, so $f_D(k) = 0$ for all $k \geq 2m$. This implies that the torsion submodule of D is bounded, and so D is bounded since A admits no nonzero torsion-free summands.

In the light of these examples, one might ask whether there are Warfield modules of arbitrary rank which are not simply presented and which do not decompose into modules of smaller rank. The answer is no, and the next theorem shows that all such examples must be of countable rank.

Theorem 23. ([4]) If A is Warfield, then $A = B \oplus C$ where B is simply presented and there is an ordinal α such that $p^{\alpha}C$ is countable and $C/p^{\alpha}C$ is torsion. In particular, C has countable rank.

We saw in Section 9 that a Warfield module can be made simply presented by 'adding' a suitable torsion module. The same result is achieved by taking the direct sum of enough copies of a Warfield module:

Theorem 24. ([4]) If A is a Warfield module then the direct sum of infinitely many copies of A is simply presented.

Let X be a decomposition basis of A. We say that X is a splitting decomposition basis of A and A splits over X if A can be written as the direct sum $\bigoplus_{x \in X} A_x$, with $x \in A_x$. Thus a module with a splitting decomposition basis is a direct sum of rank one modules. Decomposition bases which generate isomorphic direct sums of valuated cyclics are said to be isomorphic. We give an example to show that isomorphic decomposition bases in a module need not have the same splitting properties.

Example 25. Let G be the countably infinite direct sum of copies of the module A in Example 22. Then the decomposition basis X of G formed by taking the decomposition basis described in Example 22 from each copy of A is not splitting (just check the relative Ulm invariants). However, the module $H = A \oplus A_t$ is simply presented and has a splitting decomposition basis isomorphic to the decomposition basis of A (Theorem 20). Hence the countably infinite direct sum of copies of H has a splitting decomposition basis isomorphic to X. But G and H have the same Ulm and Warfield invariants and so are isomorphic.

We now set about characterizing those isomorphism classes of decomposition bases in a Warfield module which contain a splitting decomposition basis. Some preparatory definitions are needed.

Let f and g be two functions from ordinals to cardinals which vanish beyond some ordinal. If $f \geq g$, we define a function $f - g$ by

$$f(\alpha) - g(\alpha) = \begin{cases} f(\alpha) & \text{if } f(\alpha) \text{ is infinite} \\ f(\alpha) - g(\alpha) & \text{otherwise.} \end{cases}$$

Obviously $(f - g) + g = f$. Given an ordinal λ, we say that f is λ- admissible if the function f' defined by

$$f'(\alpha) = \begin{cases} f(\alpha) & \text{if } \alpha < \lambda \\ 0 & \text{otherwise} \end{cases}$$

is admissible. The length $\ell(f)$ of f is $\sup\{\alpha + 1 : f(\alpha) \neq 0\}$. We also set

$$L(f) = \sup\{\lambda \leq \ell(f) : \lambda = 0 \text{ or is a limit ordinal}\}.$$

The following result which allows us to decompose certain of these functions will also be needed.

Lemma 26. ([4,Lemma 21]) Let f and g be functions from ordinals to cardinals which vanish beyond some ordinal. If f dominates $f + g$, and $g = \sum_I g_i$, then f can be written $f = \sum_I f_i$ such that f_i dominates $f_i + g_i$ for all $i \in I$.

Theorem 27. Let A be a Warfield module with decomposition basis X. Then A has a splitting decomposition basis isomorphic to X if and only if $f_A - f_{<X>}$ is $L(g_{<X>})$-admissible and has length at least $L(g_{<X>})$.

Proof. Set $f^* = f_A - f_{<X>}$ and $\lambda = L(g_{<X>})$. Suppose A has a splitting decomposition basis isomorphic to X. Since f^* depends on X only up to isomorphism, we may assume that X itself is splitting. First assume that X contains but one element x. For each $\alpha < \lambda$, there are only finitely many elements β of $V(x)$ such that $\alpha < \beta < \alpha + \omega$ so $f_{<x>}$ contributes finitely to f_A in this range. Since f_A is admissible and $\ell(f_A) \geq \lambda$, it follows that f^* has length at least λ and is λ-admissible. Now let X be arbitrary and suppose $\alpha < \lambda$. If $\sum_{\alpha<\beta<\alpha+\omega} f_A(\beta) > \aleph_0$ then $\sum_{\alpha<\beta<\alpha+\omega} f_A(\beta) = \sum_{\alpha<\beta<\alpha+\omega} f^*(\beta)$ and there is nothing to prove. If $\sum_{\alpha<\beta<\alpha+\omega} f_A(\beta) \leq \aleph_0$ then, since there is an element x in X with $\alpha < L(g_{<x>})$ and a rank one summand B containing x, the first part of the argument shows

$$\sum_{\alpha<\beta<\alpha+\omega} f^*(\beta) \geq \sum_{\alpha<\beta<\alpha+\omega}(f_B - f_{<x>})(\beta) \geq \aleph_0 \geq \sum_{\beta\geq\alpha+\omega} f_A(\beta) \geq \sum_{\beta\geq\alpha+\omega} f^*(\beta).$$

This also proves that f^* has length at least λ.

Conversely, suppose f^* satisfies the conditions of the theorem. We claim that f^* dominates $f^* + g_{<X>}$. Since f_A dominates $f_A + g_{<X>}$ (Theorem 19), it suffices to consider only those $\alpha < \lambda$ for which $\sum_{\alpha<\beta<\alpha+\omega} f_A(\beta) \leq \aleph_0$. But then the condition on the length of f^* ensures that

$$\sum_{\alpha<\beta<\alpha+\omega} f^*(\beta) = \aleph_0 \geq \sum_{\beta\geq\alpha+\omega} (f_A + g_{<X>})(\beta) \geq \sum_{\beta\geq\alpha+\omega} (f^* + g_{<X>}(\beta))$$

and the claim is established. Now we use Lemma 26 to write $f^* = \sum f_x$ where f_x dominates $f_x + g_{<x>}$ for each $x \in X$. But then $f_x + f_{<x>}$ dominates $f_x + f_{<x>} + g_{<x>}$ for all $x \in X$, so there is a Warfield module A_x with representing sequence $0 \to <x> \to A_x \to A/<x> \to 0$ such that $f_{A_x} = f_x + f_{<x>}$. Now $\bigoplus_{x\in X} A_x$ has the same Ulm and Warfield invariants as A, and so is isomorphic to A.

Theorem 27 is useful in determining when a Warfield module is a direct sum of rank one modules. For example, the next theorem is a generalization of a result of Rotman [9].

Corollary 28. Let A be a Warfield module with decomposition basis X. If there is an ordinal α such that $\alpha \leq vx < \alpha + \omega$ for all $x \in X$ then A is a direct sum of rank one modules.

Proof. This result follows from Theorem 27 on observing that $\alpha+\omega \geq L(g_{<X>})$ and $f_A(\beta) - f_{<X>}(\beta) = f_A(\beta)$ for all $\beta < L(g_{<X>})$.

11. Decomposition bases. The notion of a nice decomposition basis is central to our development of Warfield modules. In this section we take a closer look at decomposition bases, with applications to Warfield modules in mind. First, a summand of a module with a decomposition basis has a decomposition basis - this result was first proved by Warfield [15]. Of particular interest is the question of whether a module with a decomposition basis also has a nice decomposition basis.

Since subordinates of nice decomposition bases are also nice, the following result of Warfield [15 Lemma 6], shows that it suffices to look for nice decomposition bases among the subordinates of any given decomposition basis.

Lemma 29. let X and Y be decomposition bases of a module. Then there exist subordinates X' and Y' of X and Y, respectively, such that $\langle X' \rangle = \langle Y' \rangle$.

We say that a valuated module A has finite jump type if, given an ordinal β, there are at most finitely many $\alpha < \beta$ such that $vx = \alpha$ and $vpx \geq \beta$ for some x in G. A moments reflection reveals that the condition $vpx \geq \beta$ in the preceding definition can be replaced by $vp^n x \geq \beta$ for $n < \omega$. It is important to notice that finite jump type is a condition on the number of distinct kinds of value sequences occurring in a valuated module, rather than on the number of elements with those value sequences. This point is illustrated by the fact that the (valuated) direct sum of arbitrarily many copies of the same valuated cyclic has finite jump type.

Lemma 30. Let X be a decomposition basis of the valuated module A, and let y be an element of A. If $y + \langle X \rangle$ has no element of maximum value, then there are finite subsets $S_1 \subset S_2 \subset \ldots$ of X, and elements $x_j \in \langle S_j \rangle$ such that

1) $\alpha_j = v(y - x_j)$ is the maximum value of elements in $y + \langle S_j \rangle$
2) $v(y) = \alpha_0 < \alpha_1 < \alpha_2 < \ldots$
3) every coordinate of x_j in $\langle S_j \rangle \setminus \langle S_{j-1} \rangle$ has value α_{j-1}.

Proof. Let $x_0 = 0$ and $S_0 = \emptyset$, and suppose the construction has been carried out to $j - 1$. Since $y - x_{j-1}$ is not of maximum value in $y + \langle X \rangle$, there is z in X such that $v(y - x_{j-1} + z) > \alpha_{j-1}$. We can choose z so that every nonzero coordinate of z has value α_{j-1}. Let S_j be S_{j-1} together with the elements of X where z has a nonzero coordinate. Since S_j is finite there is x_j in $\langle S_j \rangle$ maximizing $v(y - x_j)$. Then $v(x_j - x_{j-1} + z) > \alpha_{j-1}$ so, since $x_{j-1} \in \langle S_{j-1} \rangle$, every coordinate of x_j in $\langle S_j \rangle \setminus \langle S_{j-1} \rangle$ must have value α_{j-1}.

Theorem 31. Let X be a decomposition basis of the valuated module A. If X has finite jump type, then X is nice in A.

Proof. If X is not nice in A, then there is y in A such that $y + \langle X\rangle$ has no element of maximum value. Let S_j, x_j and α_j be as in Lemma 30, and suppose $p^n y \in \langle X\rangle$. For sufficiently large j, say $j \geq N$, the coordinates of $p^n y$ in $S_j \setminus S_{j-1}$ are zero. By Lemma 30, 1), we may choose $s_j \in S_j \setminus S_{j-1}$. Let z_j be the projection of x_j on $\langle s_j\rangle$. Then $vz_j = \alpha_{j-1}$. We shall show that $vp^n z_j \geq \sup \alpha_i$ for $j \geq N$, contradicting the assumption that $\langle X\rangle$ has finite jump type. Note that $p^n z_j$ is the projection of $p^n x_j$ on $\langle s_j\rangle$. If $i > j$, then $v(x_i - x_j) = v(y - x_j - y + x_i) = \alpha_j$ so the projection u_{ij} of x_i on $\langle s_j\rangle$ has value α_{j-1}. Thus $vp^n z_j = vp^n u_{ij} \geq vp^n x_i \geq \alpha_i$ for $i \geq j \geq N$, whereupon $vp^n z_j \geq \sup \alpha_i$

In particular, a finite decomposition basis generates a nice submodule. We give an application of this in:

Theorem 32. Let A be a Warfield module of finite rank. If A has a splitting decomposition basis, then A splits over every decomposition basis.

Proof. It is easy to see that if A has a decomposition basis X satisfying the conditions of Theorem 27, then every decomposition basis satisfies those conditions - the main point being that finite changes to a function do not affect its admissibility. Let X be any decomposition basis of A. By Theorem 27 there is a splitting decomposition basis Y of A which is isomorphic to X. Now A has finite rank so $f_{A,\langle X\rangle} = f_{A,\langle Y\rangle}$ while Theorem 31 shows that $\langle X\rangle$ and $\langle Y\rangle$ are nice in A. Theorem 4 extends the isomorphism $\langle Y\rangle \approx \langle X\rangle$ to an automorphism of A. This automorphism carries a splitting of A over Y into a splitting of A over X.

Theorem 33. Every countable decomposition basis of a valuated module has a nice subordinate.

Proof. Let $x_1, x_2, \ldots$ be a decomposition basis and $\alpha_1, \alpha_2, \ldots$ an enumeration of the ordinals $v(p^n x_m)$. Choose nonnegative integers $n(j)$ such that, for each $i \le j$, either

$$\alpha_i < v(p^{n(j)} x_j) ,$$

$$\text{or} \qquad \alpha_i > v(p^n x_j) \quad \text{for all } n.$$

Then $\langle p^{n(j)} x_j : j = 1, 2, \ldots \rangle$ has finite jump type and hence is nice.

Thus if A and B are direct sums of countable modules and both A and B have decomposition bases, then A and B have nice decomposition bases. It follows that A is isomorphic to B if and only if $f_A = f_B$ and $w_A = w_B$. This result was first proved by Warfield[15,Theorem 3]. We now show that there is a module which has a decomposition basis, but which has no nice decomposition basis.

Lemma 34. Let A be a valuated module with decomposition basis X. Let $\{x_j\}$ be a sequence in X such that $v(x_j) < v(x_{j+1})$ and $v(px_j) \ge \sup v(x_i)$ for all j. Then there is a valuated module B containing A as a valuated submodule, and an element y in B, such that B/A is torsion and

$$v(y + x_1 + \ldots + x_n) = v(x_{n+1})$$

for $n = 0, 1, 2, \ldots$.

Proof. First observe that it suffices to verify the inequalities $v(y + x_1 + \ldots + x_n) \ge v(x_{n+1})$, for if $v(y + x_1 + \ldots + x_n) > v(x_{n+1})$, then $v(y + x_1 + \ldots + x_n) = v(x_{n+1}) < v(x_{n+2})$. If such a y exists in A, then set B = A. Otherwise construct B as the group direct sum of A and a cyclic group of order p generated by y, valuated as follows:

$$B(\alpha) = A(\alpha) \quad \text{for } \alpha \ge \sup v(x_i);$$

$$B(\alpha) = A(\alpha) + \langle y + x_1 + \ldots + x_n \rangle \quad \text{for } \alpha \le v(x_{n+1}),$$

$$n = 0, 1, 2, \ldots .$$

Note that $B(\alpha)\cap A = A(\alpha)$ so we will be done if we show that the $B(\alpha)$ define a valuation on B. Clearly the $B(\alpha)$ are decreasing in α, and $pB(\alpha) \subset A(\alpha + 1) \subset B(\alpha + 1)$. Suppose α is a limit ordinal. If $\alpha > \sup v(x_i)$, then

$$\bigcap_{\beta<\alpha} B(\beta) = \bigcap_{\beta<\alpha} A(\beta) = A(\alpha) = B(\alpha).$$

If $\alpha \leq v(x_{n+1})$, then

$$\bigcap_{\beta<\alpha} B(\beta) = \bigcap_{\beta<\alpha} [A(\beta) + \langle y + x_1 + \dots + x_n\rangle].$$

But $A\cap\langle y + x_1 + \dots x_n\rangle = \langle px_1 + \dots + px_n\rangle \in A(\alpha)$, so

$$\bigcap_{\beta<\alpha} B(\beta) = A(\alpha) + \langle y + x_1 + \dots + x_n\rangle = B(\alpha).$$

Finally, suppose $\alpha = \sup v(x_i)$ and that $z + y \in \bigcap_{\beta<\alpha} B(\beta)$ for some z in A. If $\beta \leq v(x_{n+1})$ we can write $z + y = g_\beta + y + x_1 + \dots + x_n$ for some $g_\beta \in A(\beta)$. Therefore

$$v(-z + x_1 + \dots + x_n) \geq v(x_{n+1})$$

for each n, contrary to our initial assumption. Thus $\bigcap_{\beta<\alpha} B(\beta) \subset A$, whence

$$\bigcap_{\beta<\alpha} B(\beta) = \bigcap_{\beta<\alpha} B(\beta)\cap A = \bigcap_{\beta<\alpha} A(\beta) = A(\alpha) = B(\alpha).$$

Corollary 35. Let A be a valuated module. Then there is a module B containing A as a valuated submodule such that B/A is torsion and, whenever $\{x_j\}$ is a sequence in a decomposition basis for A such that $v(x_j) < v(x_{j+1})$ and $v(px_j) \geq \sup v(x_i)$ for all j, then there is y in B such that

$$v(y + x_1 + \dots + x_n) = v(x_{n+1})$$

for $n = 0, 1, 2, \dots$.

Proof. Transfinite application of Lemma 34 gives a valuated module B' with the desired properties. Now use Theorem 2 to put B' in a module B such that B/B' is torsion.

Theorem 36. If A is a direct sum of infinite cyclic valuated modules that does not have finite jump type, then a basis for A is a not nice decomposition basis for some module B.

Proof. Let B be as in Corollary 35. Since A does not have finite jump type, we can find a sequence $\{x_j\}$ in some subordinate decomposition basis of A such that $v(x_j) < v(x_{j+1})$ and $v(px_j) \geq \sup v(x_i)$ for all j. Let y be as in Corollary 35. Then $y + A$ has no element of maximum value, for if $v(y + g) \geq \sup vx_i$, then

$$v(x_1 + \ldots + x_n - g) \geq vx_{n+1}$$

for all n, which is impossible.

<u>Theorem</u> 37. There is a module B with a decomposition basis X such that no subordinate decomposition basis is nice in B.

<u>Proof.</u> Let S be the set of all sequences s of ordinals $s_n = \omega n + m_n(s)$ where $m_n(s) < \omega$. For $s \in S$ let x_s generate a cyclic valuated module $\langle x_s \rangle$ such that $vp^n x_s = s_n$. Let A be the direct sum of the $\langle x_s \rangle$, let $X = \{x_s : s \in S\}$, and let $B \supset A$ be as in Corollary 35. Let $f(s)$ be a nonnegative integer for each s in S, and let $X' = \{p^{f(s)}x : s \in S\}$. We shall show that $\langle X' \rangle$ is not nice in B.

The set $\{s_n : f(s) = n\}$ is infinite for some n lest there be an $s' \in S$ such that $s'_n > s_n$ whenever $f(s) = n$. So there exist x_j in X' with $vx_1 < vx_2 < \ldots$ and $vpx_j \geq \omega(n + 1) \geq \sup vx_i$. Corollary 35 and the last line in the proof of Theorem 36 show that $\langle X' \rangle$ is not nice in B.

It follows from Lemma 29 that a module of the kind described in Theorem 37 has no nice decomposition basis.

12. <u>Projective characterizations and balanced projectives.</u> Recall that a sequence $0 \to A \to B \to C \to 0$ of valuated modules is in Ext (C, A) if and only if, for all ordinals α,

$$0 \to A(\alpha) \to B(\alpha) \to C(\alpha) \to 0$$

is exact as a sequence of modules, and a valuated module is projective if and only if it is free. A sequence $0 \to A \to B \to C \to 0$ is called <u>sequentially pure</u> if the sequence of modules

$$0 \to A(\underline{\alpha}) \to B(\underline{\alpha}) \to C(\underline{\alpha}) \to 0$$

is exact for all value sequences $\underline{\alpha}$. It is readily seen that a valuated module is projective with respect to all sequentially pure exact sequences if and only if it is a direct sum of cyclics.

When attention is restricted to such sequences and their relative projectives within the category of modules, we have:

Theorem 38. (Warfield [19]) (a) A module is projective relative to all short exact sequences of modules which are in Ext in V_p if and only if it is Warfield and contains a free decomposition basis.

(b) A module is projective relative to all sequentially pure sequences of modules if and only if it is Warfield.

The modules described in part (a) of the preceding theorem are called balanced projectives. They comprise a particularly tractable class of Warfield modules which we now examine in some detail. Let A be balanced projective. Then A contains a free valuated subgroup C such that A/C has a composition series. It is immediate from Theorem 31 that C is nice in A, and hence a balanced projective module is Warfield. Thus a module A is balanced projective if and only if A has a representing sequence

$$0 \to C \to A \to A/C \to 0$$

with hom.dim.C = 0 and hom.dim.A/C ≤ 1. Hence hom.dim.A ≤ 1 for any balanced projective module A.

Lemma 39. Let A be a balanced projective module and C a free valuated submodule of A. Then A/C has a composition series.

Proof. By Theorem 32, $0 \to C \to A \to A/C \to 0$ is a V_p-extension. As hom.dim.C = 0 and hom.dim.A ≤ 1, it follows that hom.dim.A/C ≤ 1.

Balanced projective modules are in fact simply presented:

Theorem 40. Let A be a balanced projective module with free decomposition basis X. Then $A = \oplus_{x \in X} A_x$ where each A_x is rank one simply presented and contains x.

Proof. By Theorem 19, f_A dominates $f_A + g_{<X>}$ but $f_{<X>} = 0$ so $f_A = f_{A,<X>}$ and Theorem 20 completes the proof.

The Warfield invariants of a balanced projective module are particularly easy to describe. Let $\underline{\alpha} = \alpha_0, \alpha_1, \ldots$ be a value sequence. If A is a balanced projective module and $w_A(\underline{\alpha}) \neq 0$, then there is a positive integer i such that $\alpha_{i+k} = \alpha_i + k$ for $k = 0, 1, \ldots$. Thus we might as well consider only those sequences $\alpha, \alpha + 1, \alpha + 2, \ldots$ where α is a limit ordinal, and hence only limit ordinals. With this in mind, set

$$h_A(\alpha) = \begin{cases} w_A(\alpha, \alpha + 1, \alpha + 2, \ldots) & \text{for } \alpha \text{ a limit, and} \\ 0 & \text{otherwise.} \end{cases}$$

For a balanced projective module A and limit ordinal α, it is clear that $h_A(\alpha)$ is the dimension of the vector space $p^\alpha M/(p^{\alpha+1}M + (p^\alpha M)_t)$; that is, $h_A(\alpha)$ and the invariant $h(\alpha, A)$ defined by Warfield [15,18] agree on balanced projective modules. The isomorphism and existence theorems for balanced projective modules (due to Warfield[19]) now follow directly from Theorem 10 and Theorem 19, respectively:

Theorem 41. Let A and B be balanced projective modules. Then A is isomorphic to B if and only if $f_A = f_B$ and $h_A = h_B$.

Theorem 42. Let f and h be functions from ordinals to cardinals which vanish beyond some ordinal. Then there is a balanced projective module A with $f_A = f$ and $h_A = h$ if and only if f dominates $f + h$.

Of course, there is also an analogue of Theorem 11 for balanced projective modules. More interesting is the following result which was proved in [19,Corollary 46] and is essential for the localization arguments used in that paper.

Theorem 43. Let A and B be isomorphic balanced projective modules with free decomposition bases X and Y, respectively. Then any isomorphism $X \to Y$ extends to an isomorphism $A \to B$.

Proof. We have already seen that $\langle X\rangle$ and $\langle Y\rangle$ are nice submodules of A and B, respectively, such that $A/\langle X\rangle$ and $B/\langle Y\rangle$ have composition series. Now $f_{A,\langle X\rangle} = f_A = f_B = f_{B,\langle Y\rangle}$ follows from $f_{\langle X\rangle} = f_{\langle Y\rangle} = 0$ and Theorem 4 completes the proof.

Remark. The importance of the preceding theorem lies in the way it differs from Theorem 11. The requirement that $A/\langle X\rangle$ and $B/\langle Y\rangle$ be simply presented does not appear in the statement of Theorem 43 as its equivalent does in Theorem 11, and this is the main obstruction to a direct application of Theorem 4. This problem was overcome in [19] by passing to modules over a complete discrete valuation ring. However, Lemma 39 is exactly what is needed (see also the remark following the statement of [19,Theorem 4.4]).

13. *The torsion submodule of a Warfield module.* The torsion submodule of a balanced projective module is called an *S- module*. In this section we show that the torsion submodule of a Warfield module is an S-module. Hence, the larger class of Warfield modules does not, in this way at least, lead to a larger class of torsion modules. S-modules were first studied by Warfield in [17] and we refer the reader to that paper for a full account.

The torsion submodule of a rank one Warfield module is characterized in:

Theorem 44 [6]. Let A be a rank one Warfield module. Then the torsion submodule of A is an S-module. Moreover, if x is an element of infinite order in A and V(x) has infinitely many gaps or $L(g_{<x>})$ is cofinal with ω, then the torsion submodule of A is simply presented.

Now for the general case:

Theorem 45. Let A be Warfield. Then $A = B \oplus C$ where B is balanced projective and the torsion submodule of C is simply presented. In particular, the torsion submodule of A is an S-module.

Proof. It is a straightforward application of Theorems 10 and 19, together with some manipulation of Ulm and Warfield invariants, to prove that $A = B \oplus C$ where B is balanced projective and every element x of a decomposition basis for C is such that V(x) has infinitely many gaps or $L(g_{<x>})$ is cofinal with ω. Thus it is enough to show that C_t is simply presented. Let T be simply presented so that $C \oplus T$ is simply presented. By Theorem 44, $C_t \oplus T$ is simply presented and hence C_t is simply presented.

In view of the preceding theorem, it is natural to ask whether we can decompose a Warfield module A as $B \oplus C$ so that B is balanced projective and each element of a decomposition basis for C has infinitely many gaps in its value sequence. However, this is not possible, and an easy counterexample can be obtained by adjoining a new element with value sequence $\omega, \omega + 1, \omega + 2, \ldots$ to the decomposition basis of the module in example 22 and leaving all the other details of the construction unchanged.

REFERENCES

1. Crawley, P., and Hales, A. W. The structure of abelian p-groups given by certain presentations, J. Algebra. 12(1969), 10-23.
2. Fuchs, L., Infinite Abelian Groups, Vol. II, Academic Press, New York, 1973.
3. Hill, P., On the classification of abelian groups, (preprint 1967).
4. Hunter, R., Richman, F., and Walker, E., Existence theorems for Warfield groups, to appear, Trans, Amer. Math. Soc.

5. ______________________ Simply presented valuated abelian p-groups, to appear, J. Algebra.

6. Hunter, R. H., Balanced subgroups of abelian groups, Trans. Amer. Math. Soc. 215 (1976), 81-98.

7. Richman, F., and Walker, E. A., Valuated groups, (preprint 1976).

8. ______________________ Ext in pre-abelian categories, Pacific J. Math. (to appear).

9. Rotman, J., Mixed modules over valuation rings, Illinois J. Math. 10 (1960), 607-623.

10. Rotman, J., and Yen, T., Modules over a complete discrete valuation ring, Trans. Amer. Math. Soc. 98 (1961), 242-254.

11. Stanton, R. O., An invariant for modules over a discrete valuation ring, Proc. Amer. Math. Soc. 49 (1975), 51-54.

12. ______________________ Relative S-invariants, (preprint 1976).

13. Walker, C., and Warfield, R. B. Jr., Unique decomposition and isomorphic refinement theorems in an additive category, J. Pure Appl. Algebra, 7 (1976), 347-359.

14. Walker, E. A., Ulm's theorem for totally projective groups, Proc. Amer. Math. Soc. 37 (1973), 387-392.

15. Warfield, R. B. Jr., Invariants and a classification theorem for modules over a discrete valuation ring, (preprint 1971).

16. ______________________ Simply presented groups, Proc. Sem. Abelian Group Theory, University of Arizona lecture notes (1972).

17. ______________________ Classification theorems for p-groups and modules over a discrete valuation ring, Bull. Amer. Math. Soc. 78 (1972), 88-92.

18. ______________________ A classification theorem for abelian p-groups, Trans. Amer. Math. Soc. 210 (1975), 149-168.

19. ______________________ Classification theory of abelian groups I: Balanced projectives, Trans. Amer. Math. Soc. 222 (1976), 33-63.

20. ______________________ Classification theory of abelian groups II: Local theory, (preprint 1976).

FINITE VALUATED GROUPS

Yonina S. Cooper

1. Introduction. Let R be a discrete valuation ring and p the prime element in R. Let Γ denote the ordinals together with ∞. The usual convention that $\alpha < \infty$ for all $\alpha \in \Gamma$ is made. A valuation on an R-module M is a function $h : M \to \Gamma$ such that

(i) $h(x + y) \geq \min\{hx, hy\}$ for all $x,y \in M$,

(ii) $h(ux) = hx$ if u is a unit in R, and

(iii) $h(px) > hx$ for all $x \in M$.

An R-module together with a valuation h_M on M is called a valuated R-module. The category V_R of valuated R-modules is the category whose objects are valuated R-modules and whose morphisms are those $\phi \in \mathrm{Hom}_R(M, N)$ such that $h_N\phi(x) \geq h_Mx$ for all $x \in M$. Every R-module M has a valuation, $ht_M : M \to \Gamma$ where $ht_Mx = \alpha$, α an ordinal, if $x \in p^\alpha M \backslash p^{\alpha+1}M$, and $ht_Mx = \infty$ if $x \in p^\alpha M$ for all ordinals α. Let M denote the category of R-modules. Let C be a nonempty class of valuations on M, $M \in M$. Then $h_M : M \to \Gamma$ defined by $h_Mx = \min\{hx : h \in C\}$ is a valuation on M. In particular, if C is the class of all valuations on M, the h_M defined is ht_M.

The category V_R is pre-abelian, that is, additive with kernels and cokernels. Moreover, it has arbitrary products and coproducts. Let $M_i \in V_R$ have valuation h_i for the index set I. The product and coproduct are the usual modules with the product valuation h_π given by $h_\pi\{x_i\}_{i\in I} = \min\{h_ix_i : i \in I\}$, and the coproduct valuation h_Σ given by $h_\Sigma\left(\sum_{i\in I} x_i\right) = \min\{h_ix_i : i \in I\}$. If N is a submodule of M in M and $h_N = h_{M|N}$ (h_M restricted to N), then N is called a submodule of M in V_R. We say that h_N is the induced valuation from M. Let N be a submodule of M in V_R. Consider the module M/N and $h : M/N \to \Gamma$ given by $h(x + N) = \sup\{h_M(x + y) : y \in N\}$. Let $C_{M/N}$ be the class of all valuations $v : M/N \to \Gamma$ such that $v(x + N) \geq h(x + N)$. C is nonempty: $v_\infty \in C$ where $v_\infty(x + N) = \infty$ for all $x + N$. So

$h_{M/N}(x + N) = \min\{v(x + N) : v \in C\}$ is a valuation on M/N. The valuation $h_{M/N}$ is called the cokernel valuation on M/N. For $f \in \mathrm{Hom}_{V_R}(M, N)$, the kernel of f is the kernel in M with the induced valuation from M, and the cokernel of f is the cokernel in M with the cokernel valuation. The image of f is $\mathrm{im}\, f = \ker \mathrm{coker}\, f$, and the coimage of f is $\mathrm{coim}\, f = \mathrm{coker} \ker f$. For $\mathrm{coim}\, f : M \to M'$, we also call M' the coimage of f. Using this identification, we define a sequence

$$\cdots \to M_{i-1} \xrightarrow{f_{i-1}} M_i \xrightarrow{f_i} M_{i+1} \xrightarrow{f_{i+1}} \cdots$$

to be exact at M_i if $\mathrm{coim}\, f_{i-1} = \ker f_i$. Further the sequence is said to be an exact sequence if it is exact at each M_i. Hence for a sequence $0 \to A \xrightarrow{f} B \xrightarrow{g} C \to 0$ to be (short) exact requires exactness at A, B, and C. This is equivalent to requiring both $f = \ker g$ and $g = \mathrm{coker}\, f$.

If R is the ring Z_p of integers localized at the prime p, the category is denoted V_p. If G is a p-group, multiplication by a prime $q \neq p$ is an automorphism of G, so G is a Z_p-module in the obvious way. The category F_p of finite valuated groups is the full subcategory of V_p of finite p-groups G whose valuation h_G satisfies $h_G x < \omega$ if $x \neq 0$.

The main concern of this paper is the homological algebra of F_p. Since the category F_p is pre-abelian, the theory developed by Richman and Walker [4] applies. However, the group of extensions $\mathrm{Ext}_{F_p}(C, A)$ is the usual group of equivalence classes of short exact sequences. The category F_p has no projectives but enough injectives. The injectives are the valuated groups $G \in F_p$ such that the valuation on G is the height function on G (Theorem 2). The category F_p has homological dimension one (Theorem 5). The group of extensions of a cyclic by a cyclic (Theorem 11) is determined.

For a valuated p-group G, the subgroup of all elements of value greater than or equal to n is denoted $G(n)$. An exact sequence $0 \to A \to B \to C \to 0$ in F_p is called n-split exact if the sequence $0 \to A(n) \to B(n) \to C(n) \to 0$ splits in F_p. The n-split exact sequences determine a relative homological algebra of dimension one (Corollary 16). These sequences are contained in the p^n socle of $\mathrm{Ext}_{F_p}(C, A)$ (Theorem 17). It is shown (Theorem 13) that the n-split projectives are those groups in F_p in which the value of each element is greater than or equal to n. Theorem 14 shows that the n-split injectives are the valuated groups $G \in F_p$ where the cokernel valuation on G/p^nG is the height function. There are enough n-split injectives. If the sequence $0 \to A(n) \to B(n) \to C(n) \to 0$ splits in the category *Ab* of abelian groups, the exact sequence $0 \to A \to B \to C \to 0$ is called Ab n-split exact. The *Ab* n-split exact sequences determine a relative homological algebra of dimension two (Theorem 23). There are neither enough *Ab* n-split projectives nor enough *Ab* n-split injectives. The *Ab* n-split injectives are the same as the injectives, those G whose valuation is the height function (Theorem 19); the *Ab* n-split projectives are direct sums in F_p of cyclics in which the valuation h on a generator x has the property that $hp^{m+1}x = hp^mx + 1$ for every non-negative integer m and $hx = n$ (Theorem 20).

The following notation will be standard throughout this paper. The cyclic group whose generator is x is written $\langle x\rangle$. The direct sum (coproduct) of valuated groups C_i, $1 \le i \le n$, is given $\sum_{i=1}^{n} C_i$, or by $C_1 \oplus C_2$ if $n = 2$. We write $h_G[x]$ for the value sequence $\{h_G p^i x\}_{i \in N}$, where N is the set of non-negative integers. Also we write $o(x)$ for the order of x and $Z(p^n)$ for the cyclic of order p^n.

2. Injectives and Projectives in F_p. If H is a subgroup of $G \in F_p$, written $H \subseteq G$, then the cokernel valuation $h_{G/H}$ is given by $h_{G/H}(x + H) = \max\{h_G(x + h) : h \in H\}$. So any epimorphism $f : H \to G$ is a cokernel if and only if for each $y \in G$, there is $x \in H$ such that $h_H x = h_G y$ and $f(x) = y$. We denote the class of cokernels in F_p by E_e; correspondingly E_m is the class of kernels in F_p.

It is easy to see that there are no (nonzero) projectives in F_p. The following lemma is a special case of a more general result of Richman and

Walker [5; Theorem 1].

<u>Lemma 1</u>. For $G \in \mathcal{F}_p$, there exist $H \in \mathcal{F}_p$ and $f : G \to H$ such that $h_H = ht_H$ and $f \in \mathcal{E}_m$.

<u>Theorem 2</u>. The injectives in $\mathcal{F}_p$ are the valuated groups G such that $h_G = ht_G$. There are enough injectives in $\mathcal{F}_p$.

<u>Proof</u>. Let G be injective. By Lemma 1, $G \subseteq H$ where $h_H = ht_H$ and $f : G \to H$ is in $\mathcal{E}_m$. So there is $f' \in \mathrm{Hom}_{\mathcal{F}_p}(H, G)$ such that $f'f = 1_G$, the identity map on G. For $x \in G$, $ht_G x \le h_G x = ht_H f(x) \le ht_G f'f(x) = ht_G x$. So $h_G = ht_G$. Conversely, suppose that in the diagram

$$\begin{array}{ccccc} 0 & \to & G & \to & H \\ & & g \downarrow & \swarrow g' & \\ & & X & & \end{array}$$

$X \in \mathcal{F}_p$ and $h_X = ht_X$. Let $G \subseteq H$, $g \in \mathrm{Hom}_{\mathcal{F}_p}(G, X)$. To obtain g', we induct on $|H/G|$. Let $x \in H \backslash G$ with $px \in G$ and $h_H x = h_{H/G}(x + G)$. The $ht_X g(px) \ge 1 + h_H x$, so there exists $y \in X$ such that $py = g(px)$ and $ht_X y \ge h_H x$. Then we extend g by $x \to y$. Now we have the commutative diagram

$$\begin{array}{ccccc} 0 & \to & G + \langle x \rangle & \to & H \\ & & \downarrow & \swarrow & \\ & & X & & \end{array}$$

and $|H/G + \langle x \rangle| < |H/G|$, so by induction there is $g' \in \mathrm{Hom}_{\mathcal{F}_p}(H, X)$ as desired.

Lemma 1 shows there are enough injectives.

Not only is every finite valuated group G a subgroup of an injective one, but there is a unique "minimal" such injective for each $G \in \mathcal{F}_p$. This injective is called the <u>injective envelope</u> of G. A subgroup H of G, or $f : H \to G$ with $f \in \mathcal{E}_m$, is called <u>essential</u> if $K \subseteq G$ and $K + H = K \oplus H$ implies $K = 0$; additionally H, or f, is <u>proper essential</u> if $H \subsetneq G$. Essential embeddings give an alternate characterization of injectives.

<u>Theorem 3</u>. G is injective if and only if there is no proper essential $f : G \to H$.

<u>Proof</u>. The proof follows the proof of Proposition 11.2 in [3; p. 102].

A modification of the proof of Theorem 11.3 in [3; p. 103] gives the injective envelope of G as the maximal essential extension of G.

<u>Theorem 4</u>. For every $G \in \mathcal{F}_p$, there is an essential embedding $f : G \to H$ with H injective. Moreover, if $g : G \to F$ is another essential embedding with F injective, then there is an isomorphism $k : H \to F$ with $kf = g$.

3. <u>$\mathrm{Ext}_{\mathcal{F}_p}(C, A)$</u>. If $A, C \in \mathcal{F}_p$, an <u>extension</u> of A by C is a short exact sequence $E : 0 \to A \to B \to C \to 0$. A <u>morphism</u> $\Gamma : E \to E'$ of extensions is a triple $\Gamma = (\gamma, \delta, \eta)$ of category morphisms such that the diagram

$$\begin{array}{lccccccccc} E: & 0 & \to & A & \xrightarrow{\alpha} & B & \xrightarrow{\beta} & C & \to & 0 \\ & & & \downarrow{\scriptstyle\gamma} & & \downarrow{\scriptstyle\delta} & & \downarrow{\scriptstyle\eta} & & \\ E': & 0 & \to & A' & \xrightarrow{\alpha'} & B' & \xrightarrow{\beta'} & C' & \to & 0 \end{array}$$

is commutative. Suppose E is an extension of A by C and $\gamma : A \to A'$ and $\eta' : C' \to C$. Then γE and $E\eta'$ are constructed in the usual manner (see [4; §2]). Following Richman and Walker [4], an extension E is called <u>stable</u> if γE and $E\eta'$ are exact for all maps γ and η'. Then $\mathrm{Ext}_{\mathcal{F}_p}(C, A)$ is defined to be the stable extensions of A by C, where two extensions are considered equal if there is a morphism $(1_A, \delta, 1_C)$ between them with δ an isomorphism. However, all extensions of A by C in $\mathcal{F}_p$ are stable [5; Theorem 6]. Then $\mathrm{Ext}^n_{\mathcal{F}_p}(C, A)$ consists of the usual Yoneda composites (see [4; §7]). As a result of Theorem 2 and the fact that coimages of injectives are injective, we have

<u>Theorem 5</u>. $\mathrm{Ext}^n_{\mathcal{F}_p}(C, A) = 0$ for $n \geq 2$.

For $a \in A \in \mathcal{V}_p$, $c \in C \in \mathcal{V}_p$, $h_A[a] \geq h_C[c]$ means that $h_A p^i a \geq h_C p^i c$ for all non-negative integers i. We define $A[c] = \{a \in A : h_A[a] \geq h_C[c]\}$, and for a non-negative integer n, $A(n) = \{a \in A : h_A a \geq n\}$. Trivially $A[c]$ and $A(n)$ are abelian groups, and if $C = \langle c\rangle$, $\mathrm{Hom}_{V_p}(C, A) \cong A[c]$ in Ab.

Let $A \in V_p$ such that

(i) A is a finitely generated p-group,

(ii) $h_A x < \omega$ if $x \neq 0$, and

(iii) there exists a non-negative integer n such that $p^n A$ is a direct sum of cyclics $\langle x\rangle$ with $h_A p^m x = h_A x + m$.

Richman and Walker [5; Theorem 30] have characterized these valuated groups as subgroups of finitely generated groups in V_p whose valuation is the height function. Let F_p^* be the full subcategory of V_p whose objects satisfy (i), (ii), and (iii). Richman and Walker [5] also show that F_p^* is pre-abelian and that an infinite cyclic $\langle c\rangle \in F_p^*$ with $h[c] = (m, m+1, m+2, \ldots)$ is projective in F_p^*. If $0 \to A \to B \to C \to 0$ is in $\mathrm{Ext}_{F_p^*}(C, A)$ with $A, C \in F_p$, then clearly $B \in F_p$. Thus $\mathrm{Ext}_{F_p}(C, A) = \mathrm{Ext}_{F_p^*}(C, A)$.

<u>Theorem 6</u>. Let $C = \langle c\rangle \in F_p$, $h_C[c] = (\ell_0, \ell_0 + 1, \ell_0 + 2, \ldots, k_0 - 2, k_0 - 1, \ell_1, \ell_1 + 1, \ldots, k_1 - 1, \ldots, \ell_m, \ell_m + 1, \ldots, k_m - 1, \infty, \ldots)$. (Note that $\ell_i < k_i < \ell_{i+1}$.) Define $n_i = k_i - \ell_i$, $0 \leq i \leq m$. Let $A \in F_p$ and define $\alpha : \sum_{i=0}^{m} A(\ell_i) \to \sum_{i=0}^{m} A(k_i)$ on the ith component as follows:

$$\alpha(a_i) = \begin{cases} (p^{n_0}a_0, 0, \ldots, 0) & \text{if } i = 0, \\ (-a_1, p^{n_1}a_1, 0, \ldots, 0) & \text{if } i = 1, \\ \vdots & \vdots \\ (0, \ldots, 0, -a_m, p^{n_m}a_m) & \text{if } i = m. \end{cases}$$

Then $\mathrm{Ext}_{F_p}(C, A) \cong \mathrm{coker}\ \alpha$.

<u>Proof</u>. Let $X = \sum_{i=0}^{m} \langle x_i\rangle$, each x_i have infinite order and $h_X[x_i] = (k_i, k_i + 1, k_i + 2, \ldots)$; let $Y = \sum_{i=0}^{m} \langle y_i\rangle$, each y_i have infinite order and $h_Y[y_i] = (\ell_i, \ell_i + 1, \ell_i + 2, \ldots)$. Then the sequence

$$0 \to X \xrightarrow{\gamma} Y \xrightarrow{\beta} C \to 0, \tag{1}$$

where $\gamma(x_i) = p^{n_i} y_i - y_{i+1}$ for $0 \le i < m$, $\gamma(x_m) = p^{n_m} y_m$, $\beta(y_i) = p^{r_i} c$, $r_i = \sum_{j=0}^{i-1} n_j$, $1 \le i \le m$, and $r_0 = 0$, is exact. Since $\mathrm{Ext}_{F_p^*}(Y, A) = 0$, $\mathrm{Hom}_{F_p^*}(Y, A) \to \mathrm{Hom}_{F_p^*}(X, A) \to \mathrm{Ext}_{F_p^*}(C, A) \to 0$ is exact [4; Theorem 12]. $\mathrm{Hom}_{F_p^*}(Y, A) \cong \sum_{i=0}^{m} A(\ell_i)$ by $\phi_Y(f) = \sum_{i=0}^{m} f(y_i)$, and $\mathrm{Hom}_{F_p^*}(X, A) \cong \sum_{i=0}^{m} A(k_i)$ by $\phi_X(f) = \sum_{i=0}^{m} f(x_i)$. Now the diagram

$$\begin{array}{ccc}
\mathrm{Hom}_{F_p^*}(Y, A) & \xrightarrow{\gamma^*} & \mathrm{Hom}_{F_p^*}(X, A) \\
\downarrow{\scriptstyle \phi_Y} & & \downarrow{\scriptstyle \phi_X} \\
\sum_{i=0}^{m} A(\ell_i) & \xrightarrow{\alpha} & \sum_{i=0}^{m} A(k_i)
\end{array}$$

is commutative. So $\mathrm{Ext}_{F_p}(C, A) \cong \mathrm{coker}\ \alpha$.

<u>Corollary 7</u>. Let $C = \langle c \rangle \in F_p$, $o(c) = p^n$, $h_C[c] = (m, m+1, m+2, \ldots, m+n-1, \infty \ldots)$, and $A \in F_p$. Then

$$\mathrm{Ext}_{F_p}(C, A) \cong \frac{A(m+n)}{p^n A(m)} .$$

<u>Proof</u>. $\mathrm{Ext}(C, A) \cong \mathrm{coker}\ \alpha$, where $\alpha : A(m) \to A(m+n)$ is given by $\alpha(a) = p^n a$.

So in particular, if C is injective, $C = \sum_{i=0}^{m} \langle x_i \rangle$ with $o(x_i) = p^{n_i}$, then $\mathrm{Ext}_{F_p}(C, A) = \sum_{i=0}^{m} \mathrm{Ext}_{F_p}(\langle x_i \rangle, A) = \sum_{i=0}^{m} \frac{A(n_i)}{p^{n_i} A}$.

<u>Corollary 8</u>. For $C = \langle c \rangle \in F_p$, $\mathrm{Ext}_{F_p}(C, C) = 0$.

<u>Proof</u>. Coker $\alpha = 0$ (in Theorem 6) since $p^{n_i} C(\ell_i) = C(k_i)$.

<u>Corollary 9</u>. Let $C \in F_p$ be as in Theorem 6 with $m = 1$. Let $A \in F_p$ be cyclic. Then

$$\operatorname{Ext}_{F_p}(C, A) = \begin{cases} \dfrac{A(k_0)}{p^{n_0}A(\ell_0) + A(\ell_1)} \oplus \dfrac{A(k_1)}{p^{n_1}(A(\ell_1) \cap p^{n_0}A(\ell_0))} & \text{if } p^{n_1}A(k_0) \subseteq A(k_1), \\ \\ \dfrac{A(k_0)}{p^{n_0}A(\ell_0) + A(\ell_1)[p^{n_1}]} \oplus \dfrac{A(k_1)}{p^{n_1}A(\ell_1)} & \text{if } A(k_1) \subsetneq p^{n_1}A(k_0). \end{cases}$$

Proof. The proof is an easy verification that the sequence

$$A(\ell_0) \oplus A(\ell_1) \xrightarrow{\alpha} A(k_0) \oplus A(k_1) \xrightarrow{\beta} \frac{A(k_0)}{p^{n_0}A(\ell_0) + A(\ell_1)} \oplus \frac{A(k_1)}{p^{n_1}(A(\ell_1) \cap p^{n_0}A(\ell_0))} \to 0$$

is exact if $p^{n_1}A(k_0) \subseteq A(k_1)$, where β is defined as follows:

$$\beta(a_0, 0) = \begin{cases} (\bar{a}_0, p^{n_1}\bar{a}_0) & \text{if } p^{n_0}A(\ell_0) \subsetneq A(\ell_1), \\ \\ (\bar{a}_0, 0) & \text{if } A(\ell_1) \subseteq p^{n_0}A(\ell_0), \text{ and} \end{cases}$$

$$\beta(0, a_1) = (0, \bar{a}_1),$$

where $\bar{a}$ is the appropriate coset of a, and that the sequence

$$A(\ell_0) \oplus A(\ell_1) \xrightarrow{\alpha} A(k_0) \oplus A(k_1) \xrightarrow{\beta} \frac{A(k_0)}{p^{n_0}A(\ell_0) + A(\ell_1)[p^{n_1}]} \oplus \frac{A(k_1)}{p^{n_1}A(\ell_1)} \to 0$$

is exact if $A(k_1) \subsetneq p^{n_1}A(k_0)$, where β is defined as follows:

$$\beta(a_0, 0) = (\bar{a}_0, 0),$$

$$\beta(0, a_1) = \begin{cases} (0, \bar{a}_1) & \text{if } p^{n_1}A(\ell_1) = 0, \text{ and} \\ \\ (\bar{a}'_0, \bar{a}_1) & \text{if } p^{n_1}A(\ell_1) \neq 0, \end{cases}$$

where $p^{n_1}a'_0 = a_1$, $a_0 \in A(k_0)$.

Definition 10. Let C, ℓ_i, k_i, n_i, m be as in Theorem 6. Let $A \in F_p$ be cyclic. We then define inductively the subgroups C_i of A. $C_m = 0$, and for $0 \leq i < m$, if $r_j = \sum_{k=i+1}^{j} n_k$,

$$C_i = \begin{cases} A(\ell_{i+1}) & \text{if } p^{n_{i+1}}A(k_i) \subseteq A(k_{i+1}), \\ \{x \in A(\ell_{i+1}) : p^{r_j}x \in C_j,\ i+1 \leq j < t < m\} & \text{if } p^{r_t}A(k_i) \subseteq A(k_t) \text{ and} \\ & A(k_{s-1}) \subsetneq p^{r_{s-1}}A(k_i) \text{ for all } s,\ i+1 < s \leq t \leq m, \text{ and} \\ \{x \in A(\ell_{i+1}) : p^{r_j}x \in C_j \text{ for all } j,\ i+1 \leq j \leq m\} & \text{if } A(k_t) \subsetneq p^{r_t}A(k_i) \\ & \text{for all } t,\ i+1 < t \leq m. \end{cases}$$

We also inductively define the subgroups B_i. $B_0 = p^{n_0}A(\ell_0)$, and for $0 < i \leq m$, if $s_j = \sum_{k=j}^{i} n_k$,

$$B_i = \begin{cases} p^{n_i}A(\ell_i) & \text{if } A(k_i) \subseteq p^{n_i}A(k_{i-1}), \\ p^{n_i}(A(\ell_i) \cap p^{n_{i-1}}(A(\ell_{i-1}) \cap p^{n_{i-2}}(\ldots \cap p^{n_j}A(\ell_j)))) & \\ \quad \text{if } p^{s_{j+1}}A(k_j) \subseteq A(k_i) \text{ for all } j,\ 0 < j \leq i-1 \text{ and} & \\ \quad A(k_i) \subsetneq p^{s_j}A(k_{j-1}), \text{ and} & \\ p^{n_i}(A(\ell_i) \cap p^{n_{i-1}}(\ldots \cap p^{n_0}A(\ell_0))) & \text{if } p^{s_{j+1}}A(k_j) \subseteq A(k_i) \\ \quad \text{for all } j,\ 0 \leq j \leq i-1. & \end{cases}$$

Theorem 11. Let A and C be cyclics in F_p, C as in Theorem 6. Then $\mathrm{Ext}_{F_p}(C, A) \cong \sum_{i=0}^{m} \frac{A(k_i)}{B_i + C_i}$, where B_i and C_i are as in Definition 10.

Proof. The proof is by defining β and showing the sequence $\sum_{i=0}^{m} A(\ell_i) \xrightarrow{\alpha} \sum_{i=0}^{m} A(k_i) \xrightarrow{\beta} \sum_{i=0}^{m} \frac{A(k_i)}{B_i + C_i}$ 0 is exact, where α is defined in Theorem 6. Proceed by induction on m; Corollary 9 is the initial step. Extending β' for $n = m - 1$ to β for the case $n = m$ and verifying that β is the cokernel of α is a simple but tedious exercise.

For any $A \in F_p$, there is an exact sequence $0 \to A \to \overline{A} \xrightarrow{\pi} \overline{A}/A \to 0$ with $h_{\overline{A}} = ht_{\overline{A}}$. Then if $C = \langle c \rangle \in F_p$, $\mathrm{Hom}_{F_p}(C, \overline{A}/A) = \frac{\overline{A}}{A}[c]$ and $\mathrm{Hom}(C, \overline{A}) = \overline{A}[c]$. Furthermore $\pi_*(\mathrm{Hom}(C, \overline{A})) = \frac{\overline{A}[c] + A}{A}$, so that $\mathrm{Hom}_{F_p}(C, \overline{A}) \xrightarrow{\pi_*} \mathrm{Hom}_{F_p}(C, \overline{A}/A) \to \mathrm{Ext}_{F_p}(C, A) \to 0$ exact implies

$$\mathrm{Ext}_{F_p}(C, A) \cong \frac{\frac{\overline{A}}{A}[c]}{\frac{\overline{A}[c] + A}{A}} . \tag{2}$$

If $A = \langle a \rangle$, $h[a] = (\ell_0, \ldots, \ell_0 + n_0 - 1, \ell_1, \ldots, \ell_1 + n_1 - 1, \ldots, \ell_m, \ldots, \ell_m + n_m - 1, \infty, \ldots)$, $k_i = \ell_i + n_i$, then $\overline{A} = \sum_{i=0}^{m} \langle x_i \rangle$, $o(x_i) = p^{k_i}$ and $\overline{A}/A = \sum_{i=0}^{m} \langle y_i \rangle$, $o(y_i) = p^{\ell_i}$. If $m = 0$, then $\mathrm{Ext}_{F_p}(C, A)$ is easily computed by (2) and is seen to be cyclic; in fact, $\mathrm{Ext}_{F_p}(C, A) = 0$ unless $h_A a < h_C c$. If $m > 0$, computing by (2) is similar to computing by Theorem 6 since $\pi : \overline{A} \to \overline{A}/A$ is given by $\pi(x_i) = y_i - p^{\ell_{i+1} - k_i} y_{i+1}$, $0 \le i < m$, and $\pi(x_m) = y_m$. If A and C are both cyclic, then (2) together with Theorem 6 shows that $\mathrm{Ext}_{F_p}(C, A) = \sum_{j=0}^{n} \langle x_i \rangle$, where $n = \min\{m_1, m_2\}$, with m_1 and m_2 the number of "jumps" in the value sequence of the generators of C and A, respectively.

Hunter, Richman, and Walker [1] have characterized the groups in F_p which are direct sums of cyclics and, consequently, for which Theorem 11 computes $\mathrm{Ext}_{F_p}(C, A)$. In particular, p^2-bounded groups are direct sums of cyclics [1; Theorem 4], but Hunter, Richman, and Walker [1] give a p^3-bounded group in F_p which is not a direct sum of cyclics. In fact, the indecomposables in F_p are not necessarily simply presented (as defined by Hunter, Richman, and Walker [2]). For example, if $C = \sum_{i=1}^{3} \langle c_i \rangle$, where $o(c_1) = p^2$, $o(c_2) = p^4$, $o(c_3) = p^5$, and $h_C = ht_C$, then the subgroup D of C generated by $x = c_1 + pc_2 + p^2c_3$ and $y = -pc_1 - p^3c_3$ is indecomposable and not simply-presented.

4. Relative homological algebras. Let $\mathcal{E}$ denote a class of short exact sequences in $\mathcal{F}_p$. If $A \subseteq B$ and the sequence $0 \to A \to B \to B/A \to 0$ belongs to $\mathcal{E}$, we write $A \square B$. A class $\mathcal{E}$ of short exact sequences is a proper class if it has the following six closure properties.

(i) $\mathcal{E}$ contains all split short exact sequences.

(ii) $\mathcal{E}$ is closed under congruence.

(iii) If $A \square C$ and $A \subseteq B \subseteq C$, then $A \square B$.

(iv) If $B \square C$ and $A \subseteq B$, then $B/A \square C/A$.

(v) If $A \square B$ and $B \square C$, then $A \square C$.

(vi) If $A \square C$ and $B/A \square C/A$, then $B \square C$.

Richman and Walker [4; §8] point out that in any pre-abelian category a proper class of exact sequences $\mathrm{Pext}(C, A)$ forms a subgroup of $\mathrm{Ext}(C, A)$, and moreover that if $\mathrm{Pext}^n(C, A)$ is defined in the usual way, we get the usual long exact sequences of abelian groups.

Let $E : 0 \to A \xrightarrow{\alpha} B \xrightarrow{\beta} C \to 0$ be exact. Then it is easy to see that $E_n : 0 \to A(n) \to B(n) \to C(n) \to 0$ is exact for every non-negative integer n. Let $n\text{-SPLIT}_{\mathcal{F}_p}(C, A)$ denote the subset of $\mathrm{Ext}_{\mathcal{F}_p}(C, A)$ consisting of those sequences E such that E_n is split exact in $\mathcal{F}_p$. If $E \in n\text{-SPLIT}_{\mathcal{F}_p}(C, A)$, we say that E is an n-split exact sequence. A routine check verifies

Theorem 12. The class of n-split exact sequences is a proper class.

The following is an easy result.

Theorem 13. P is an n-split projective if and only if $P = P(n)$. There are not enough.

Theorem 14. G is an n-split injective if and only if G/p^nG is injective.

Proof. Let G be an n-split injective and $x \in G\backslash p^nG$. Suppose $h_G x \geq n$. Then the sequence

$$\begin{array}{ccccccccc} 0 & \to & \langle x\rangle & \xrightarrow{f} & \langle y\rangle & \to & \langle y\rangle/\langle x\rangle & \to & 0 \\ & & {\scriptstyle i}\downarrow & \swarrow{\scriptstyle g} & & & & & \\ & & G & & & & & & \end{array} \qquad (3)$$

is n-split exact where $h[y] = (0, 1, \ldots, n - 1, h_G x, h_G px, \ldots)$. Then there exists $g \in \mathrm{Hom}_{F_p}(\langle y\rangle, G)$ such that $gf = i$ since G is n-split injective. But this implies $ht_G x \geq n$, so $x \in p^n G$. Hence $h_G x < n$. Let $h[y] = (0, 1, \ldots, h_G x - 1, h_G x, h_G px, \ldots)$. Then (3) is n-split exact, and the existence of g implies $ht_G x \geq h_G x$. Thus $ht_G x = h_G x$ for $x \in G\backslash p^n G$. Hence $G/p^n G$ is injective.

Conversely, suppose $h_{G/p^nG} = ht_{G/p^nG}$, $0 \to A \xrightarrow{\alpha} B \to C \to 0$ is n-split, and $\gamma \in \mathrm{Hom}_{F_p}(A, G)$. For simplification we identify $\alpha(A)$ with A. Let $\beta \in \mathrm{Hom}_{F_p}(B(n), A(n))$ such that $\beta\alpha = 1_{A(n)}$. Choose $x \in B\backslash A$ such that $h_B x$ is maximal. Then $px \in A$ and $h_B(a + x) = \min\{h_B a, h_B x\}$ for $a \in A$. If $h_B x \geq n$, define $\alpha'(x) = \gamma\beta(x)$; then $\alpha' \in \mathrm{Hom}_{F_p}(A + \langle x\rangle, G)$. If $\gamma(px) = 0$, define $\alpha'(x) = 0$. Thus we may assume $h_B x < n$ and $\gamma(px) \neq 0$. Now $h_G\gamma(px) \geq h_A px = h_B px \geq ht_B px \geq 1$. Also $G/p^n G$ injective implies that if $x \in G\backslash p^n G$, that is, $ht_G x < n$, then $h_G x = ht_G x$, and hence that if $h_G x \geq n$, then $ht_G x \geq n$. Thus if $h_G\gamma(px) \geq n$, there exists $x' \in G$ with $h_G x' \geq n - 1 \geq h_B x$ and $px' = \gamma(px)$; if $h_G\gamma(px) < n$, then $h_G\gamma(px) = ht_G\gamma(px) = m$ and there exists $x' \in G$ with $\gamma(px) = px'$ and $ht_G x' = ht_G x' = m - 1 \geq h_B x$. So define $\alpha'(x) = x'$. Then $h_G\alpha'(x + a) \geq \min\{h_G\alpha'(x), h_G\alpha'(a)\} \geq \min\{h_B x, h_B a\} = h_B(x + a)$. Hence (by finiteness) we can extend γ to $\alpha' \in \mathrm{Hom}_{F_p}(B,G)$ such that $\alpha'\alpha = \gamma$. So G is an n-split injective.

In order to show that there are enough injectives, we need the following generalization of Lemma 1.

<u>Theorem 15</u>. Let $A \in F_p$ and let n be non-negative integer. Then there exists $G \in F_p$ such that

(i) $A \subseteq G$,

(ii) $p^n G = A(n)$, and

(iii) $h_{G/p^nG} = ht_{G/p^nG}$.

<u>Proof</u>. The proof is by induction on $|A|$. Let $x \in A$ with $h_A x$ maximal. We

may assume $h_A x \geq n$ since otherwise the assertion is Lemma 1. By induction, there exists $H \in F_p$ with $A/\langle x\rangle \subseteq H$, $p^n H = \frac{A}{\langle x\rangle}(n)$, and $h_{H/p^n H} = ht_{H/p^n H}$. There is an exact sequence $E : 0 \to A/\langle x\rangle \to H \to H' \to 0$. But $\mathrm{Ext}_{F_p}(H', A) \xrightarrow{\pi_*} \mathrm{Ext}_{F_p}(H', A/\langle x\rangle \to 0$ is exact, so there is an exact sequence $E' : 0 \to A \to G' \to H' \to 0$ and $\pi_*(E') = E$. Hence $H \cong G'/\langle x\rangle$. Let $G = \dfrac{G' \oplus \langle x'\rangle}{\langle (x, -p^n x')\rangle}$ where $h[x'] = (0, 1, \ldots, n-1, h_G x, \infty, \ldots)$. Then $A \subseteq G$ by $a \to (a, 0) + \langle (x, -p^n x')\rangle = \bar{a}$ since trivially $h_A a = h_G \bar{a}$. Also

$$p^n G = \frac{p^n G' \oplus \langle p^n x'\rangle + \langle x, -p^n x'\rangle}{\langle (x, -p^n x')\rangle} = \frac{(p^n G' + \langle x\rangle) \oplus \langle p^n x'\rangle}{\langle (x, -p^n x')\rangle} \cong A(n) \quad \text{since}$$

$$\frac{A}{\langle x\rangle}(n) = \frac{A(n)}{\langle x\rangle} \cong \frac{p^n G' + \langle x\rangle}{\langle x\rangle}. \quad \text{Furthermore,}$$

$$G/p^n G = \frac{(G' \oplus \langle x\rangle)/\langle (x, -p^n x')\rangle}{((p^n G' + \langle x\rangle) \oplus \langle p^n x'\rangle)/\langle (x, -p^n x')\rangle} \cong \frac{G' \oplus \langle x\rangle}{(p^n G' + \langle x\rangle) \oplus \langle p^n x'\rangle}$$

since x has maximal value in A. Now $G'/p^n G' + \langle x\rangle \cong \dfrac{G'/\langle x\rangle}{(p^n G' + \langle x\rangle)/\langle x\rangle}$ which has the height valuation by induction, and $\langle x\rangle/\langle p^n x'\rangle$ has the height valuation by construction. Thus $h_{G/p^n G} = ht_{G/p^n G}$.

Corollary 16. For $A, C \in F_p$ and any non-negative integer n, $n\text{-SPLIT}^2_{F_p}(C, A) = 0$.

Proof. There are enough n-split injectives, and coimages of n-split injectives are n-split injectives, so the relative homological dimension ≤ 1.

Theorem 17. $n\text{-SPLIT}_{F_p}(C, A) \subseteq \mathrm{Ext}_{F_p}(C, A)[p^n]$.

Proof. By Theorem 15, $A \subseteq G$ with $p^n G = A(n)$ and G n-split injective. So $G' = G/A$ is such that $p^n G' = 0$ and $\mathrm{Hom}_{F_p}(C, G') \to n\text{-SPLIT}_{F_p}(C, A) \to 0$ is exact. Hence $p^n(n\text{-SPLIT}_{F_p}(C, A)) = 0$ since $p^n \mathrm{Hom}_{F_p}(C, G') = 0$.

There is another family of relative homological algebras which is similar to the family of n-split relative homological algebras. For each non-negative integer n, let $n\text{-SPLIT}_{Ab}(C, A)$ be the exact sequences $0 \to A \to B \to C \to 0$ such that $0 \to A(n) \to B(n) \to C(n) \to 0$ splits in Ab. These sequences are called the <u>*Ab* n-split exact sequences</u>; if $n = 0$, we say the sequence is <u>*Ab* split</u>.

<u>Theorem 18</u>. The *Ab* n-split exact sequences form a proper class.

<u>Theorem 19</u>. The *Ab* n-split injectives are the injectives. There are not enough.

<u>Proof</u>. Let G be an *Ab* n-split injective. Choose $s < \omega$ such that $s > h_G z$ for all $0 \neq z \in G$. Let $x \in G$, say $o(x) = p^r$. Consider $\langle y_1 \rangle$ with $h[y_1] = (0, 1, \ldots, h_G x - 1, h_G x, h_G px, \ldots)$, and $\langle y_2 \rangle$ where $y_2 = 0$ if $h_G x \leq n$, and $h_G[y_2] = (s, s + 1, \ldots, s + r - 1, \infty, \ldots)$ if $h_G x > n$. Let $\alpha(x) = p^m y_1 + y_2$ where $h_G x = m$, and let $Z = \operatorname{coker} \alpha$. Then the sequence

$$\begin{array}{ccccccc} 0 \to & \langle x \rangle & \xrightarrow{\alpha} & \langle y_1 \rangle \oplus \langle y_2 \rangle & \to Z \to 0 \\ & {\scriptstyle i}\downarrow & \swarrow & & \\ & G & & & \end{array}$$

is *Ab* n-split exact. Since G is *Ab* n-split injective, there exists α' with $\alpha'\alpha = i$ where i is the inclusion map $i : \langle x \rangle \to G$. Then since $\alpha'(y_2) = 0$, we have $ht_G x \geq h_G x$. Hence $ht_G x = h_G x$ and G is injective. To see that there are not enough injectives, consider $\langle x \rangle$ with $o(x) = p$ and $hx = n + 1$. It is impossible to embed $\langle x \rangle$ in G with $h_G = ht_G$ <u>and</u> have $0 \to \langle x \rangle \to G \to G/\langle x \rangle \to o$ be *Ab* n-split.

The corresponding calculation for the projectives yields

<u>Theorem 20</u>. P is an *Ab* n-split projective if and only if $P = \sum_{i=0}^{m} \langle x_i \rangle$ where $h_P[x_i] = (n, n + 1, n + 2, \ldots, n + m_i, \infty, \ldots)$, $m_i \geq 0$. There are not enough.

<u>Corollary 21</u>. For the *Ab* split sequences, the *Ab* split projectives are the *Ab* split injectives which are the injectives in F_p.

Since we do not have enough *Ab* n-split injectives nor enough *Ab* n-split projectives in F_p, we consider the *Ab* n-split exact sequences in F_p^*, written $n^*\text{-SPLIT}_{Ab}(C, A)$. In order to distinguish the category, we call these

Ab n^*-split exact sequences. Recall that the infinite cyclics $\langle c\rangle$ where $h[c] = (m, m+1, m+2, \ldots)$ are projective in F_p^*. It is easy to see that the Ab n-split projectives in F_p are also Ab n^*-split projectives.

Let $C = \langle c\rangle \in F_p$, $h[c] = (\ell_0, \ell_0 + 1, \ldots, \ell_0 + n_0 - 1, \ell_1, \ldots, \ell_1 + n_1 - 1, \ldots, \ell_m, \ell_m + 1, \ldots, \ell_m + n_m - 1, \ldots)$ where $\ell_i + n_i < \ell_{i+1}$, $n_i \geq 1$, $m \geq 0$. Let $r = \sum_{i=0}^{m} n_i$. Then $p^r = o(c)$. Define $k_i = \ell_i + n_i$. Note that if $\ell_m + n_m \leq n$, the sequence (1) of Theorem 6 is an Ab n^*-split projective resolution of C. Then for all $A \in F_p$, $n\text{-SPLIT}^2_{Ab}(C, A) = n^*\text{-SPLIT}^2_{Ab}(C, A) = 0$. An induction on the rank of B shows that for any $B \in F_p$ such that $B(n) = 0$, $n\text{-SPLIT}^2_{Ab}(B, A) = 0$ for $A \in F_p$. Thus for $B \in F_p$, the sequence $0 \to B(n) \to B \to B/B(n) \to 0$ is Ab n-split exact. So $0 = n^*\text{-SPLIT}^t_{Ab}(B/B(n), A) \to n^*\text{-SPLIT}^t_{Ab}(B, A) \to n^*\text{-SPLIT}^t_{Ab}(B(n), A) \to n^*\text{-SPLIT}^{t+1}_{Ab}(B/B(n), A) = 0$ is exact for $t \geq 2$. Thus $n^*\text{-SPLIT}^t(B, A) \cong n^*\text{-SPLIT}^t_{Ab}(B(n), A)$ if $t \geq 2$. Thus we may assume $\ell_0 \geq n$. Consider $P_0 = B \oplus \sum_{i=0}^{m} C_i$, where $B = \langle b\rangle$, $o(b) = p^r$, $h[b] = (n, n+1, \ldots, n+r-1, \infty, \ldots)$, and each C_i is an infinite cyclic $\langle c_i\rangle$ with $h[c_i] = (\ell_i, \ell_i + 1, \ell_i + 2, \ldots)$. Then define $\gamma_0 : P_0 \to C$ by $\gamma_0(b) = c$ and $\gamma_0(c_i) = p^{r_i}c$, where $r_0 = 0$ and $r_i = \sum_{j=0}^{i-1} n_j$, $0 < i \leq m$. Note that $\ker \gamma_0 = \sum_{i=0}^{m+1} \langle d_i\rangle$, where $h[d_{m+1}] = (n, n+1, \ldots, n+r-1, \ell_0 + r, \ell_0 + r + 1, \ell_0 + r + 2, \ldots)$, and $h[d_i] = (k_i, k_i + 1, k_i + 2, \ldots)$ for $0 \leq i \leq m$. Now let $P_1 = \left(\sum_{i=0}^{m} \langle d_i\rangle\right) \oplus \langle f_1\rangle \oplus \langle f_2\rangle$, where $\langle f_i\rangle$ are infinite cyclics with $h[f_1] = (n, n+1, n+2, \ldots)$, and $h[f_2] = (\ell_0 + r, \ell_0 + r + 1, \ell_0 + r + 2, \ldots)$. Define $\gamma_1 : P_1 \to P_0$ by $\gamma_1(d_i) = d_i$, $0 \leq i \leq m$, $\gamma_1(f_1) = d_{m+1}$, and $\gamma_1(f_2) = p^r d_{m+1}$. Then $\ker \gamma_1 = G = \langle g\rangle$, where $h[g] = (n+r, n+r+1, n+r+2, \ldots)$. Let $P_2 = G$. Then $0 \overset{\gamma_2}{\to} P_2 \overset{\gamma_1}{\to} P_1 \overset{\gamma_0}{\to} P_0 \to C \to 0$ is an Ab n^*-split projective resolution for C. Hence $n^*\text{-SPLIT}^3_{Ab}(C, A) = 0$ for a cyclic $C \in F_p$. If $C(n) = C = \Sigma\langle x_i\rangle$ in Ab, $0 \to \langle x_1\rangle \to C \to C/\langle x_1\rangle \to 0$ is Ab n-split. Now an induction on the rank of C gives $n^*\text{-SPLIT}^3_{Ab}(C, A) = 0$ for all $C \in F_p$.

Theorem 22. For $A, C \in F_p$, $n\text{-SPLIT}^3_{Ab}(C, A) = 0$.

Proof. We need only observe $n\text{-SPLIT}^3_{Ab}(C, A) = n^*\text{-SPLIT}^3_{Ab}(C, A)$.

Consider $A = \langle a \rangle$, $o(a) = p^3$, $h_A[a] = (n, n+2, n+4, \infty, \ldots)$, $B = \langle b_1 \rangle \oplus \langle b_2 \rangle$, $o(b_1) = p^3$, $o(b_2) = p^2$, $h_B[b_1] = (n, n+3, n+4, \infty, \ldots)$, $h_B[b_2] = (n+1, n+2, \infty, \ldots)$, and $C = \langle c \rangle$, $o(c) = p^2$, $h_C[c] = (n+1, n+3, \infty, \ldots)$. Then $0 \to A \overset{\alpha}{\to} B \overset{\beta}{\to} C \to 0$ is *Ab* n-split exact if $\alpha(a) = b_1 + b_2$, $\beta(b_1) = -c$ and $\beta(b_2) = c$. So $n\text{-SPLIT}_{Ab}(C, C) \to n\text{-SPLIT}_{Ab}(B, C) \to n\text{-SPLIT}_{Ab}(A, C) \to n\text{-SPLIT}^2_{Ab}(C, C)$ is exact. Now $n\text{-SPLIT}_{Ab}(C, C) = 0$, $n\text{-SPLIT}_{Ab}(B, C) = Z(p)$, and $n\text{-SPLIT}_{Ab}(A, B) = Z(p) \oplus Z(p)$. Hence $n\text{-SPLIT}^2_{Ab}(C, C) \neq 0$, and we have the following theorem.

Theorem 23. The class of *Ab* n-split exact sequences has relative homological dimension 2.

REFERENCES

[1] R. Hunter, F. Richman, and E. A. Walker, Finite direct sums of cyclic valuated p-groups, Pac. J. Math. 69(1977) 97-104.

[2] R. Hunter, F. Richman, and E. A. Walker, Simply presented valuated abelian p-groups, to appear in J. Alg.

[3] S. MacLane, "Homology", Springer-Verlag, New York 1967.

[4] F. Richman and E. A. Walker, Ext in pre-abelian categories, to appear in Pac. J. Math.

[5] F. Richman and E. A. Walker, Valuated groups, to appear.

CRITERIA FOR FREENESS IN GROUPS AND VALUATED VECTOR SPACES

Paul Hill

1. Introduction. For more than forty years there has been a very useful criterion, due to Pontryagin [14], for the freeness of countable torsion-free groups. This well-known criterion is the following. It is understood throughout that all groups are Abelian.

Criterion 1. A countable torsion-free group is free if all of its subgroups of finite rank are free.

Another criterion for freeness of countable groups is also well known and has been around for a little over a quarter of a century. In 1950 Specker [18] proved the following.

Criterion 2. A countable torsion-free group is free if it can be embedded in a product of infinite cyclic groups.

Let $P = \prod Z$ denote an infinite product of infinite cyclic groups; the exact cardinality of the index set for the infinite product is not important. We shall let B denote the subgroup of P consisting of the "bounded sequences". That is, if I is the index set for the product P, then the element $\{n_i\}$ in P belongs to B if there exists a positive integer k such that $n_i \leq k$ for all i in I. Specker also established the following criterion, which is of a restricted nature, but does apply to certain uncountable groups.

Criterion 3. A group of cardinality not exceeding $\aleph_1$ is free if it can be embedded in B.

Through the fifties and even through the early and middle sixties (when there was great progress in Abelian groups) there were no further outstanding developments on criteria for freeness; at least

I do not recall any. Certainly at the time of the first New Mexico State conference on Abelian groups in 1962 there was not available a good general test for the freeness of an uncountable group. As a matter of fact, at the time of the first New Mexico State conference, three of the most prestigious open problems in Abelian groups all concerned criteria for freeness but that point was not emphasized; see Topics in Abelian Groups [21] , the proceedings of that conference. The three big problems to which I have referred were: Baer's problem, Whitehead's problem, and Specker's problem. It is well known that each of the above problems remained open for a long time. With the advantage of hindsight, I can give a partial explanation of the extreme difficulty in finding solutions. It seems that in each case the problem was naturally and elegantly formulated but did not attack directly the real structural problem involved. This viewpoint will be explained further in the next section, where these problems are discussed briefly case by case.

2. Survey of Contemporary Results. Apparently Griffith [5] was the first to realize that a question really more basic to the structure of mixed Abelian groups than Baer's problem is the following. How much can the "torsion-free part" of a group be changed by collapsing the torsion? It was found that the torsion-free part can be transformed so completely that it is possible for a group G to have only free torsion-free subgroups and yet for G/T to be divisible where T is the torsion of G. This fact enabled Griffith to solve Baer's problem immediately and thereby to establish the following criterion for freeness.

Criterion 4. If $\mathrm{Ext}(G,T) = 0$ for all torsion groups T, then G is free.

The preceding criterion was established by Griffith in June, 1967 during or about the time of the conference on Abelian groups in Montpellier, France.

Although progress on Whitehead's problem was made by Stein [19], Rotman [16], Chase [2], Nunke [13], and Griffith [6], by far the most important development is due to Shelah [17]. The main breakthrough was the consideration of how the problem depends on certain set-theoretical axioms consistent with ZFC. Shelah [17] has established the following.

<u>Criterion 5.</u> Under the assumption $V = L$, the axiom of constructibility, a group G of cardinality not exceeding $\aleph_1$ is free provided that $\mathrm{Ext}(G,Z) = 0$.

For a complete exposition of Shelah's result on Whitehead's problem (and related results), we refer to Eklof [3].

We now turn to Specker's problem: Is B free, where B is the subgroup of $P = \prod Z$ consisting of the bounded sequences? The first solution to Specker's problem (without the aid of the continuum hypothesis) was given by Nöbeling [12]. This provided the following criterion.

<u>Criterion 6.</u> If G can be embedded in B, the group of bounded sequences, then G is free.

Shortly after the preceding criterion was established by Nöbeling, a stronger result was proved by Bergman [1]: any torsion-free commutative ring generated by idempotents is free (as far as its additive structure is concerned). In [8], I gave an elementary proof that a subset of an idempotent generating set of a commutative ring always generates a pure subring. The latter result is all that is needed to prove Bergman's theorem and to obtain the solution of Specker's problem. I suspect that this result can be improved by replacing the condition that the ring is generated by idempotents with a more general condition on the generators, but in any case we have the following criterion due to Bergman.

Criterion 7. If G is torsion free and can be embedded in the additive group of a commutative ring generated by idempotents, then G is free.

It is, of course, well known that a difficult question can sometime be changed to an easier one by making it more general. Certainly the freeness of R^+ for a torsion-free, commutative ring generated by idempotents is simpler than the direct solution to Specker's original problem even though the first result yields the second. The procedure that I use next is of an entirely different nature. The basic idea here is that one takes a remarkable and elegant result and redresses it so that it is no longer charming, and then he refines the theorem because something must be done now to improve it. It was only seven or eight years ago that I observed that Pontryagin's theorem can be stated as follows.

Criterion 1'. Let G be a countable group. If G is the union of an ascending sequence of pure subgroups G_n, then G is free if G_n is free for each positive integer n.

A genuine acquaintance with the preceding version of Pontryagin's theorem leads to an awareness that it has an awkward feature. The countability condition on G is not essential. This fact was proved in [9], and establishes the following criterion.

Criterion 8. If G is the union of an ascending sequence of pure subgroups G_n, then G is free if G_n is free for each positive integer n.

One step leads to another. In the case at hand, we notice that the significance of G_n being pure is that the small (finite rank) subgroups of G/G_n are free. The idea is developed further in [10] and [11], and criteria for freeness of G are given in terms of a development of G as a smooth chain of free subgroups G_α where the quotients $G_{\alpha+1}/G_\alpha$ are almost free. In particular, the following criterion is given in [11].

Criterion 9. If G is the union of a smooth chain

$$G_0 \subseteq G_1 \subseteq \cdots \subseteq G_\alpha \subseteq \cdots$$

of length not exceeding $\aleph_n$ $(n < \omega)$ of free groups G_α, then G is free provided that $G_{\alpha+1}/G_\alpha$ is $\beth_n$-free for each α.

Our final criterion is due to Griffith [7].

Criterion 10. If G is the extension of a free group by a totally projective primary group, then G is free.

I would like to emphasize at this point that it has not been my purpose in this paper to give a complete and detailed survey of presently known criteria for freeness. Rather I have limited the discussion to ten criteria for freeness. Seven of the ten criteria have been established in the last ten years. There is some duplication among the ten in the sense that one criterion may generalize or extend a previous one, but in such case the first criterion carries considerable historical significance. At any rate, it should be apparent that great strides have been made recently concerning the freeness of abelian groups, and many criteria are now available. The state of the art for freeness in valuation vector spaces is not so advanced. We shall investigate this subject in the next section, where some original results will be presented on criteria for freeness in valuation vector spaces.

3. Valuation Vector Spaces. Recently two important papers have appeared on valuated groups and vector spaces. One is by Fuchs [4] and the other is by Richman and Walker [15]. By a valuation vector

space, we shall mean a vector space with a valuation satisfying the conditions of [4]; in particular, $|x + y| \geq \min\{|x|, |y|\}$, where $|x|$ denotes the value of x. The terminology here, however, shall be abbreviated to "valuation space". We mention also that the dimension of a valuation space is the same, by definition, as it would be if the vector space had no valuation.

We should perhaps stress at the outset that there are some fundamental differences between the structure of groups and the structure of valuation spaces although there are some similarities, too. For example, a subspace of a free space need not itself be free. Moreover, for valuation spaces in general the isomorphism

$$(*) \qquad B/B\cap A \cong \langle A,B\rangle/A$$

does not hold. We shall see, however, that the isomorphism (*) is valid under a certain condition specified below. We have been somewhat too vague about the meaning of (*). Actually, when we refer to the isomorphism (*), we mean that the natural map from $B/B\cap A$ onto $\langle A,B\rangle/A$ is an isometry. Thus (*) refers to a particular isomorphism.

<u>Definition 1</u>. If A and B are subspaces of a valuation space V, then A is said to be compatible with B if for each pair (a,b) in $A\times B$ there exists $c\in A\cap B$ such that $|a+c| \geq |a+b|$.

It is easily demonstrated that the relation of being compatible is symmetric, and we write $A \parallel B$ if A and B are compatible.

<u>Proposition 1</u>. If $A\parallel B$, the isomorphism (*) holds.

We omit the formality of a proof of Proposition 1 since the only thing to verify is that values are preserved, and the compatibility of A and B ensures this.

The next concept is most important in the study of valuation spaces.

<u>Definition 2.</u> A subspace A of a valuation space V is called a

separable subspace if for each $x \in V$ there exists a sequence $\{a_n\} \in A$, depending on the choice of x, such that

$$|x + A| = \sup_{n<\omega}\{|x + a_n|\}.$$

If the sequence above is replaced by a single element $a \in A$, the subspace is called a nice subspace. Nice subspaces were studied in [4] and [15], but apparently the significance of separable subspaces was overlooked. We now use the preceding concepts to define certain classes of spaces.

<u>Definition 3.</u> A valuation space V is called an SF-space if V can be embedded in a free space. Likewise, V is called an SSF-space if it can be embedded as a separable subspace of a free space, and finally V is called an NSF-space if it can be embedded as a nice subspace of a free space.

The above definitions identify three problems.

<u>Problem 1.</u> Characterize the SF-spaces.

<u>Problem 2.</u> Characterize the SSF-spaces.

<u>Problem 3.</u> Characterize the NSF-spaces.

We do not have solutions to the preceding problems. It is easy to construct examples of SF-spaces that are not free. It is much more difficult to construct examples of SSF-spaces or NSF-spaces that are not free. In fact, it was asserted in [4] that NSF-spaces are free. However, a counterexample was presented by Richman and Walker in [15]. The next result puts a strong restraint on an SSF-space.

<u>Theorem 1.</u> Any SSF-space must be absolutely separable, that is, it must be separable in every containing space.

Proof. First we demonstrate that a free space is absolutely separable. Let F be a free space with a fixed basis, $F = \sum_{i \in I} \langle b_i \rangle$.

Suppose that F is contained in an arbitrary valuation space V. Denote by $F(\beta)$ the subspace of F defined as follows:

$$F(\beta) = \sum_{|b_i| < \beta} \langle b_i \rangle .$$

Now assume that for some fixed $x \in V$ and every sequence $\{f_n\} \in F$ the inequality $|x + F| > \sup_{n<\omega} \{|x + f_n|\}$ prevails. Let $|x + F| = \beta$ and let $|x| = \alpha_0 < \beta$. There exist $\alpha_1 > \alpha_0$ and an element $f_1 \in F(\alpha_1)$ such that $|x + f_1| = \alpha_1$. Suppose that we have already chosen a finite ascending sequence of values $\alpha_0 < \alpha_1 < \cdots < \alpha_n$ and elements $f_i \in F(\alpha_i)$ so that $|x + f_i| = \alpha_i$. There exist $\alpha_{n+1} > \alpha_n$ and an element $f_{n+1} \in F(\alpha_{n+1})$ such that $|x + f_{n+1}| = \alpha_{n+1}$. Therefore, there exist an ascending sequence of values α_n and elements $f_n \in F(\alpha_n)$ such that $|x + f_n| = \alpha_n$. Let $\alpha = \sup_{n<\omega}\{\alpha_n\}$. There exists $f \in F(\alpha)$ such that $|x + f| \geq \alpha$. We observe that $|f - f_n| = \alpha_n$, which implies that $f \notin F(\alpha_n)$ for any positive integer. However, this is impossible since $F(\alpha) = \bigcup_{n<\omega} F(\alpha_n)$. Hence F must be separable in V, which shows that a free space is absolutely separable.

Now let A be a separable subspace of a free space F, and let A be contained in an arbitrary space V. Construct the amalgamated sum of F and V over A,

$$S = F \underset{A}{\oplus} V = \{(f,v) + A^* : f \in F, \ v \in V\},$$

where $A^* = \{(a,-a) : a \in A\}$. There are natural embeddings i_F and i_V of F and V, respectively, into S. Suppose that A is not separable in V, and let $\beta = |x+A| > \sup\{|x+a_n|\}$ for every sequence $\{a_n\}$ in A. It follows immediately that β is not cofinal with ω. Moreover, since $i_F(F) \supseteq i_F(A) = i_V(A)$, we know that $|(0,x)+A^* + i_F(F)| \geq \beta$. But $i_F(F)$ is free and hence separable in E. Therefore, there exists $f \in F$ such that $|(f,x)+A^*| \geq \beta$ since β is not cofinal with ω. Since A is separable in F, there exists $a \in A$ such that $|f + a| \geq \beta$. This implies that $|x - a| \geq \beta$, contrary to the choice of

$x \in V$. We conclude that A is separable in V, and the theorem is proved.

Lemma 1. Let A be a separable subspace of a free space $F = \sum_{i \in I} \langle x_i \rangle$. Let B be any subspace of F having infinite dimension. There exists a subset J of I such that the following conditions are satisfied, where F_J denotes the subspace $\sum_{j \in J} \langle x_j \rangle$ of V:

(i) $B \subseteq F_J$

(ii) $\dim(B) = \dim(F_J)$, and

(iii) $F_J \parallel A$.

Proof. It is enough to prove the theorem for the case that $B = F_{J(0)}$, where $J(0)$ is some subset of I. Since A is separable, there exists a subset $J(1) \supseteq J(0)$ having the same cardinality as $J(0)$ such that for each pair $(x,a) \in F_{J(0)} \times A$ there exists $c \in F_{J(1)} \cap A$ with the property that $|x + c| \geq |x + a|$. Inductively, choose $F_{J(n)}$ for each positive integer n, with the property that for each pair $(x,a) \in F_{J(n)} \times A$ there exists $c \in F_{J(n+1)} \cap A$ such that $|x + c| \geq |x + a|$. The set $J(n)$ can be selected so that it has the same cardinality as $J(n-1)$. The desired set J is obtained by letting $J = \bigcup J(n)$.

Theorem 2. Let A be a separable subspace of a free space F. If $\dim(A) \leq \aleph_1$, then A is free.

Proof. We may assume that $\dim(F) = \aleph_1$. If we write $F = \sum_{i \in I} \langle x_i \rangle$, there exists by Lemma 1 a smooth ascending chain of countable subsets $J(\alpha)$ of I whose union is I with the property that $F_{J(\alpha)} \parallel A$ for each α. Letting $A \cap F_{J(\alpha)} = A_\alpha$, we can show that A_α is nice in A. Since $A_{\alpha+1}$ has countable dimension over A_α, we conclude that A_α splits out of $A_{\alpha+1}$ and that A is free.

We need a generalization of Lemma 1 in order to prove our next theorem. First, however, we observe the following.

Lemma 2. Suppose that A is any subspace of V and that B is a finite dimensional subspace of V not contained in A. There exist elements $b_1, b_2, \cdots, b_n$ of B such that:

(1) $\{b_1+A, b_2+A, \cdots, b_n+A\}$ is a basis of $\langle B,A\rangle/A$, and

(2) if $|a + \sum t_i b_i| = \alpha$ where t_i is a scalar and $a \in A$, there exist elements $a_i \in A$ such that $|a_i + t_i b_i| \geq \alpha$ for each $i \leq n$.

Proof. The proof is by induction on $\dim(\langle B,A\rangle/A)$. Let $n = \dim(\langle B,A\rangle/A)$. If $n = 1$, there is nothing to prove. Assume $n > 1$, and choose b_1 so that $b_1 + A$ has the largest value among the nonzero elements of $\langle B,A\rangle/A$. Moreover, if possible, we choose b_1 so that $|b_1 + A| = |b_1 + a_1|$ for some $a_1 \in A$. Let

$$\langle B,A\rangle/A = \langle b_1 + A\rangle + \sum_{i=2}^{n} \langle b_i + A\rangle$$

where, by the induction hypothesis, the set $\{b_2, b_3, \cdots, b_n\}$ is chosen so that, for each $a \in A$ and each choice of n scalars t_i, there exist elements $a_2, a_3, \cdots, a_n$ in A with the property that $|a + \sum_{i=2}^{n} t_i b_i| \leq |a_i + t_i b_i|$ if $2 \leq i \leq n$. Now suppose that $a \in A$ and t_i is a scalar for $i \leq n$. Let $|a + \sum_{i=1}^{n} t_i b_i| = \alpha$. By the choice of b_1, we know that $|b_1 + A| \geq \alpha$, and by the induction hypothesis we may assume $t_1 \neq 0$. Let $|b_1 + A| = \beta$. If $\alpha = \beta$, there exists, by the choice of b_1, an element $a_1 \in A$ such that $|a_1 + t_1 b_1| = \alpha$. Otherwise, we could replace b_1 by $\sum_{i=1}^{n} t_i b_i$. From the fact that $|a_1 + t_1 b_1| = \alpha$, we conclude that $|a - a_1 + \sum_{i=2}^{n} t_i b_i| \geq \alpha$, and the induction hypothesis produces the desired result. Therefore, assume $\alpha < \beta$. Since $|b_1 + A| = \beta > \alpha$, $|t_1 b_1 + A| = \beta > \alpha$ and there exists $a_1 \in A$ such that $|t_1 b_1 + a_1| \geq \alpha$. The rest of the proof is the same as before.

Lemma 3. Let A be a separable subspace of the valuation space V and let B be any subspace of V. There exists a subspace C of V such that the following conditions are satisfied:

(i) $B \subseteq C$,

(ii) $\dim(C) \leq \aleph_0 \dim(B)$, and

(iii) $C \parallel A$.

Proof. The proof is rather trivial by induction on $\dim(B)$ once the proof survives finite dimensions. Thus it is enough to prove the lemma for the case that B has finite dimension n. By Lemma 2, we can choose $b_1, b_2, \cdots, b_n$ in B so that conditions (1) and (2) of Lemma 2 hold. For each b_i, there exists a sequence $\{a_{j,i}\}$ with the property that $|b_i + A| = \sup\{|b_i + a_{j,i}|\}$. Set $C_0 = B$ and define $C_1 = \langle C_0, a_{j,i}\rangle$. Inductively, define C_{n+1} in the indicated manner. Observe that if $(x,a) \in C_n \times A$, there exists $c \in C_{n+1} \cap A$ such that $|x + c| \geq |x + a|$. The space $C = \bigcup C_n$ has the desired properties, and the lemma is proved.

We are now prepared to prove one of our main results. The special case of the following theorem where V has a countable number of values was proved by Thomas [20].

Theorem 3. If the valuation space V is the set-theoretical union of a countable number of free subspaces, then V must be free.

Proof. Suppose that $V = \bigcup A_n$, where A_n is free for each positive integer n. Let $A_n = \sum_{i \in I(n)} \langle x_{n,i}\rangle$. The proof is by induction on $\dim(V)$. If $\dim(V) \leq \aleph_0$, the conclusion of the theorem is trivial since any space having countable dimension is free. Thus we shall assume that V has uncountable dimension. It follows from the separability of the free subspaces A_n and Lemma 3 that there exists a smooth chain

$$0 = B_0 \subseteq B_1 \subseteq B_2 \subseteq \cdot\cdot\cdot \subseteq B_\alpha \subseteq \cdot\cdot\cdot$$

of subspaces B_α of V with the following properties for each α

and each positive integer n:

(1) $\dim(B_\alpha) < \dim(V)$,

(2) $B_\alpha \| A_n$, and

(3) $B_\alpha \cap A_n = \sum_{i \in J(n,\alpha)} \langle x_{n,i} \rangle$, where $J(n,\alpha)$ is a subset of $I(n)$.

It is routine to demonstrate that B_α is nice in $\langle B_\alpha, A_n \rangle$ and therefore B_α is nice in V. Observe that $B_{\alpha+1}/B_\alpha = \bigcup \langle B_{\alpha+1} \cap A_n, B_\alpha \rangle / B_\alpha$ and that $B_{\alpha+1} \cap A_n$ is compatible with B_α. Thus $\langle B_{\alpha+1} \cap A_n, B_\alpha \rangle / B_\alpha$ is isomorphic to $B_{\alpha+1} \cap A_n / B_\alpha \cap A_n$ and is therefore free in view of condition (3). The induction hypothesis implies that $B_{\alpha+1}/B_\alpha$ is free since $\dim(B_{\alpha+1}/B_\alpha) < \dim(V)$. Since B_α is nice in $B_{\alpha+1}$, we know that $B_{\alpha+1} = B_\alpha + C_\alpha$ where C_α is free. It follows that $V = \sum C_\alpha$ is free.

Theorem 2 and Theorem 3 give criteria for freeness for valuation spaces. One of the fundamental problems concerning the structure of valuation vector spaces is the following.

<u>Main Problem.</u> Find necessary and sufficient conditions for a subspace of a free space to be free.

In order to solve the preceding problem we need the following concept.

<u>Definition 4.</u> A valuation space is called a QFF-space if it is the quotient of a free space by a free space.

<u>Proposition 2.</u> Every valuation space V of dimension not exceeding $\aleph_1$ is a QFF-space.

Proof. There is an exact sequence

$$K \rightarrowtail F \twoheadrightarrow V$$

where F is free and K is nice in F. Since $\dim(V) \leq \aleph_1$, we can take $\dim(F) \leq \aleph_1$. By Theorem 1, K is free, and V is a QFF-space.

Richman and Walker [15] were the first to find an example of a space that is not, in our terminology, a QFF-space. The specific example constructed by Richman and Walker involves the continuum hypothesis. By using König's theorem, we can eliminate this assumption. Thus the next result does not use the continuum hypothesis.

Theorem 4. There exist valuation spaces that are not QFF-spaces having the countable ordinals for values.

Proof. Choose an increasing chain

$$m_0 < m_1 < \cdots < m_\alpha < \cdots \qquad (\alpha < \omega_1)$$

of cardinals m_α so that

$$m_{\alpha+1} > m_\alpha^{\aleph_0}.$$

It follows that

$$m_0 m_1 m_2 \cdots m_\lambda \leq m_\lambda^{\aleph_0} < m_{\lambda+1}.$$

Letting $\rho = \prod_{\alpha<\omega_1} m_\alpha$ and $\sigma = \sum_{\alpha<\omega_1} m_\alpha$, we know that $\rho > \sigma$ by König's theorem. Define H_α to be a homogeneous space of value α and dimension m_α. Let $P = \prod_{\alpha<\omega_1} H_\alpha$, and set

$$K = \bigcup_{\beta<\omega_1} (\prod_{\alpha\leq\beta} H_\alpha).$$

It should be observed that $\dim(\prod_{\alpha\leq\beta} H_\alpha) < m_{\beta+1}$. Therefore, $\dim(K) = \sigma < \rho$. Obviously, P/K is homogeneous with value ω_1.

Assume that $P = A/B$ where A and B are both free. There is no loss of generality in assuming that A has no elements with value ω_1 since P has no elements having value ω_1. Let $A = \sum_{i\in I} \langle a_i \rangle$ and $B = \sum_{j\in J} \langle b_j \rangle$. There exists a subset I^* of I satisfying the following conditions.

(1) $|I^*| \leq \sigma$.

(2) $K \subseteq \langle \sum_{i\in I^*} \langle a_i \rangle, B \rangle / B$.

(3) $\sum_{i\in I^*} \langle a_i \rangle \parallel B$.

(4) $B \cap \sum_{i\in I^*} \langle a_i \rangle$ is a summand B^* of $B = \sum_{j\in J} \langle b_j \rangle$.

For simplicity, let $A^* = \sum_{i\in I^*} \langle a_i\rangle$, and let $B^* = B \cap A^*$. Now observe that $(A/B)/\langle A^*,B\rangle/B \cong (A/A^*)/\langle A^*,B\rangle/A^*$. Moreover, by (3) and (4), $\langle A^*,B\rangle/A^* \cong B/B^*$ is free. Thus $\langle A^*,B\rangle/A^*$ is separable in A/A^*, and therefore $\langle A^*,B\rangle/B$ is separable in A/B. Since B is separable in A, it follows that $\langle A^*,B\rangle$ is separable in A. But this is impossible since $A/\langle A^*,B\rangle$ is homogeneous having value ω_1 and since A has no elements having value ω_1. Therefore, P is not a QFF-space.

The proof of the preceding result can be generalized to establish the next theorem.

<u>Theorem 5.</u> Any QFF-space has a separable development. That is, every QFF-space V having infinite dimension m is the union of a smooth chain of subspaces V_α having the following properties for each α.

(1) $\dim(V_\alpha) < m$.

(2) V_α is separable in V.

(3) V_α is a QFF-space.

<u>Proof.</u> Since any space having countable dimension is free, we may assume that m is uncountable. Let $V = A/B$ where A and B are free. Set $A = \sum_{i\in I}\langle a_i\rangle$ and $B = \sum_{j\in J}\langle b_j\rangle$. For convenience of notation, let $A_\alpha = \sum_{i\in I(\alpha)}\langle a_i\rangle$ for a subset $I(\alpha)$ of I. Similarly, define $B_\alpha = \sum_{j\in J(\alpha)}\langle b_j\rangle$ for a subset $J(\alpha)$ of J. There exists a smooth chain of summands A_α of A that lead up to A and satisfy the following conditions.

(i) $\dim(A_\alpha) < m$.

(ii) $A_\alpha \parallel B$.

(iii) $A_\alpha \cap B = B_\alpha$, a summand of B.

In connection with the above, we simply comment that we use Lemma 3 and the fact that a free subspace is separable in order to obtain condition (ii). Define $V_\alpha = \langle A_\alpha, B\rangle/B$. Then $V_\alpha \cong A_\alpha/B_\alpha$, and V_α is a QFF-space. Thus conditions (1) and (3) are already established for

V_α. In order to establish (2), we need to show that $\langle A_\alpha, B\rangle/B$ is separable in $A/B = V$. However, this is equivalent to showing that $\langle A_\alpha, B\rangle$ is separable in A and, in turn, is equivalent to showing that $\langle A_\alpha, B\rangle/A_\alpha$ is separable in A/A_α. Since $\langle A_\alpha, B\rangle/A_\alpha \cong B/B_\alpha$ is free, the proof that $\langle A_\alpha, B\rangle/B$ is separable in A/B is complete.

Our final result is a solution to the Main Problem, posed above.

Theorem 6. A subspace C of a free space F is itself free if and only if:

(1) C is separable.

(2) F/C is a QFF-space.

Proof. Since conditions (1) and (2) are obviously necessary in order for the subspace C to be free, we need to concern ourselves only with the sufficiency. Thus suppose that C is a separable subspace of the free space F and that $\varphi: F/C \rightarrowtail\!\!\!\twoheadrightarrow A/B$ is an isomorphism from F/C onto A/B where A and B are free. We shall show that C is free by induction on $\dim(F)$. If F has countable dimension, then C must be free because C, too, has countable dimension. Thus we assume that $\dim(F)$ is uncountable. Let

$$F = \sum_{i\in I}\langle f_i\rangle,$$
$$A = \sum_{j\in J}\langle a_j\rangle, \text{ and}$$
$$B = \sum_{k\in K}\langle b_k\rangle.$$

If $I(\alpha)$, $J(\alpha)$, and $K(\alpha)$ are subsets of I, J and K, respectively, we let $F_\alpha = \sum_{i\in I(\alpha)}\langle f_i\rangle$, $A_\alpha = \sum_{j\in J(\alpha)}\langle a_j\rangle$, and $B_\alpha = \sum_{k\in K(\alpha)}\langle b_k\rangle$. Using this notation, we claim that there exists a smooth chain of summands of F

$$0 = F_0 \subseteq F_1 \subseteq \cdots \subseteq F_\alpha \subseteq \cdots$$

which satisfy the folling conditions for each α.

(1) $\dim(F_\alpha) < \dim(F)$.

(2) $F = \bigcup F_\alpha$.

(3) $F_\alpha \parallel C$.

(4) $\varphi(\langle F_\alpha, C\rangle/C) = \langle A_\alpha, B\rangle/B$ (for some subset $J(\alpha)$ of J).

(5) $A_\alpha \parallel B$.

(6) $A_\alpha \cap B = B_\alpha$ (for some subset $K(\alpha)$ of K).

Finally, we let $C_\alpha = C \cap F_\alpha$ for each α, and observe that C_α is a smooth chain leading up to C. Moreover, C_α is nice in C since F_α is nice in F. To prove that C is free, it suffices to prove that $C_{\alpha+1}/C_\alpha$ is free. We propose to show that $C_{\alpha+1}/C_\alpha$ is free by using the induction hypothesis. First, we observe that

$$C_{\alpha+1}/C_\alpha = (C \cap F_{\alpha+1})/(C \cap F_\alpha) \cong \langle C \cap F_{\alpha+1}, F_\alpha\rangle/F_\alpha ,$$

so all we need is to prove that $\langle C \cap F_{\alpha+1}, F_\alpha\rangle/F_\alpha$ is free. Moreover, $\langle C \cap F_{\alpha+1}, F_\alpha\rangle/F_\alpha$ is a subspace of the free space $F_{\alpha+1}/F_\alpha$. We intend to show that $\langle C \cap F_{\alpha+1}, F_\alpha\rangle$ is separable in $F_{\alpha+1}$ and that the space $F_{\alpha+1}/\langle C \cap F_{\alpha+1}, F_\alpha\rangle$ is QFF. In demonstrating that $\langle C \cap F_{\alpha+1}, F_\alpha\rangle$ is separable in $F_{\alpha+1}$, we need to observe that $C \cap F_{\alpha+1}$ is separable in $F_{\alpha+1}$. Therefore, the desired result can be obtained by showing that $\langle C \cap F_{\alpha+1}, F_\alpha\rangle/(C \cap F_{\alpha+1})$ is separable in $F_{\alpha+1}/(C \cap F_{\alpha+1})$, which is equivalent to showing that $\langle F_\alpha, C\rangle/C$ is separable in $\langle F_{\alpha+1}, C\rangle/C$. Under the isomorphism φ, this can be accomplished by showing that $\langle A_\alpha, B\rangle/B$ is separable in $\langle A_{\alpha+1}, B\rangle/B$, which is equivalent to showing that $\langle A_\alpha, B\rangle/A_\alpha$ is separable in $\langle A_{\alpha+1}, B\rangle/A_\alpha$ since A_α and B are both separable in A. However, $\langle A_\alpha, B\rangle/A_\alpha$ is absolutely separable since $\langle A_\alpha, B\rangle/A_\alpha \cong B/B_\alpha$ is free. This completes the proof that $\langle C \cap F_{\alpha+1}, F_\alpha\rangle/F_\alpha$ is a separable subspace of $F_{\alpha+1}/F_\alpha$.

We shall now demonstrate that $F_{\alpha+1}/\langle C \cap F_{\alpha+1}, F_\alpha\rangle$ is a QFF-space. First, observe that

$$F_{\alpha+1}/\langle C \cap F_{\alpha+1}, F_\alpha\rangle = F_{\alpha+1}/(\langle C, F_\alpha\rangle \cap F_{\alpha+1}) \cong \langle F_{\alpha+1}, C\rangle/\langle F_\alpha, C\rangle \cong$$

$$(\langle F_{\alpha+1}, C\rangle/C)/(\langle F_\alpha, C\rangle/C) \cong (\langle A_{\alpha+1}, B\rangle/B)/(\langle A_\alpha, B\rangle/B) \cong$$

$\langle A_{\alpha+1},B\rangle/\langle A_{\alpha},B\rangle \cong A_{\alpha+1}/(A_{\alpha+1}\cap\langle A_{\alpha},B\rangle) \cong (A_{\alpha+1}/A_{\alpha})/\langle A_{\alpha},A_{\alpha+1}\cap B\rangle/A_{\alpha}$ which is QFF since $A_{\alpha+1}/A_{\alpha}$ is free and since $\langle A_{\alpha}, A_{\alpha+1}\cap B\rangle/A_{\alpha}$ is isomorphic to $(A_{\alpha+1}\cap B)/(A_{\alpha}\cap B) = B_{\alpha+1}/B_{\alpha}$, which is free. This completes the proof of Theorem 6.

Corollary. Let V be a valuation space. Then V = A/B where A and B are free and B is a nice subspace of A if and only if V is QFF.

REFERENCES

1. G. Bergman, Boolean rings of projection maps, J. London Math. Soc. 4 (1972), 593-598.

2. S. Chase, On group extensions and a problem of J. H. C. Whitehead, Topics in Abelian groups, Chicago, Illinois (1963), 173-193.

3. P. Eklof, Independence results in algebra, Yale lecture notes, 1975.

4. L. Fuchs, Vector spaces with valuations, J. Algebra 35 (1975), 23-38.

5. P. Griffith, A solution to the splitting mixed problem of Baer, Trans. Amer. Math. Soc. 139 (1969), 261-269.

6. ________ , Separability of torsion free groups and a problem of J. H. C. Whitehead, Illinois J. Math. 12 (1968), 654-659.

7. ________ , Extensions of free groups by torsion groups, Proc. Amer. Math. Soc. 24 (1970), 677-679.

8. P. Hill, The additive group of commutative rings generated by idempotents, Proc. Amer. Math. Soc. 38 (1973), 499-502.

9. _______ , On the freeness of abelian groups: a generalization of Pontryagin's theorem, Bull. Amer. Math. Soc. 76 (1970), 1118-1120.

10. _______ , New criteria for freeness in abelian groups, Trans. Amer. Math. Soc. 182 (1973), 201-209.

11. ______ , New Criteria for freeness in abelian groups, II, Trans.

Amer. Math. Soc. 196 (1974), 191-201.

12. G. Nöbeling, Verallgemeinerung eines Satzes von Herrn E. Specker, Invent. Math. 6 (1968), 41-55.

13. R. Nunke, Modules of extensions over Dedekind rings, Illinois J. Math. 3 (1959), 222-241.

14. L. Pontryagin, The theory of topological commutative groups, Ann. Math. 35 (1934), 361-388.

15. F. Richman and E. Walker, Valuated groups, to appear.

16. J. Rotman, On a problem of Baer and a problem of Whitehead in abelian groups, Acta. Math. Acad. Sci. Hungar. 12 (1961), 245-254.

17. S. Shelah, Infinite abelian groups, Whitehead problem and some constructions, Israel J. Math. 18 (1974), 243-256.

18. E. Specker, Additive Gruppen von Folgen ganzer Zahlen, Portugaliae Math. 9 (1950), 131-140.

19. K. Stein, Analytiche Funktionen mehrerer komplexes Veranderlichen zu vorgegebenen Periodizitatsmoduln und das zweite Cousinsche Problem, Math. Ann. 123 (1951), 201-222.

20. D. Thomas, On the freeness of valued vector spaces, preprint.

21. J. Irwin and E. Walker, Topics in Abelian groups, Chicago, Illinois, 1963.

SUBFREE VALUED VECTOR SPACES

Laszlo Fuchs

1. Introduction. It was Prüfer who first discovered (in 1923) the relevance of the socle in the theory of abelian p-groups. He used the socle to trace certain properties of the p-groups or subgroups, like divisibility or purity.

A construction for basic subgroups, based on socles, has been given by Charles in 1955. Also, Honda in 1964 pointed out the relevance of socles. Hill in 1966 proved that direct sums of torsion-complete p-groups are completely determined, up to isomorphism of course, by their socles furnished with the height function. In 1968, Hill and Megibben studied summable p-groups. The Ulm invariants as well as Hill's relative invariants are computed in the socle. The importance of socles became more apparent in various proofs, too.

In 1975, I started a systematic study of valued vector spaces in general (with trivially valued base field) in order to obtain general information about the behavior of socles. What I had in mind was to accept valued vector spaces as new invariants (in a sense, they are easier to survey than p-groups) and to reduce the structure problem of certain classes of p-groups to valued vector spaces. I was--and still am--convinced that a lot of information can be obtained by a systematic use of valued vector spaces.

Here my main purpose is to study certain valued vector spaces which occur as socles of totally projective p-groups. These S are subfree in the sense that they are s-dense subspaces in some free valued vector space F. We wish to show that there is a great deal of difference between the cases where the values of vectors in F/S are ordinals cofinal with ω or not.

Our Theorem 5 has been obtained, in a different form, by P. Hill too.

For unexplained terminology and notation, we refer to our paper [1].

2. Basic Definitions. Let Γ be the class of ordinals and Φ an arbitrary field.

We consider vector spaces V over Φ.

By a valuation v of V is meant a function $v: V \to \Gamma^*$ (where Γ^* denotes Γ with a symbol ∞ adjoined that is considered larger than every ordinal) such that

(i) $v(a) = \infty$ if and only if $a = 0$;

(ii) $v(\alpha a) = v(a)$ for all $a \in V$ and non-zero $\alpha \in \Phi$;

(iii) $v(a+b) \geqq \min(v(a), v(b))$ for all $a, b \in V$.

A valued vector space is a pair (V,v) where V is a Φ-vector space and v is a valuation of V. The category of valued vector spaces has its objects the valued vector spaces and its morphisms those Φ-linear maps $\mu: V \to W$ which do not decrease values:

$$v(\mu a) \geqq v(a) \quad \text{for all } a \in V.$$

Let U be a subspace of the valued vector space (V,v). The restriction of v to U makes U into a valued vector space, while

$$v(a+U) = \sup_{u \in U} v(a+u)$$

defines a valuation of the quotient vector space V/U.

If $v(a) \geqq v(a+u)$ for all $u \in U$, then we say a is orthogonal to U, in notation: $a \perp U$. If every coset of V mod U contains an element orthogonal to U, we say: U is nice in V.

A subspace U of V is s-dense in V if 0 is the only element of V orthogonal to U. On the other hand, U is s-closed in V if it is not s-dense in any subspace of V properly containing U; s-closure is equivalent to niceness.

Two valued vector spaces are isometric if there is a vector space isomorphism between them that preserves values; such a map is called an isometry. A vector space map $U \to V$ is called an injection or embedding if it is one-to-one and value-preserving.

If (V_i, v_i) are valued vector spaces, then their coproduct (V,v) is $V = \oplus V_i$ (vector space direct sum) furnished with the valuation $v(\Sigma a_i) = \min v_i(a_i)$ where Σa_i is a finite sum. A free valued vector space F is simply the coproduct of 1-dimensional valued vector spaces. All 1-dimensional and all homogeneous (i.e. all nonzero vectors carry the same valuation) valued vector spaces are free.

If for a valued vector space V and an ordinal $\sigma \in \Gamma$, we write

$$V^{\sigma} = \{a \in V \mid v(a) \geqq \sigma\},$$

$$V_{\sigma} = \{a \in V \mid v(a) > \sigma\},$$

then there is a σ-homogeneous subspace $B(\sigma)$ such that

$$V^{\sigma} = B(\sigma) \oplus V_{\sigma}$$

(coproduct representation). The coproduct

$$B = \bigoplus_{\sigma} B(\sigma)$$

is an s-dense, free subspace of V. It is called a basic subspace of V and is unique up to isometry.

Every valued vector space V can be embedded in an injective valued vector space; among them there are minimal ones, they are all isometric over V. We call them the injective hull of V, in notation: $\hat{V}$. Both V and its basic subspaces are s-dense in $\hat{V}$, and if $B = \bigoplus_{\sigma} B(\sigma)$ is a basic subspace of V, then

$$\hat{V} \cong \prod_{\sigma} B(\sigma)$$

where the value of an $a \in \hat{V}$ is the minimum of the coordinate-values of a. The support of V is defined as $\operatorname{supp} V = \{v(a) \mid a \neq 0 \text{ in } V\}$. It follows that $\operatorname{supp} V = \operatorname{supp} B = \operatorname{supp} \hat{V}$.

3. Subfree Valued Vector Spaces. Our goal is to study certain valued vector spaces that occur as socles of totally projective p-groups.

A valued vector space S will be called subfree if it is an s-dense subspace in some free valued vector space F. Note that such an F is unique up to isometry; in fact, any basic subspace of S is basic in F, so necessarily isometric to F.

Manifestly, the following are true:

(a) An s-dense subspace of a subfree valued vector space is again subfree.

(b) Coproducts of subfree valued vector spaces are subfree.

There is an abundance of subfree spaces that fail to be free; let us now give an explicit example.

<u>Example</u>. Let $F = \oplus_\sigma \Phi a_\sigma$ be a free valued vector space with $v(a_\sigma) = \sigma$, where σ runs over all ordinals less than some limit ordinal λ not cofinal with ω. Define S as the subspace of F spanned by all $a_\sigma - a_\rho$ (for all $\rho, \sigma < \lambda$). Since S, together with any a_σ, spans F, but $S \neq F$, it is clear that F/S is one-dimensional of value λ. As F has no vector of value λ, S is s-dense in F. Suppose S is free; then necessarily $S \cong F$, so $S = \oplus_\sigma \Phi b_\sigma$ with $v(b_\sigma) = \sigma$. Since $v(a_0 + (a_\sigma - a_0)) = \sigma$, $v(a_0 + (a_\rho - a_0)) = \rho$, the coordinates of $a_\rho - a_0$ and $a_\sigma - a_0$ $(\sigma < \rho)$ are the same in every Φb_α with $\alpha < \sigma$. Since $a_0 \notin S$, there must exist a sequence $\rho_1 < \ldots < \rho_n < \ldots$ such that $a_{\rho_{n+1}} - a_0$ has a non-zero coordinate in some $\Phi b_{\alpha_{n+1}}$ in which all of $a_{\rho_1} - a_0, \ldots, a_{\rho_n} - a_0$ have 0 coordinates. Now $\operatorname{cof} \lambda \neq \omega$ implies that some $\rho_0 < \lambda$ exists with $\sup \rho_n < \rho_0$. But then $a_{\rho_0} - a_0$ has the same non-zero coordinate in V_{α_n} as $a_{\rho_n} - a_0$, in contradiction to $a_{\rho_0} - a_0 \in S$. Thus S cannot be free.

4. <u>Subfree Spaces that are Free</u>. In a number of cases, subfree spaces are free; the most obvious examples are the basic subspaces of free valued vector spaces. Here we want to present a sufficient condition for a subfree space to be free.

First of all, let us note that if S is a subfree space in a free valued vector space F, then all the values of vectors in F/S are limit ordinals. For, if for some $x \in F$, $v(x+S) = \sigma$ were an isolated ordinal, then there would be an $s \in S$ with $v(x+s) = \sigma$, i.e. $x + s \perp S$, in contradiction to the s-density of S in F.

We shall require a couple of lemmas. F will denote a free valued vector space, $\Sigma = \{b_i\}_{i\in J}$ a fixed basis of F and S an s-dense subspace of F.

<u>Lemma 1</u>. Suppose $F = F_1 \oplus F_2$ and $C = S \cap F_1$ satisfies the following condition:

(i) $v(x+C) = v(x+S)$ for all $x \in F_1$.

Then C satisfies:

(ii) C is s-dense in F_1;

Richman and Walker [15] were the first to find an example of a space that is not, in our terminology, a QFF-space. The specific example constructed by Richman and Walker involves the continuum hypothesis. By using König's theorem, we can eliminate this assumption. Thus the next result does not use the continuum hypothesis.

Theorem 4. There exist valuation spaces that are not QFF-spaces having the countable ordinals for values.

Proof. Choose an increasing chain

$$m_0 < m_1 < \cdots < m_\alpha < \cdots \qquad (\alpha < \omega_1)$$

of cardinals m_α so that

$$m_{\alpha+1} > m_\alpha^{\aleph_0}.$$

It follows that

$$m_0 m_1 m_2 \cdots m_\lambda \leq m_\lambda^{\aleph_0} < m_{\lambda+1}.$$

Letting $\rho = \prod_{\alpha<\omega_1} m_\alpha$ and $\sigma = \sum_{\alpha<\omega_1} m_\alpha$, we know that $\rho > \sigma$ by König's theorem. Define H_α to be a homogeneous space of value α and dimension m_α. Let $P = \prod_{\alpha<\omega_1} H_\alpha$, and set

$$K = \bigcup_{\beta<\omega_1} (\prod_{\alpha\leq\beta} H_\alpha).$$

It should be observed that $\dim(\prod_{\alpha\leq\beta} H_\alpha) < m_{\beta+1}$. Therefore, $\dim(K) = \sigma < \rho$. Obviously, P/K is homogeneous with value ω_1.

Assume that $P = A/B$ where A and B are both free. There is no loss of generality in assuming that A has no elements with value ω_1 since P has no elements having value ω_1. Let $A = \sum_{i\in I} \langle a_i\rangle$ and $B = \sum_{j\in J} \langle b_j\rangle$. There exists a subset I^* of I satisfying the following conditions.

(1) $|I^*| \leq \sigma$.

(2) $K \subseteq \langle \sum_{i\in I^*} \langle a_i\rangle, B\rangle / B$.

(3) $\sum_{i\in I^*} \langle a_i\rangle \parallel B$.

(4) $B \cap \sum_{i\in I^*} \langle a_i\rangle$ is a summand B^* of $B = \sum_{j\in J} \langle b_j\rangle$.

If ρ is a limit ordinal, then by smoothness we have no choice other than setting

$$C_\rho = \bigcup_{\sigma<\rho} C_\sigma \quad \text{and} \quad \Sigma_\rho = \bigcup_{\sigma<\rho} \Sigma_\sigma.$$

Then (1) is evident for $\sigma \leq \rho$ and (2) for $\sigma + 1 \leq \rho$. To check (3), let $x \in \langle \Sigma_\rho \rangle$, i.e. $x \in \langle \Sigma_\sigma \rangle$ for some $\sigma < \rho$. Application of (3) to this σ gives

$$v(x+S) = v(x+C_\sigma) \leqq v(x+C_\rho) \leqq v(x+S)$$

so that (3) holds true for $\sigma \leq \rho$.

If ρ is an isolated ordinal, then a more elaborate proof is required. Write $\rho = \mu + 1$, and consider C_μ, Σ_μ. If $C_\mu \neq S$, then select a subspace A_1 of S that contains C_μ properly such that $\dim A_1/C \leqq \aleph_0$. Clearly, there is a countable subset Δ_1 of $\Sigma \setminus \Sigma_\mu$ such that $A_1 \leqq \langle \Sigma_\mu \cup \Delta_1 \rangle$. Since A_1 can be replaced by $\langle \Sigma_\mu \cup \Delta_1 \rangle \cap S$, without loss of generality

$$A_1 = \langle \Sigma_\mu \cup \Delta_1 \rangle \cap S$$

can be assumed. Consider those $x \in \langle \Delta_1 \rangle$ which fail to belong to A_1. If $v(x+A_1) < v(x+S)$, then because of $v(x+S)$ is cofinal with ω we can find a countable subset $s_1, \ldots, s_n, \ldots$ of S such that $\sup v(x+s_n) = v(x+S)$. For every x in a basis of the free valued vector space $\langle \Sigma_\mu \cup \Delta_1 \rangle / A_1$ we adjoin these s_n to A_1 to obtain a subspace A_2 of S such that A_2 contains A_1, A_2/A_1 is countable dimensional, and for every $x \in \langle \Sigma_\mu \cup \Delta_1 \rangle$, the equality $v(x+A_2) = v(x+S)$ holds. Next we choose a countable subset Δ_2 of $\Sigma \setminus \Sigma_\mu$ that contains Δ_1 and satisfies $A_2 \leqq \langle \Sigma_\mu \cup \Delta_1 \rangle$; here again, there is no loss of generality in assuming

$$A_2 = \langle \Sigma_\mu \cup \Delta_2 \rangle \cap S.$$

For every x in a basis of the free space $\langle \Sigma_\mu \cup \Delta_2 \rangle / A_2$ such that $v(x+A_2) < v(x+S)$, we select a countable subset $t_1 \ldots, t_n, \ldots$ of S with $\sup v(x+t_n) = v(x+S)$ and adjoin it to A_2. In this way, a subspace A_3 of S is obtained that contains A_2 such that $\dim A_3/A_2 \leqq \aleph_0$ and $v(x+A_3) = v(x+S)$ for every $x \in \langle \Sigma_\mu \cup \Delta_2 \rangle$. Continuing this process, we obtain a chain $A_1 \leqq \ldots \leqq A_n \leqq \ldots$ of subspaces of S, and a chain $\Delta_1 \subseteq \ldots \subseteq \Delta_n \subseteq \ldots$ of countable subsets of $\Sigma \backslash \Sigma_\mu$. With their aid, we define

$$C_\rho = \bigcup_n A_n \quad \text{and} \quad \Sigma_\rho = \Sigma_\mu \cup (\bigcup_n \Delta_n).$$

Condition (1) obviously holds for $\sigma \leq \rho$. That (2) holds for $\sigma + 1 \leq \rho$ is clear from the construction. That (3) holds for $\sigma \leq \rho$ follows in the same way as in the case of limit ordinals, using the fact that $x \in C_\rho$ means that $x \in A_n$ for some n.

This transfinite process comes to an end when $C_\tau = S$ for some ordinal τ. Then $\Sigma_\tau = \Sigma$ must hold owing to the s-density of S in F. This completes the proof of Lemma 2.

We can now formulate and prove:

<u>Theorem 3</u>. Let S be an s-dense subspace of the free valued vector space F. If all the values in F/S are cofinal with ω, then S is free.

<u>Proof</u>. From the preceding lemma we conclude the existence of a smooth well-ordered ascending chain of subspaces $C_\sigma (\sigma < \tau)$ of S with the properties (1) - (3). By Lemma 1, (1) and (3) guarantee that C_σ is s-closed in S, and hence in $C_{\sigma+1}$, for every σ. By (2), $C_{\sigma+1}/C_\sigma$ is countable dimensional, so it is free. Free implies projective, thus C_σ is a summand of $C_{\sigma+1}$, and we can write

$$C_{\sigma+1} = C_\sigma \oplus S_\sigma \quad \text{for every } \sigma$$

for a suitable subspace S_σ of $C_{\sigma+1}$. It is a standard argument to prove that

$$S = \bigoplus_{\sigma<\tau} S_\sigma.$$

Since each S_σ is free, S is free, as claimed.

We know that if S is s-dense in F and the values of vectors in F/S are not all cofinal with ω, then S need not be free. In this general case we can still find a smallest subspace V of F that contains S and the support of F/V contains only ordinals cofinal with ω.

<u>Theorem 4</u>. Let S be an s-dense subspace of a free valued vector space F. Then there exists a subspace V such that:

(a) $S \subseteq V \subseteq F$;

(b) the non-zero vectors in V/S have values not cofinal with ω;

(c) the non-zero vectors in F/V have values cofinal with ω.

Proof. Choose a basic subspace $\overline{B} = \oplus \overline{B}(\sigma)$ of F/S; here $\overline{B}(\sigma)$ denotes the σ-homogeneous component of $\overline{B}$. Consider an s-closure $\overline{C}$ of the direct sum of those $\overline{B}(\sigma)$ where σ is not cofinal with ω, and define V by $V/S = \overline{C}$. Then (a) and (b) hold true. Note that no vector $\overline{a} \in F/S$ orthogonal to $\overline{C}$ can have a value ρ not cofinal with ω, because then there exists a $\overline{b} \in B(\rho)$ with $v(\overline{a-b}) > \rho$, in contradiction to $\overline{a} \perp \overline{C}$. For $x \in F$, consider $V(x+V)$; without loss of generality, $x + S \perp \overline{C}$ may be assumed, i.e. $v(v+S) \geqq v(x+u+S)$ for all $u \in V$. Clearly, $v(x+u+S) \geqq v(x+u)$ whence $v(x+S) \geqq \sup v(x+u) = v(x+V)$. The converse inequality is trivial, thus $v(x+S) = v(x+V)$. As we have noticed, $v(x+S)$ is cofinal with ω, proving the assertion.

Let me take this opportunity to correct an erroneous statement in my paper [1]. As it was pointed out by F. Richman and E. A. Walker, Theorem 3 fails to hold in general, i.e. s-closed subspaces of free valued vector spaces are not necessarily free. We want to add a condition to the hypotheses in order to get a correct statement.

We say that the valued vector space V satisfies the countability condition if there is no limit ordinal which is not cofinal with ω and which is the supremum of values of vectors in V. It is readily checked that subspaces and quotients of V likewise satisfy the countability condition.

Now Theorem 7 of [1] should read: Quotients of injective valued vector spaces with the countability condition are again injective. The proof given there simplifies, since every nest of balls contains a cofinal countable subnest.

From this, we can derive a correct form of Theorem 3 in [1]. If an s-closed subspace C of a free valued vector space F satisfies the countability condition, then C is free. The proof given there is correct, only the reference should be to the corrected Theorem 7.

We can combine this result with Theorem 3 above and state:

Theorem 5. Let H be a subspace of a free valued vector space F. If H satisfies the countability condition, then it is free.

Proof. Let C be an s-closure of H in F; it is readily seen that supp C = supp H. Hence C is an s-closed subspace of F with the countability condition, so C is free. The countability condition implies that all values of vectors in C/H are cofinal with ω. We can apply Theorem 3 to the s-dense subspace H of C to verify the freeness of H.

5. Basic subspaces of subfree valued vector spaces. It is natural to inquire what can be said about the basic subspaces B of subfree spaces S. We have already noticed that if S is s-dense in the free valued vector space F, then $B \cong F$. What we want to show now is that the dimension of S/B cannot be small unless S is free itself.

In our discussion, we require two lemmas.

Lemma 6. If $B = \oplus B(\sigma)$ is a basic subspace of a valued vector space V and if V/B has dimension 1, then $v(V/B) = \lambda$ is cofinal with ω and under the canonical embedding $\phi: V \to \Pi B(\sigma)$, every $a \in V \setminus B$ is represented by a vector with countably many non-zero coordinates.

Proof. Without loss of generality, we can assume that supp V = supp B contains only values $< \gamma$. Let $a \in V \setminus B$ be represented, under the Conrad embedding $\phi: V \to \hat{B} = \prod_{\sigma} B(\sigma)$, by the vector $\phi a = (\ldots, b_\sigma, \ldots)$ where $b_\sigma \in B(\sigma)$. For every $\mu < \lambda$, there exists a $b \in B$ such that $v(a-b) > \mu$; therefore if we write $b = b_{\sigma_1} + \ldots + b_{\sigma_n}$ with $b_{\sigma_i} \in B(\sigma_i)$, then clearly $\phi(a-b)$ has no non-zero coordinates for indices $\leqq \mu$, i.e., ϕa has only a finite number of non-zero coordinates for indices $\leqq \mu$ for any ordinal $\mu < \lambda$. Hence λ is the sup of a countable subset in supp B and our claim follows.

Lemma 7. If V is a valued vector space that contains a free valued vector space B with dim $V/B \leqq \aleph_0$, then V is free.

Proof. It suffices to prove this in case B is s-dense in V. In fact, an s-closure C of B in V will have countable codimension in V, so C s-closed in V and V/C free imply $V = C \oplus F$ for some free valued vector space F.

Thus let B be as stated and s-dense in V, i.e. B is basic in V. Choose a vector space basis $\{a_n+B\}_{n<\omega}$ of V/B. By Lemma 6, each a_n is a countable vector $a_n = (\ldots,b_{\sigma_i},\ldots)$, so there is a countable dimensional summand B' of B, say $B = B' \oplus B''$, such that all coordinates b_{σ_i} for all a_n $(n<\omega)$ are contained in B'. Then B' is s-dense in $G = \langle B', a_1,\ldots,a_n,\ldots\rangle$, so $B' \perp B''$ implies $G \perp B''$. Hence and from $G + B'' = V$ we conclude $V = G \oplus B''$. Here G is countable dimensional and so free.

From the last lemma it is now easy to derive:

<u>Theorem 8</u>. If S is a subfree, but not free valued vector space and B is a basic subspace of S, then the dimension of S/B is uncountable.

<u>Proof</u>. If S/B was countable, then by Lemma 7, S would be free.

Applications to totally projective p-groups will be given elsewhere.

Reference

[1] L. Fuchs, Valued vector spaces, <u>Journal of Algebra</u> 35(1975), 23-38.

ON CLASSIFYING TORSION FREE MODULES OVER DISCRETE VALUATION RINGS

E. L. Lady

Let V be a discrete valuation ring with quotient field Q, let V^* be the completion of V and Q^* its quotient field. Let K be an intermediate field between Q and Q^* and let $R = V^* \cap K$. A finite rank torsion free V-module G is called <u>K-decomposable</u> if $R \otimes G$ is the direct sum of a free R-module and a divisible R-module. We also say that K is a <u>splitting field</u> for G. Every finite rank torsion free V-module has a splitting field which is a finitely generated extension of Q.

In this paper we consider the following question: For what finite algebraic extensions K of Q can the K-decomposable modules be classified up to quasi-isomorphism?

1. <u>Concepts and Results</u>. If G is a torsion free V-module, we let QG denote its divisible hull. By a <u>quasi-homomorphism</u> from G to H is meant a Q-linear map $\phi : QG \to QH$ such that for some $0 \neq n \in V$, $n\phi(G) \subseteq H$. The group of quasi-homomorphisms from G to H can be identified with $Q \otimes \mathrm{Hom}(G, H)$, which we will write simply as $\mathrm{QHom}(G, H)$.

Since $G = (Q \otimes G) \cap (R \otimes G)$, we easily see the following:

<u>Basic Fact</u>: Let G and H be K-decomposable modules, let $U_1 = QG$, $U_2 = QH$, and let D_1 and D_2 be the divisible submodules of $R \otimes G$ and $R \otimes H$. Then a Q-linear map $\phi : U_1 \to U_2$ is a quasi-homomorphism if and only if $\phi(D_1) \subseteq D_2$. (Here ϕ also denotes the unique extension $K \otimes U_1 \to K \otimes U_2$.)

It follows that if we are interested in quasi-homomorphisms, then a K-decomposable module can be identified with an ordered pair (U, D), where U is a Q-vector space of dimension r (= rank of the module) and D is a K-subspace of $K \otimes U$. Actually, U and D determine G up to quasi-equality. If $G = (U, D)$, then $\dim_K D$ is called the <u>corank</u> of G. A morphism $(U_1, D_1) \to (U_2, D_2)$ is then a Q-linear map $\phi : U_1 \to U_2$ such that $\phi(D_1) \subseteq D_2$.

The category described in the above paragraph will be denoted by C.

In particular, if $G = (U_1, D_1)$ and $H = (U_2, D_2)$, then G and H are quasi-isomorphic if and only if there is an isomorphism $\phi : U_1 \to U_2$ such that $\phi(D_1) = D_2$. Since U_1 and U_2 can then be identified with Q^r, where $r = \text{rank } G$, we can restate this as follows:

Theorem 1. There is a one-to-one correspondence between the quasi-isomorphism classes of K-decomposable modules with rank r and corank c and the set of orbits of $GL(r, Q)$ acting on $\text{Grass}(r, c; K)$ in the obvious way, where $\text{Grass}(r, c; K)$ is the set of c-dimensional subspaces of K^r.

Corollary. If K is not algebraic over Q or if $[K : Q] \geq 4$, then there are infinitely many pairwise non-quasi-isomorphic K-decomposable modules with rank 2 and corank 1.

Proof. $\text{Grass}(2, 1; K)$ is simply projective 1-space over K and hence is an n-dimensional Q-variety, where $n = [K : Q]$. Now $GL(2, Q)$ is a 4-dimensional Q-algebraic group, with a 1-dimensional center which acts trivially on $\text{Grass}(2, 1; K)$, and the orbits of $GL(2, Q)$ have dimension 3 over Q. Since $n > 3$, there must be an infinite number of orbits.

Now suppose that $[K : Q] = n < \infty$. The following theorem answers the question raised at the beginning of this paper.

Theorem 2. 1) If $n = 2$, then there are exactly three strongly indecomposable K-decomposable V-modules, namely V, Q and R.

2) If $n = 3$, then there are exactly five strongly indecomposable K-decomposable V-modules.

3) If $n = 4$, there are infinitely many strongly indecomposable K-decomposable V-modules. These are classifiable, but a detailed classification does not as yet exist. The difficulty occurs in classifying those modules whose rank is twice the corank.

4) If $n \geq 5$, then Corner's Theorem is valid. Namely, for every finite dimensional Q-algebra A, there exists a K-decomposable module G such that A is the quasi-endomorphism ring of G. Since the classification of all finite

dimensional Q-algebras is generally regarded as hopeless, we should not expect a classification in this case.

2. Methodology. If K is a finite algebraic extension of Q, then the study of K-decomposable modules can be reduced to the study of finite length modules over the ring $\Lambda = \begin{pmatrix} Q & 0 \\ K & K \end{pmatrix}$.

We begin by abelianizing the category C. We define C_a to consist of triples (U, D, α) where U is a finite dimensional Q-vector space, D a finite dimensional K-vector space and $\alpha : D \to U$ is Q-linear. A morphism $(U_1, D_1, \alpha_1) \to (U_2, D_2, \alpha)$ consists of a Q-linear map $\phi : U_1 \to U_2$ and a K-linear map $\psi : D_1 \to D_2$ such that $\phi\alpha_1 = \alpha_2\psi$.

Now let $\tau : K \to Q$ be a fixed Q-linear splitting for the inclusion map $Q \to K$. Then since $[K : Q] < \infty$, it is routine to check that any Q-linear map $\alpha : D \to U$ can be written as $(U \otimes \tau)\tilde{\alpha}$ for a unique $\tilde{\alpha} \in \mathrm{Hom}_K(D, K \otimes U)$. We now see that we can identify C with the subcategory of C_a consisting of those (U, D, α) such that $\alpha = (U \otimes \tau)\tilde{\alpha}$ where $\tilde{\alpha}$ is a monomorphism.

On the other hand, C_a can easily be identified with the category of finite length right Λ-modules. For Λ is a hereditary Q-algebra having exactly two simple right modules, namely $\tilde{V} = \begin{pmatrix} Q & 0 \\ 0 & 0 \end{pmatrix}$ and $S = \Lambda / \begin{pmatrix} Q & 0 \\ K & 0 \end{pmatrix}$. Since $\tilde{V}$ is projective and S injective, any finite length right Λ-module M can be put into an exact sequence

$$0 \to U \to M \to D \to 0$$

where U is a direct sum of copies of $\tilde{V}$ and D is a direct sum of copies of S. Now U is essentially just a Q-vector space and D just a K-vector space. If $j = \begin{pmatrix} 0 & 0 \\ 1 & 0 \end{pmatrix}$, then multiplication on M by j induces a Q-linear map $\alpha : D \to U$, and the triple (U, D, α) determines M, thus giving an equivalence between C_a and mod-Λ. Furthermore, one can see that under this equivalence, C corresponds to the full subcategory of mod-Λ consisting of modules not having a summand isomorphic to S.

Finally, if we define the rank of a finite length right Λ-module M to be the number of times $\tilde{V}$ occurs as a composition factor for M, and the <u>corank</u> to be the number of times S occurs as a composition factor, then the embedding functor from C into mod-Λ preserves rank and corank.

From here on, of course, everything is straight ring theory. The necessary tools can be found in [2] and [3]. We will simply draw attention to the most important of these tools. Auslander [1] has noted the importance in mod-Λ of the functors $D = \mathrm{Hom}_Q(\_, Q)$ (Matlis duality) and $Tr = \mathrm{Ext}^1_\Lambda(\_, \Lambda)$ (transpose). Dlab and Ringel [2] define the Coxeter functors C^+ and C^- on C_a. It is not terribly hard to show that in fact, $C^+ = DTr$ and $C^- = TrD$. Furthermore, if M has no projective (injective) summand, then M is indecomposable if and only if $C^+(M)$ is indecomposable ($C^-(M)$ is indecomposable.) Furthermore, with the same restrictions, if M has rank r and corank c, then $C^+(M)$ has rank $nc - r$ and corank $(n - 1)c - r$, and $C^-(M)$ has rank $(n - 1)r - nc$ and corank $r - c$. Now starting from the two indecomposable projective Λ-modules and the two indecomposable injective Λ-modules, we can apply the Coxeter functors in an iterated way to get a whole sequence of indecomposable finite length right Λ-modules, which for convenience we will call singular Λ-modules. An important fact is that if M is an (indecomposable) singular Λ-module with rank r and corank c, then it is (up to isomorphism) the only indecomposable Λ-module with rank r and corank c. It is shown in [2] that if these sequences terminate (i.e. if there are only finitely many singular modules), then in fact all indecomposable Λ-modules are singular. On the other hand, if there are infinitely many singular modules, then there will also be infinitely many non-singular indecomposable modules. (In the terminology of [2], if $n = 4$, these are called regular.) The non-singular indecomposables are the ones which are difficult to classify. Apparently, if there exists a non-singular indecomposable with rank r and corank c, then there exist infinitely many of this rank and corank, although at the moment I do not have a proof for this.

From the formulas for rank and corank already given, it is easy to see that the sequences of singular modules terminate precisely when $n \leq 3$. Parts

3) and 4) of Theorem 2 are a special case of results in [3].

REFERENCES

[1] M. Auslander, Representation theory of Artin algebras III-Almost split sequences, Comm. Algebra 3(1975), 239-294.

[2] V. Dlab and C. M. Ringel, Indecomposable representations of graphs and algebras, Mem. Amer. Math. Soc. 6(1976), no. 173.

[3] C. M. Ringel, Representations of K-species and bi-modules, J. Algebra 41(1976), 269-302.

A SHEAF - THEORETIC INTERPRETATION OF THE KUROŠ THEOREM

Mary Turgi

1. Introduction. This paper offers a reinterpretation of the Kuroš-Mal'cev-Derry classification theorem for torsion-free modules of finite rank over a principal ideal domain in the language of sheaves and sheaf cohomology. Specifically, we identify the equivalence classes of matrix sequences arising in the traditional theorem as cohomology classes; this identification enables us to give a sheaf-theoretic proof of the theorem.

For the reader's convenience, we review the definitions and some elementary properties of sheaves and non-abelian sheaf cohomology in section 2. Basic references for this material are [3], [4], [5], and [6]. Section 3 contains a detailed account of the results for modules over a discrete valuation ring; section 4 consists of a sketch of the global version. In section 5, we state some theorems which may be proved using the sheaf-theoretic approach. (These will appear elsewhere.)

2. Sheaves and sheaf cohomology.

Presheaves and sheaves. Let τ be a topology defined on a set X. Since the sets of τ are quasi-ordered by inclusion, τ can be regarded as a category:

the objects are the open sets;

$$\mathrm{Hom}(U,V) = \begin{cases} \phi, & \text{if } U \not\subset V \\ \{i_V^U\}, & \text{if } i_V^U : U \longrightarrow V \text{ is the inclusion.} \end{cases}$$

Definition 2.1. If $\mathcal{C}$ is a category, a presheaf on X with values in $\mathcal{C}$ is a contravariant functor $\mathcal{F} : \tau \longrightarrow \mathcal{C}$.

For example, if $\mathcal{C}$ is the category of sets, a presheaf on X is a family of sets $\mathcal{F}(U)$ and set maps $\rho_U^V : \mathcal{F}(V) \longrightarrow \mathcal{F}(U)$ defined whenever $U \subset V$, subject to the following conditions:

1) The map $\rho_U^U = 1_{\mathcal{F}(U)}$;

2) Whenever $U \subset V \subset W$, there is a commutative diagram

$$\begin{array}{ccc} \mathcal{F}(W) & \xrightarrow{\rho_U^W} & \mathcal{F}(U) \\ & {\scriptstyle \rho_V^W} \searrow \quad \nearrow {\scriptstyle \rho_U^V} & \\ & \mathcal{F}(V) & \end{array} .$$

The maps ρ_U^V are called restriction maps. For simplicity, if $s \in \mathcal{F}(V)$ and $U \subset V$, we shall denote $\rho_U^V(s)$ by $s|U$.

Although sheaves may be defined for more general categories, in this paper we shall restrict our attention to categories which are subcategories of the category of sets.

Definition 2.2. Let $\mathcal{C}$ be the category of sets. A presheaf $\mathcal{F}$ on X with values in $\mathcal{C}$ is called a sheaf of sets on X if it satisfies the additional properties:

1) If $\{U_i : i \in I\}$ is an open cover of a set U in X, and if $x, y \in \mathcal{F}(U)$ satisfy $x|U_i = y|U_i$, for all $i \in I$, then $x = y$;

2) If $\{U_i : i \in I\}$ is as above, and if $x_i \in \mathcal{F}(U_i)$, $i \in I$, satisfy $x_i|U_i \cap U_j = x_j|U_i \cap U_j$, for all i,j, then there is an element $x \in \mathcal{F}(U)$ such that $x|U_i = x_i$, for all i.

It is clear that if $\mathcal{C}$ is any subcategory of the category of sets, we can define a sheaf with values in $\mathcal{C}$. For example, a sheaf of groups on X is a presheaf $\mathcal{F}$ on X with values in the category of groups which satisfies the sheaf axioms in Definition 2.2. In this case, observe that condition (2) implies that $\mathcal{F}(\phi)$ is the trivial group.

If $\mathcal{F}$ is a sheaf on X, and if U is any open set in X, then the restriction of $\mathcal{F}$ to the open sets $V \subset U$ is a sheaf on U, denoted $\mathcal{F}|U$.

If $x \in X$, then the family of all open sets containing x is a directed family under reverse inclusion. If $\mathcal{C}$ is a category with direct limits, and $\mathcal{F}$ is a sheaf on X with values in $\mathcal{C}$, then the stalk of $\mathcal{F}$ at x, denoted $\mathcal{F}_x$, is defined as $\varinjlim_{U \ni x} \mathcal{F}(U)$.

If $\mathcal{F}$ and $\mathcal{G}$ are sheaves on X with values in $\mathcal{C}$, then a morphism $f: \mathcal{F} \longrightarrow \mathcal{G}$ is a natural transformation, i.e., a family of maps $\{f_U : \mathcal{F}(U) \longrightarrow \mathcal{G}(U)$, for each open U in X$\}$ such that, whenever $U \subset V$, there is a commutative diagram

$$\begin{array}{ccc} \mathcal{F}(V) & \xrightarrow{f_V} & \mathcal{G}(V) \\ \text{res} \downarrow & & \downarrow \text{res} \\ \mathcal{F}(U) & \xrightarrow{f_U} & \mathcal{G}(U) \end{array} .$$

It is clear that every morphism $f: \mathcal{F} \longrightarrow \mathcal{G}$ induces, for each $x \in X$, a morphism $f_x : \mathcal{F}_x \longrightarrow \mathcal{G}_x$ between stalks.

Definition 2.3. Let $\mathcal{A}$ be a sheaf of commutative rings on X. Assume that $\mathcal{M}$ is a sheaf of abelian groups on X such that $\mathcal{M}(U)$ is an $\mathcal{A}(U)$ - module for each U. If $U \subset V$, the ring map $\mathcal{A}(V) \longrightarrow \mathcal{A}(U)$ allows us to consider $\mathcal{M}(U)$ as an $\mathcal{A}(V)$ - module. If the restriction map $\mathcal{M}(V) \longrightarrow \mathcal{M}(U)$ is an $\mathcal{A}(V)$ - map, then $\mathcal{M}$ is called a sheaf of $\mathcal{A}$ -Modules.

The class of $\mathcal{A}$ -Modules can be shown to form an abelian category; exact sequences are defined in the usual way. A useful characterization of exact sequences is given in the following proposition.

Proposition 2.4. A sequence $\mathcal{F} \xrightarrow{\varphi} \mathcal{G} \xrightarrow{\psi} \mathcal{H}$ of $\mathcal{A}$ -Modules is exact if and only if $\mathcal{F}_x \xrightarrow{\varphi_x} \mathcal{G}_x \xrightarrow{\psi_x} \mathcal{H}_x$ is an exact sequence of $\mathcal{A}_x$-modules for each $x \in X$. [9, p. 56].

Gluing of sheaves. Assume that $\mathcal{C}$ is a category having inverse limits. Let X be a topological space and $\{U_i : i \in I\}$ an indexed open cover of X. For each $i \in I$, let $\mathcal{F}_i$ be a sheaf on U_i with values in $\mathcal{C}$. Suppose further that for each ordered pair of indices (i,j), there is a sheaf isomorphism $\theta_{ij} : \mathcal{F}_j|U_i \cap U_j \longrightarrow \mathcal{F}_i|U_i \cap U_j$ such that, for each ordered triple (i, j, k), we have $\theta'_{ij}\,\theta'_{jk} = \theta'_{ik}$. Here θ'_{ij}, θ'_{jk}, and θ'_{ik} denote the restrictions of θ_{ij}, θ_{jk}, and θ_{ik} to $U_i \cap U_j \cap U_k$. Then there is a sheaf $\mathcal{F}$ on X with values in $\mathcal{C}$ and an isomorphism $\eta_i : \mathcal{F}|U_i \longrightarrow \mathcal{F}_i$, for each $i \in I$. The sheaf $\mathcal{F}$ is unique up to isomorphism and is called the sheaf obtained by gluing the $\mathcal{F}_i$ by means of the θ_{ij} ; the θ_{ij} are called gluing data. The explicit construction of $\mathcal{F}$ is described in [6, p. 77].

With the same notation as above, suppose that, for each $i \in I$, $\mathcal{G}_i$ is a second sheaf defined on U_i and that for each ordered pair (i, j), gluing data $\omega_{ij} : \mathcal{G}_j|U_i \cap U_j \longrightarrow \mathcal{G}_i|U_i \cap U_j$ are given. Suppose further that, for each $i \in I$, there is a morphism of sheaves $f_i : \mathcal{F}_i \longrightarrow \mathcal{G}_i$ such that the diagrams

$$\begin{array}{ccc} \mathcal{F}_j|U_i \cap U_j & \xrightarrow{f_j} & \mathcal{G}_j|U_i \cap U_j \\ \theta_{ij}\downarrow & & \downarrow \omega_{ij} \\ \mathcal{F}_i|U_i \cap U_j & \xrightarrow{f_i} & \mathcal{G}_i|U_i \cap U_j \end{array}$$

commute. If $\mathcal{G}$ is the sheaf obtained by gluing the $\mathcal{G}_i$ by means of the ω_{ij}, then there is a unique morphism $f: \mathcal{F} \longrightarrow \mathcal{G}$ such that the diagrams

$$\begin{array}{ccc} \mathcal{F}|U_i & \xrightarrow{f|U_i} & \mathcal{G}|U_i \\ \downarrow & & \downarrow \\ \mathcal{F}_i & \xrightarrow{f_i} & \mathcal{G}_i \end{array}$$

commute for all $i \in I$. If each f_i is an isomorphism, then f is an isomorphism.

Non-abelian cohomology. Let $\mathcal{G}$ be a sheaf of groups on a space X; let $\mathcal{U} = \{U_i : i \in I\}$ be an open cover of X. We shall write $U_{i_0 i_1 \ldots i_p}$ for the intersection $U_{i_0} \cap U_{i_1} \cap \ldots \cap U_{i_p}$ for any $(p+1)$ - tuple in I^{p+1}.

Definition 2.5. The group of p - cochains of the covering $\mathcal{U}$ with coefficients in the sheaf of groups $\mathcal{G}$ is the product group

$$C^p(\mathcal{U}, \mathcal{G}) = \prod_{I^{p+1}} \mathcal{G}\,(U_{i_0 i_1 \ldots i_p}) \quad .$$

Since $\mathcal{G}(\phi)$ is the trivial group $\{e\}$, we may restrict our attention to $(p+1)$ - tuples for which $U_{i_0 i_1 \ldots i_p}$ is non-empty.

<u>Definition 2.6.</u> A 1-cochain in $C^1(\mathcal{U}, \mathcal{G})$ is called a <u>1 - cocycle of the covering $\mathcal{U}$ with coefficients in $\mathcal{G}$</u> if it satisfies the "cocycle conditions":

$$g_{ik} = g_{ij}\ g_{jk} \quad \text{in } \mathcal{G}(U_{ijk})$$

whenever $U_{ijk} \neq \phi$.

Observe that the cocycle conditions imply that $g_{ii} = e$ and $g_{ji} = g_{ij}^{-1}$.

Two 1 - cocycles (g_{ij}) and (g'_{ij}) in $C^1(\mathcal{U}, \mathcal{G})$ are said to be <u>cohomologous</u> if there is a 0 - cochain (h_i) in $C^o(\mathcal{U}, \mathcal{G})$ such that

(2.7) $$g'_{ij} = h_i\, g_{ij}\, h_j^{-1}, \quad \text{for all pairs } i, j.$$

It is clear that this relation is an equivalence relation on $C^1(\mathcal{U}, \mathcal{G})$ which respects the set of cocycles; the equivalence classes are called <u>cohomology classes</u>.

<u>Definition 2.8.</u> The set <u>$H^1(\mathcal{U}, \mathcal{G})$</u> is the set of cohomology classes of the covering $\mathcal{U}$ with coefficients in $\mathcal{G}$.

If $\mathcal{V} = \{V_j : j \in J\}$ is another open cover of X which is finer than $\mathcal{U}$, it can be shown that there is a canonical map $\psi_{\mathcal{V},\mathcal{U}} : H^1(\mathcal{U}, \mathcal{G}) \longrightarrow H^1(\mathcal{V}, \mathcal{G})$ which is always an injection. Moreover, if $\mathcal{W} = \{W_k : k \in K\}$ is a third open cover of X finer than $\mathcal{V}$, then $\psi_{\mathcal{W},\mathcal{U}} = \psi_{\mathcal{W},\mathcal{V}} \circ \psi_{\mathcal{V},\mathcal{U}}$. This enables us to define the <u>first cohomology set of X with coefficients in $\mathcal{G}$</u> :

$$H^1(X, \mathcal{G}) = \varinjlim_{\mathcal{U}} H^1(\mathcal{U}, \mathcal{G})$$

where the limit is taken over arbitrarily fine coverings of X. Details may be found in [5].

3. <u>The Kuroš invariants as cohomology classes: local case</u>. Throughout this section, V will denote a discrete valuation ring with maximal ideal (p). Let Q be the quotient field of V, V^* the p-adic completion of V, and Q^* the quotient field of V^*. Regard the set $X = \{x_1, x_2, x_3\}$ as the topological space with open sets $X_o = X$, $X_1 = \{x_1, x_3\}$, $X_2 = \{x_2, x_3\}$, $X_3 = \{x_3\}$, and ϕ .

Since we shall be considering sheaves on the space X, we summarize several useful observations regarding such sheaves in the following proposition.

<u>Proposition 3.1.</u> 1) Every presheaf $\mathcal{F}$ on X_1 (or X_2) is a sheaf.

2) If A, B, C are groups with

$$\begin{array}{ccc} & & B \\ & & \downarrow \\ A & \hookrightarrow & C \end{array},$$

then the presheaf $\mathcal{F}$ on X defined by setting $\mathcal{F}(X) = A \cap B$, $\mathcal{F}(X_1) = A$, $\mathcal{F}(X_2) = B$, and $\mathcal{F}(X_3) = C$, and restriction maps the given inclusions, is a sheaf.

3) If $\mathcal{F}$ is a sheaf on X, then the stalk $\mathcal{F}_{x_i} = \mathcal{F}(X_i)$, for $i = 1, 2, 3$.

4) If $\mathcal{F}$ is a sheaf on X, then $H^1(\{X_1, X_2\}, \mathcal{F}) = H^1(X, \mathcal{F})$.

<u>Proof.</u> 1) An open covering of an open set $U \subset X_1$ must contain the set U itself. Since the restriction map ρ_U^U is $1_{\mathcal{F}(U)}$, the sheaf axioms are trivially satisfied.

2) The presheaves $\mathcal{F}|X_1$ and $\mathcal{F}|X_2$ are sheaves on X_1 and X_2 by part (1), and $\mathcal{F}$ is the sheaf obtained by gluing these sheaves via identity maps.

3) The stalk $\mathcal{F}_{x_i} = \varinjlim_{U \ni x_i} \mathcal{F}(U)$, where U is open in X. But, for each i, $\varinjlim_{U \ni x_i} \mathcal{F}(U) = \mathcal{F}(X_i)$.

4) The set $H^1(X, \mathcal{F}) = \varinjlim_{\mathcal{U}} H^1(\mathcal{U}, \mathcal{F})$, where $\mathcal{U}$ ranges over arbitrarily fine coverings of X. Since the cover $\{X_1, X_2\}$ refines every open cover of X, $\varinjlim_{\mathcal{U}} H^1(\mathcal{U}, \mathcal{F}) = H^1(\{X_1, X_2\}, \mathcal{F})$.

We apply Proposition 3.1 (2) to two important situations. By first letting $A = V^*$, $B = Q$, and $C = Q^*$, we obtain a sheaf $\check{V}$ of rings on X with $\check{V}(X) = V$. Secondly, if G is a torsion-free (hence flat) V - module, we may obtain a $\check{V}$ - Module by setting $A = V^* \otimes G$, $B = Q \otimes G$, and $C = Q^* \otimes G$ ($\otimes = \otimes_V$). Since $(V^* \otimes G) \cap (Q \otimes G) = V \otimes G \cong G$

(the intersection occurring within $Q^* \otimes G$), we may define a sheaf $\check{G}$ on X as indicated in the following diagram:

$$
(3.2)\qquad
\begin{array}{ccccc}
 & & G = \check{G}(X) & & \\
 & \swarrow^{\rho_1} & & \searrow^{\rho_2} & \\
\check{G}(X_1) = V^* \otimes G & & & & Q \otimes G = \check{G}(X_2) \\
 & \searrow & & \swarrow & \\
 & & Q^* \otimes G = \check{G}(X_3) & &
\end{array}
\quad ;
$$

here ρ_i is the canonical map $\rho_i : g \longmapsto 1 \otimes g$, for $i = 1, 2$. Furthermore, if G and H are two torsion-free V-modules and $f : G \longrightarrow H$ is a V-homomorphism, then setting $\check{f}_o = f$ and $\check{f}_i = 1_{\check{V}(X_i)} \otimes f$, for $i = 1, 2, 3$, we obtain a morphism of sheaves $\check{f} : \check{G} \longrightarrow \check{H}$. It is easily seen that the correspondence $G \longmapsto \check{G}$ defines a functor S from the category of torsion-free V-modules into the category of $\check{V}$-Modules.

<u>Proposition 3.3.</u> The functor S is additive, exact, and is a full embedding of the category of torsion-free V-modules into the category of $\check{V}$-Modules.

<u>Proof</u>. The functor S is additive since tensor product is additive. If A is a torsion-free V-module, then A is V-flat and $A \otimes \_\_$ is an exact functor. This, together with Propositions 2.4 and 3.1 (3) gives the exactness of S. Since S is additive and takes non-zero morphisms to non-zero morphisms, S is faithful. Clearly, S is one-to-one on objects. Finally, S is full for if $h : \check{G} \longrightarrow \check{H}$ is any sheaf morphism, then h is a family h_i, $i = 0, 1, 2, 3$, of $\check{V}(X_i)$ - homomorphisms such that the diagrams below commute:

$$
\begin{array}{ccc}
G & \xrightarrow{\;h_o\;} & H \\
\Big\downarrow{\scriptstyle \text{res}} & & \Big\downarrow{\scriptstyle \text{res}} \\
\check{V}(X_i) \otimes G & \xrightarrow{\;h_i\;} & \check{V}(X_i) \otimes H
\end{array}
\quad .
$$

Commutativity of the above diagrams, together with the fact that each h_i is a $\check{V}(X_i)$ - homomorphism, ensures that $h_i = 1_{\check{V}(X_i)} \otimes h_o$ for $i = 1, 2, 3$, and thus $h = \check{h}_o$.

Fix a positive integer n, and let k and ℓ be non-negative integers with $k + \ell = n$. We shall define three sheaves which will be used throughout the remainder of this section.

Let $\Gamma(n, k)$ denote the group of $n \times n$ matrices of the form

$$\begin{bmatrix} Z & Y \\ 0 & W \end{bmatrix}$$

where $W \in GL(\ell, Q^*)$, $Z \in GL(k, V^*)$, and Y is a $k \times \ell$ matrix with entries in Q^*. Using Proposition 3.1(2), we define a sheaf $\mathcal{G}\ell\,(n, k)$ on X as indicated below:

$$\begin{array}{ccccc} & & GL(n, Q) \cap \Gamma(n, k) = \mathcal{G}\ell(n, k)(X) & & \\ & \swarrow & & \searrow & \\ \mathcal{G}\ell(n, k)(X_1) = GL(n, Q) & & & & \Gamma(n, k) = \mathcal{G}\ell(n, k)(X_2) \\ & \searrow & & \swarrow & \\ & & GL(n, Q^*) = \mathcal{G}\ell(n, k)(X_3) & & \end{array}$$

where all the restriction maps are inclusions.

The next proposition follows directly from Definition 2.6 and equations (2.7).

<u>Proposition 3.4</u>. 1) A 1-cocycle of the covering $\{X_1, X_2\}$ with coefficients in the sheaf $\mathcal{G}\ell(n, k)$ is a 4-tuple $(M_{11}, M_{12}, M_{21}, M_{22})$ of matrices in in $GL(n, Q^*)$ satisfying $M_{21} = M_{12}^{-1}$, and $M_{11} = M_{22} = I$, the $n \times n$ identity matrix.

2) Two 1-cocycles (I, M, M^{-1}, I) and (I, N, N^{-1}, I) in $C^1(\{X_1, X_2\}, \mathcal{G}\ell(n, k))$ are cohomologous if and only if there are matrices B in $GL(n, Q)$ and C in $\Gamma(n, k)$ such that $BMC^{-1} = N$.

(3.5) Sheaves $\mathcal{F}_1$ and $\mathcal{F}_2$ will be defined on X_1 and X_2, respectively. Choose a fixed set of symbols $\{z_1, \ldots, z_n\}$, and set

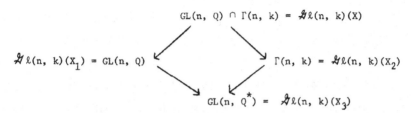

$$F_1 = \sum_{i=1}^{k} V^* z_i \oplus \sum_{j=k+1}^{n} Q^* z_j \quad ,$$

$$F_2 = \sum_{i=1}^{n} Q z_i \ , \qquad \text{and}$$

$$F_3 = \sum_{i=1}^{n} Q^* z_i \ .$$

Let $\mathcal{A}_1$ be the sheaf defined on X_1 by

$$\begin{array}{c} \mathcal{A}_1(X_1) = F_1 \\ \Big\downarrow i_1 \ ; \\ \mathcal{A}_1(X_3) = F_3 \end{array}$$

Similarly, define $\mathcal{A}_2$ on X_2 by

$$\begin{array}{c} \mathcal{A}_2(X_2) = F_2 \\ \Big\downarrow i_2 \ . \\ \mathcal{A}_2(X_3) = F_3 \end{array}$$

Observe that $\mathrm{Aut}_Q(F_2) \cong GL(n, Q)$, $\mathrm{Aut}_{Q^*}(F_3) \cong GL(n, Q^*)$, and $\mathrm{Aut}_{V^*}(F_1) \cong \Gamma(n, k)$, since the maximal divisible submodule of F_1 is fully invariant.

In the remainder of this section, we reinterpret the Kuroš theorem in the language of sheaves. In particular, we shall show that if G is a torsion-free V-module of rank n and p-rank k, and if M is a Kuroš matrix representative for G, then M determines gluing data for the sheaves $\mathcal{A}_1$ and $\mathcal{A}_2$. If $\mathcal{A}$ is the sheaf obtained by gluing $\mathcal{A}_1$ and $\mathcal{A}_2$ via this data, then $\mathcal{A} \cong \check{G}$ so that, in particular, $\mathcal{A}(X) \cong \check{G}(X) = G$.

If f_{11}, f_{12}, f_{21}, and f_{22} are gluing data for the sheaves $\mathcal{A}_1$ and $\mathcal{A}_2$, it is clear from the description of gluing given in [6] that the maps f_{11}, f_{22}, and f_{21} are superfluous in the actual construction of $\mathcal{A}$, i.e., the sheaf $\mathcal{A}$ can be obtained by using the map f_{12} alone. Choosing

$\mathcal{F}(X_1) = \mathcal{F}_1(X_1)$, $\mathcal{F}(X_3) = \mathcal{F}_1(X_3)$ and $\mathcal{F}(X_2) = \mathcal{F}_2(X_2)$, $\mathcal{F}$ is the sheaf described below:

$$\begin{array}{ccccc} & & \mathcal{F}_1(X_1) \cap \rho(\mathcal{F}_2(X_2)) = \mathcal{F}(X) & & \\ & \swarrow & & \searrow & \\ \mathcal{F}(X_1) = \mathcal{F}_1(X_1) & & & & \mathcal{F}_2(X_2) = \mathcal{F}(X_2) \\ & i_1 \searrow & & \swarrow \; \rho = f o i_2 & \\ & & \mathcal{F}_1(X_3) = \mathcal{F}(X_3) & & \end{array} \tag{3.6}$$

where $f : \mathcal{F}_2(X_3) \longrightarrow \mathcal{F}_1(X_3)$ is the Q^* - isomorphism given by the sheaf map f_{12} .

<u>Definition 3.7</u>. If $f : \mathcal{F}_2(X_3) \longrightarrow \mathcal{F}_1(X_3)$ is the Q^* - isomorphism given by the sheaf map f_{12} as above, and if the glued sheaf $\mathcal{F}$ is isomorphic to $\check{G}$, for a V-module G, then f is a <u>gluing map for G</u>. A matrix M is a <u>matrix for a gluing map</u> $f : \mathcal{F}_2(X_3) \longrightarrow \mathcal{F}_1(X_3)$ if M is the matrix for f with respect to the basis $\{z_1,\ldots,z_n\}$.

Let $\mathcal{V}$ denote the class of torsion-free V-modules of rank n, p-rank k. If G belongs to $\mathcal{V}$, then $V^* \otimes G$ is the direct sum of a free V^*-module of rank k and a divisible V^*-module of rank n - k [7]. A subset $\{v_1,\ldots,v_n\}$ of $V^* \otimes G$ will be called a <u>special basis</u> for $V^* \otimes G$ if

$$V^* \otimes G = \sum_{i=1}^{k} V^* v_i \oplus \sum_{j=k+1}^{n} Q^* v_j \quad .$$

To simplify notation, let C^1 denote $C^1(\{X_1, X_2\}, \mathcal{G}\ell(n, k))$ and H^1 denote $H^1(\{X_1, X_2\}, \mathcal{G}\ell(n, k))$. Let $J = \{1,2\}$.

<u>Theorem 3.8</u>. Let G belong to $\mathcal{V}$; let $\{v_1,\ldots,v_n\}$ be a special basis for $V^* \otimes G$, and $\{a_1,\ldots,a_n\}$ a basis for $Q \otimes G$. Then G, together with this choice of bases, determines a cocycle (I, M_{12}, M_{21}, I) in C^1, where M_{12} is the matrix for a gluing map for G.

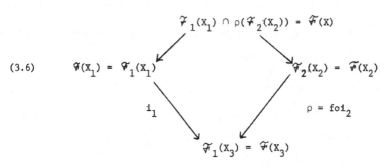

Proof. Recall that $F_1 = \sum_{i=1}^{k} V^*z_i \oplus \sum_{j=k+1}^{n} Q^*z_j$, $F_2 = \sum_{i=1}^{n} Qz_i$, and $F_3 = \sum_{i=1}^{n} Q^*z_i$. Corresponding to the given choice of bases for $V^*\otimes G$ and $Q\otimes G$, there are isomorphisms

$$g_1 : F_1 \longrightarrow V^*\otimes G \quad \text{and}$$

$$g_2 : F_2 \longrightarrow Q\otimes G$$

given by $g_1 : z_i \longmapsto v_i$ and $g_2 : z_i \longmapsto a_i$, for all i. These maps induce sheaf isomorphisms

$$\gamma_1 : \mathcal{F}_1 \longrightarrow \check{G}|X_1 \quad , \text{ and}$$

$$\gamma_2 : \mathcal{F}_2 \longrightarrow \check{G}|X_2$$

where $\mathcal{F}_1$ and $\mathcal{F}_2$ are the sheaves defined in (3.5), and $\check{G}$ is as defined in (3.2). Let $\check{G}_j$ denote $\check{G}|X_j$ and $f_j = \gamma_j|X_3$, for $j\in J$.

On $X_3 = X_1 \cap X_2$, we have

$$\mathcal{F}_j|X_3 \xrightarrow{f_j} \check{G}_j|X_3 = \check{G}_i|X_3 \xleftarrow{f_i} \mathcal{F}_i|X_3 , \quad \text{for } i, j \in J.$$

Setting $f_{ij} = f_i^{-1} \circ f_j$ we obtain sheaf isomorphisms

$$f_{ij} : \mathcal{F}_j|X_3 \longrightarrow \mathcal{F}_i|X_3 , \quad \text{for all } i, j \in J .$$

Observe that $f_{21} = f_{12}^{-1}$, $f_{11} = 1_{\mathcal{F}_1|X_3}$, and $f_{22} = 1_{\mathcal{F}_2|X_3}$.

Since X_3 is a singleton space and $\mathcal{F}_j(\phi) = \{0\}$, $j \in J$, a sheaf isomorphism $h : \mathcal{F}_j|X_3 \longrightarrow \mathcal{F}_i|X_3$ is just a pair of homomorphisms h_3 in $\mathrm{Aut}_{Q^*}(F_3)$ and the zero map h_ϕ . Each such sheaf isomorphism thus determines a Q^*-automorphism h_3 of F_3, hence a matrix M (with respect to the basis $\{z_1,\ldots,z_n\}$) in $GL(n, Q^*)$. Conversely, a matrix M_{ij} in $GL(n, Q^*)$ determines a sheaf isomorphism $h : \mathcal{F}_j|X_3 \longrightarrow \mathcal{F}_i|X_3$, $i, j \in J$. In particular, the 4-tuple $(f_{11}, f_{12}, f_{21}, f_{22})$ of sheaf isomorphisms determines a 1 - cocycle (I, M_{12}, M_{21}, I). Hence G, together with the choice of bases $\{v_1,\ldots,v_n\}$ and $\{a_1,\ldots,a_n\}$ of $V^*\otimes G$ and $Q\otimes G$, respectively, determines the cocycle (I, M_{12}, M_{21}, I) in C^1.

Since the sheaf isomorphisms $f_{ij} : \mathcal{F}_j|X_3 \longrightarrow \mathcal{F}_i|X_3$ satisfy the the compatibility relations $f_{ik} = f_{ij} \circ f_{jk}$ on $X_i \cap X_j \cap X_k$, for $i, j, k \in J$, the f_{ij} may be used as gluing data. Let $\mathcal{F}$ be the sheaf on X obtained by gluing $\mathcal{F}_1$ and $\mathcal{F}_2$ via the f_{ij}, for $i, j \in J$. The diagrams

$$\begin{array}{ccc} \mathcal{F}_j|X_3 & \xrightarrow{f_j} & \check{G}_j|X_3 \\ {\scriptstyle f_{ij}}\downarrow & & \downarrow{\scriptstyle 1} \\ \mathcal{F}_i|X_3 & \xrightarrow{f_i} & \check{G}_i|X_3 \end{array}$$

commute for all $i, j \in J$, and both f_1 and f_2 are isomorphisms; thus, we have $\mathcal{F} \cong \check{G}$. Since the matrix M_{12} is the matrix determined by the sheaf map f_{12}, M_{12} is the matrix of a gluing map for G.

<u>Theorem 3.9</u>. The correspondence $\mathcal{V} \longrightarrow H^1$ given by $G \longmapsto [(I, M_{12}, M_{21}, I)]$, where M_{12} is the matrix for a gluing map for G, is well defined.

<u>Proof</u>. In Theorem 3.8, we saw that G, together with the particular choice of bases $\{v_1,\ldots,v_n\}$ and $\{a_1,\ldots,a_n\}$ of $V^* \otimes G$ and $Q \otimes G$, respectively, determines a 4-tuple $(f_{11}, f_{12}, f_{21}, f_{22})$ of gluing data for the sheaves $\mathcal{F}_1$ and $\mathcal{F}_2$. If $\{v_1',\ldots,v_n'\}$ is a different special basis for $V^* \otimes G$ and $\{a_1',\ldots,a_n'\}$ is a different basis for $Q \otimes G$, let

$$g_1' : F_1 \longrightarrow V^* \otimes G, \quad \text{and}$$

$$g_2' : F_2 \longrightarrow Q \otimes G$$

be given by $g_1' : z_i \longmapsto v_i'$ and $g_2' : z_i \longmapsto a_i'$ for all i. The g_j' induce sheaf isomorphisms $\gamma_j' : \mathcal{F}_j \longrightarrow \check{G}_j$, for $j = 1, 2$. As in

Theorem 3.8, we set $f_j' = \gamma_j'|X_3$ and $f_{ij}' = (f_i')^{-1} \circ f_j'$, for $i, j \in J$.

Using the notation of Theorem 3.8, g_1 and g_2 are the isomorphisms defined by $g_1 : z_i \longmapsto v_i$ and $g_2 : z_i \longmapsto a_i$ for all i. Thus the maps $(g_1')^{-1} \circ g_1$ and $(g_2')^{-1} \circ g_2$ are automorphisms of F_1 and F_2, respectively, which induce sheaf automorphisms $(\gamma_1')^{-1} \circ \gamma_1$ and $(\gamma_2')^{-1} \circ \gamma_2$ of $\mathcal{A}_1$ and $\mathcal{E}_2$, respectively. Let $h_j = [(\gamma_j')^{-1} \circ \gamma_j]|X_3$, for $j \in J$.

If the maps f_{ij} are the gluing data arising from the choice of bases $\{v_1,\ldots,v_n\}$ and $\{a_1,\ldots,a_n\}$, and f_{ij}' the data arising from the choice $\{v_1',\ldots,v_n'\}$ and $\{a_1',\ldots,a_n'\}$, then

$$
\begin{aligned}
f_{ij}' &= (f_i')^{-1} \circ f_j' \\
&= [(\gamma_i')^{-1} \circ \gamma_j']|X_3 \\
&= [(\gamma_i')^{-1} \circ (\gamma_i \circ \gamma_i^{-1}) \circ (\gamma_j \circ \gamma_j^{-1}) \circ (\gamma_j')]|X_3 \qquad (3.10)\\
&= \{[(\gamma_i')^{-1} \circ \gamma_i] \circ [\gamma_i^{-1} \circ \gamma_j] \circ [\gamma_j^{-1} \circ \gamma_j']\}|X_3 \\
&= h_i \circ f_{ij} \circ h_j^{-1} ,
\end{aligned}
$$

for all $i, j \in J$.

Let (a_i'), (a_i), (v_i), and (v_i') be n-dimensional column vectors. The automorphism $(g_2')^{-1} \circ g_2$ of F_2 has matrix N in $GL(n, Q)$, where $(a_i) = N(a_i')$; the automorphism $(g_1')^{-1} \circ g_1$ of F_1 has matrix P in $\Gamma(n, k)$ where $(v_i) = P(v_i')$. If $(f_{11}, f_{12}, f_{21}, f_{22})$ corresponds to the cocycle (I, M_{12}, M_{21}, I), and $(f_{11}', f_{12}', f_{21}', f_{22}')$ corresponds to the cocycle (I, M_{12}', M_{21}', I), then the computations above (equations (3.10)) show that $M_{12}' = N^{-1}M_{12}P$. By Proposition 3.4 (2), the cocycles (I, M_{12}, M_{21}, I) and (I, M_{12}', M_{21}', I) are cohomologous.

Let $\mathcal{M}$ denote the set of isomorphism classes of torsion-free V-modules of rank n, p-rank k. The correspondence of Theorem 3.9 induces a correspondence $\psi : \mathcal{M} \longrightarrow H^1$. It is easy to show that ψ is well-defined, i.e. if $G \cong K$, then G and K determine the same cohomology class in H^1. We shall show that ψ is a bijection.

<u>Lemma 3.11</u>. If $\mathcal{F}$ and $\mathcal{F}'$ are sheaves on X, determined respectively by cohomologous cocycles (I, M_{12}, M_{21}, I) and (I, M'_{12}, M'_{21}, I) in C^1, then $\mathcal{F} \cong \mathcal{F}'$.

<u>Proof.</u> Since (I, M_{12}, M_{21}, I) and (I, M'_{12}, M'_{21}, I) are cohomologous, there are matrices B in $GL(n, Q)$ and C in $\Gamma(n, k)$ such that $M'_{12} = B\, M_{12}\, C^{-1}$. Let g_1 be the automorphism of F_1 given by C, g_2 the automorphism of F_2 given by B and γ_j the sheaf automorphism of $\mathcal{F}_j$ induced by g_j, $j \in J$. Finally, for $j \in J$, let $h_j = \gamma_j | X_3$.

If $(f_{11}, f_{12}, f_{21}, f_{22})$ is the 4-tuple of gluing data determined by (I, M_{12}, M_{21}, I), and $(f'_{11}, f'_{12}, f'_{21}, f'_{22})$ that determined by (I, M'_{12}, M'_{21}, I), then the cohomology relation ensures that the following diagrams commute for all $i, j \in J$:

$$\begin{array}{ccc} \mathcal{F}'_j | X_3 & \xrightarrow{h_j} & \mathcal{F}_j | X_3 \\ {\scriptstyle f'_{ij}} \downarrow & & \downarrow {\scriptstyle f_{ij}} \\ \mathcal{F}'_i | X_3 & \xrightarrow{h_i} & \mathcal{F}_i | X_3 \end{array} .$$

Thus, if $\mathcal{F}$ and $\mathcal{F}'$ are the sheaves on X obtained by gluing $\mathcal{F}_1$ and $\mathcal{F}_2$ via the f_{ij} and f'_{ij}, respectively, then $\mathcal{F} \cong \mathcal{F}'$.

<u>Theorem 3.12</u>. The function $\psi : \mathcal{M} \longrightarrow H^1$ is one-to-one.

<u>Proof</u>. Let G and K belong to $\mathcal{V}$. Suppose that G, together with a choice of bases for $V^* \otimes G$ and $Q \otimes G$ determines a cocycle (I, M_{12}, M_{21}, I),

and K, together with a choice of bases, determines a cocycle (I, M'_{12}, M'_{21}, I) which is cohomologous to (I, M_{12}, M_{21}, I). By Lemma 3.11, the sheaves $\mathcal{F}$ and $\mathcal{F}'$ determined by these cocycles are isomorphic so, in particular, $\mathcal{F}(X) \cong \mathcal{F}'(X)$. But by Theorem 3.8, $\mathcal{F}(X) \cong \check{G}(X) = G$ and $\mathcal{F}'(X) \cong \check{K}(X) = K$; hence, $G \cong K$.

<u>Lemma 3.13</u>. Every cohomology class in H^1 contains a cocycle (I, N_{12}, N_{21}, I), where N_{12} is a matrix with entries in V^*.

<u>Proof</u>. If (I, M_{12}, M_{21}, I) is a cocycle in C^1, then (I, M_{12}, M_{21}, I) is cohomologous to $(I, p^k M_{12}, p^{-k} M_{12}, I)$, for all $k \in Z$. Since $M_{12} \in GL(n, Q^*)$, $p^k M_{12}$ has entries in V^* for sufficiently large k.

<u>Theorem 3.14</u>. The function $\psi : \mathcal{M} \longrightarrow H^1$ is onto.

<u>Proof</u>. By Lemma 3.12, a cohomology class in H^1 contains a cocycle (I, N_{12}, N_{21}, I) where N_{12} is a V^*-matrix. Let $(f_{11}, f_{12}, f_{21}, f_{22})$ be the gluing data determined by (I, N_{12}, N_{21}, I), and f the Q^*-automorphism of F_3 given by N_{12}. The sheaf $\mathcal{F}$ obtained by gluing $\mathcal{F}_1$ and $\mathcal{F}_2$ via $(f_{11}, f_{12}, f_{21}, f_{22})$ is as shown in (3.6). If $\rho = f \circ i_2$ is the restriction map $\mathcal{F}(X_2) \longrightarrow \mathcal{F}(X_3)$, let $\rho(z_i) = a_i$, for all i. The proof that $\mathcal{F}(X)$ has rank n and p-rank k, that $\{a_1, \ldots, a_n\}$ is a basis for $Q \otimes \mathcal{F}(X)$, and that $\{z_1, \ldots, z_n\}$ is a special basis for $V^* \otimes \mathcal{F}(X)$ may be found in [2, p. 157]. Since $(a_i) = N_{12}(z_i)$, $\mathcal{F}(X)$ determines the cohomology class $[(I, N_{12}, N_{21}, I)]$.

We summarize the results of this section in the following theorem.

<u>Theorem 3.15</u>. The function $\psi : \mathcal{M} \longrightarrow H^1(X, \mathcal{GL}(n, k))$ given by $[G] \longmapsto [(I, M_{12}, M_{21}, I)]$, where M_{12} is the matrix of a gluing map for G, is a one-to-one correspondence.

If M is the matrix of a gluing map for G and $\mathcal{F}$ the sheaf obtained by gluing $\mathcal{F}_1$ and $\mathcal{F}_2$ via the data given by the cocycle (I, M, M^{-1}, I), then $\mathcal{F}(X) \cong G$. From the proof of Theorem 3.14, we see that M is a Kuroš matrix representative for G. Conversely, if M is a Kuroš matrix representative for G, then M determines the 1-cocycle (I, M, M^{-1}, I) in C^1 which, in turn, determines gluing data for the sheaves $\mathcal{F}_1$ and $\mathcal{F}_2$. If $\mathcal{F}$ is the sheaf obtained by gluing via this data, then $\mathcal{F} \cong \check{G}$; hence a Kuroš matrix representative for G is the matrix for a gluing map for G. By Proposition 3.4(2), two matrices in $GL(n, Q^*)$ are Kuroš-equivalent if and only if the corresponding 1-cocycles are cohomologous. Thus, the Kuroš equivalence classes of matrices may be identified with the cohomology classes in $H^1(X, \mathcal{G}\ell(n, k))$, and the traditional classification theorem may be viewed as a restatement of Theorem 3.15.

4. <u>The global version</u>. The results of the previous section easily globalize to torsion-free modules of finite rank over a principal ideal domain R.

Let $\mathcal{P}$ be the set of non-zero prime ideals $\mathfrak{p}$ in R; choose a fixed generator p for each such ideal. For a prime ideal (p) in R, let R_p be the localization of R at (p), R_p^* the p-adic completion of R_p, and Q_p^* the quotient field of R_p^*. Let Q be the quotient field of R. To the ring R, we associate a sheaf of rings $\check{R}$ defined on X as indicated below:

$$\begin{array}{ccccc} & & R = \check{R}(X) & & \\ & \stackrel{\Delta'}{\swarrow} & & \stackrel{i}{\searrow} & \\ \check{R}(X_1) = \prod_p R_p^* & & & & Q = \check{R}(X_2) \\ & \stackrel{(i_p)}{\searrow} & & \stackrel{\Delta}{\swarrow} & \\ & & \prod_p Q_p^* = \check{R}(X_3) & & \end{array}$$

where Δ and Δ' are diagonal maps, $i_p : R_p^* \longrightarrow Q_p^*$ is the inclusion for each p, and $i : R \longrightarrow Q$ is the inclusion.

Similarly, if G is any torsion-free R-module, we may associate to G a sheaf $\check{G}$ of $\check{R}$-Modules:

$$\begin{array}{ccccc} & & G = \check{G}(X) & & \\ & \stackrel{\Delta'}{\swarrow} & & \stackrel{i}{\searrow} & \\ \check{G}(X_1) = \prod_p (R_p^* \otimes G) & & & & Q \otimes G = \check{G}(X_2) \\ & \stackrel{(i_p)}{\searrow} & & \stackrel{\Delta}{\swarrow} & \\ & & \prod_p (Q_p^* \otimes G) = \check{G}(X_3) & & \end{array}$$

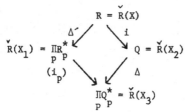

where $\Delta' : g \longmapsto (1\otimes g)$,

$i : g \longmapsto 1\otimes g$, and

$i_p : R_p^*\otimes G \longrightarrow Q_p^*\otimes G$ is the inclusion for all p.

($\otimes = \otimes_R$ throughout this section.)

As before, this association is functorial, and the functor $G \longmapsto \check{G}$ is additive, exact, and a full embedding of the category of torsion-free R-modules into the category of $\check{R}$-Modules.

As in the local case, we next define three sheaves. Let n be a fixed positive integer. For each prime p, let ℓ_p and k_p be non-negative integers such that $k_p + \ell_p = n$. Let $\kappa = (k_p)$ be the family of all the k_p. We define the sheaf $\mathcal{A}\ell(n, \kappa)$ on X by

$$\begin{array}{ccccc}
 & & \mathcal{A}\ell(n, \kappa)(X) & & \\
 & \swarrow^{\rho_1} & & \searrow^{\rho_2} & \\
\mathcal{A}\ell(n, \kappa)(X_1) = GL(n, Q) & & & & \prod_p \Gamma_p(n, k_p) = \mathcal{A}\ell(n, \kappa)(X_2) \\
 & \searrow_{\Delta} & & \swarrow_{(i_p)} & \\
 & & \prod_p GL(n, Q_p^*) = \mathcal{A}\ell(n, \kappa)(X_3) & &
\end{array}$$

where $\Gamma_p(n, k_p)$ is defined as in the local case, $i_p : \Gamma_p(n, k_p) \longrightarrow GL(n, Q_p^*)$ is the inclusion for each p, Δ is the diagonal map, and $(\mathcal{A}\ell(n, \kappa)(X), \rho_1, \rho_2)$ is a pullback.

To define sheaves $\mathcal{F}_1$ and $\mathcal{F}_2$ on X_1 and X_2, respectively, we choose a fixed set of symbols $\{z_1,\ldots,z_n\}$, and set

$$F_1^p = \sum_{i=1}^{k_p} R_p^* z_i \oplus \sum_{j=k_p+1}^{n} Q_p^* z_j, \quad \text{for all } p,$$

$$F_2 = \sum_{i=1}^{n} Q z_i, \quad \text{and}$$

$$F_3^p = \sum_{i=1}^{n} Q_p^* z_i \quad .$$

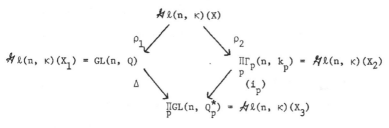

The sheet $\mathcal{F}_1$ is then defined on X_1 by

$$\begin{array}{c} \mathcal{F}_1(X_1) = \prod_p F_1^p \\ \big\downarrow (i_p) \\ \mathcal{F}_1(X_3) = \prod_p F_3^p \end{array}$$

where $i_p : F_1^p \longrightarrow F_3^p$ is the inclusion for all p.

Similarly, $\mathcal{F}_2$ is defined on X_2 by

$$\begin{array}{c} \mathcal{F}_2(X_2) = F_2 \\ \big\downarrow \Delta \\ \mathcal{F}_2(X_3) = \prod_p F_3^p \end{array} .$$

Observe that $\mathrm{Aut}_{\check{R}(X_2)}(F_2) \cong GL(n, Q)$, $\mathrm{Aut}_{\check{R}(X_1)}(\prod_p F_1^p) \cong \prod_p \Gamma_p(n, k_p)$, and $\mathrm{Aut}_{\check{R}(X_3)}(\prod_p F_3^p) \cong \prod_p GL(n, Q_p^*)$.

The following proposition is the global version of Proposition 3.4.

<u>Proposition 4.1</u>. 1) A 1-cocyle of the covering $\{X_1, X_2\}$ with coefficients in $\mathcal{G}\ell(n, \kappa)$ is a 4-tuple $((I), (M_p)_{12}, (M_p^{-1})_{21}, (I))$ of elements of $\prod_p GL(n, Q_p^*)$, where (I) denotes the element with an $n \times n$ identity matrix in each component.

2) Two 1-cocycles $((I), (M_p)_{12}, (M_p^{-1})_{21}, (I))$ and $((I), (N_p)_{12}, (N_p^{-1})_{21}, (I))$ in $C^1(\{X_1, X_2\}, \mathcal{G}\ell(n, \kappa))$ are cohomologous if and only if there is a matrix B in $GL(n, Q)$, and an element (C_p) of $\prod_p \Gamma_p(n, k_p)$ such that $BM_pC_p^{-1} = N_p$, for each prime p.

We define a gluing map for G as in the local case. Since $\mathrm{Aut}_{\check{R}(X_3)}(\prod_p F_3^p) \cong \prod_p GL(n, Q_p^*)$, a gluing map $f = (f_p)$ for G has a matrix representation (M_p), where M_p is the matrix for f_p in the basis $\{z_1, \ldots, z_n\}$.

Let $\mathcal{R}$ be the class of torsion-free R-modules of rank n and p-ranks k_p, let C^1 denote $C^1(\{X_1, X_2\}, \mathcal{G}\ell(n, \kappa))$, and let H^1 denote $H^1(\{X_1, X_2\}, \mathcal{G}\ell(n, \kappa))$.

<u>Theorem 4.2</u>. Let G belong to $\mathcal{R}$. For each p, let $\{v_1^p,\ldots,v_n^p\}$ be a special basis for $R_p^* \otimes G$, and $\{a_1,\ldots,a_n\}$ a basis for $Q \otimes G$. Then G, together with this choice of bases, determines a cocycle $((I), (M_p)_{12}, (M_p^{-1})_{21}, (I))$, where $(M_p)_{12}$ is the matrix representation of a gluing map for G.

<u>Proof</u>. Corresponding to the choice of bases for $R_p^* \otimes G$ and $Q \otimes G$, there are isomorphisms

$$g_1^p : F_1^p \longrightarrow R_p^* \otimes G, \quad \text{for each } p, \text{ and}$$

$$g_2 : F_2 \longrightarrow Q \otimes G$$

given by $g_1^p : z_i \longmapsto v_i^p$ and $g_2 : z_i \longmapsto a_i$, for all i. These yield sheaf isomorphisms

$$\gamma_1 : \mathcal{F}_1 \longrightarrow \check{G}|X_1 = \check{G}_1, \quad \text{and}$$

$$\gamma_2 : \mathcal{F}_2 \longrightarrow \check{G}|X_2 = \check{G}_2 .$$

Let $f_j = \gamma_j|X_3$, for $j \in J = \{1,2\}$, and $f_{ij} = f_i^{-1} \circ f_j$, for $i, j \in J$. Since $f_{ij} = (f_{ij}^p)$, where f_{ij}^p is the pth-component of $f_i^{-1} \circ f_j$, the 4-tuple $(f_{11}, f_{12}, f_{21}, f_{22})$ determines a 1-cocycle $((I), (M_p)_{12}, (M_p^{-1})_{21}, (I))$ in C^1. Here M_p is the matrix of f_{12}^p in the basis $\{z_1,\ldots,z_n\}$, for each p.

As in the local case, the f_{ij} may be used as gluing data. The diagrams

$$\begin{array}{ccc} \mathcal{F}_j|X_3 & \xrightarrow{f_j} & \check{G}_j|X_3 \\ {\scriptstyle f_{ij}}\downarrow & & \downarrow{\scriptstyle 1} \\ \mathcal{F}_i|X_3 & \xrightarrow{f_i} & \check{G}_i|X_3 \end{array}$$

commute for all $i, j \in J$; thus if $\mathcal{F}$ is the sheaf obtained by gluing via this data, then $\mathcal{F} \cong \check{G}$. Since $(M_p)_{12}$ is determined by the sheaf map f_{12}, $(M_p)_{12}$ is the matrix representative of a gluing map for G.

Theorems 4.3, 4.4, and 4.6 globalize Theorems 3.9, 3.12, and 3.14. The proofs are analogous and will be omitted.

Theorem 4.3. The correspondence $\mathcal{R} \longrightarrow H^1$ given by $G \longmapsto [(I), (M_p)_{12}, (M_p^{-1})_{21}, (I))]$, where (M_p) is the matrix representation of a gluing map for G, is well-defined.

Let $\mathcal{N}$ be the set of isomorphism classes of torsion-free R-modules of rank n, p-ranks k_p. The correspondence of Theorem 4.3 induces a correspondence $\psi : \mathcal{N} \longrightarrow H^1$ which is easily seen to be well-defined.

Theorem 4.4. The function $\psi : \mathcal{N} \longrightarrow H^1$ is one-to-one.

Definition 4.5. Let $\mathcal{S}$ be the subset of H^1 comprised of cohomology classes which contain a cocycle $((I), (M_p)_{12}, (M_p^{-1})_{21}, (I))$ where each M_p is an R_p^*-matrix.

It is easy to show that $\mathcal{S}$ is a proper subset of H^1; for example, the class $[(I), (p^{-1}I)_{12}, (pI)_{21}, (I)]$ is not in $\mathcal{S}$.

Theorem 4.6. The function $\psi : \mathcal{N} \longrightarrow \mathcal{S}$ is onto.

We summarize the above results.

Theorem 4.7. The function $\psi : \mathcal{N} \longrightarrow \mathcal{S} \subset H^1$ given by $[G] \longmapsto [(I), (M_p)_{12}, (M_p^{-1})_{21}, (I))]$, where (M_p) is the matrix representation of a gluing map for G, is a one-to-one correspondence.

As before, one easily sees that the matrix representation of a gluing map for G is a Kuroš-Mal'cev-Derry (KMD) matrix sequence for G. Conversely, any KMD matrix sequence (M_p) for a module G determines the 1-cocycle $((I), (M_p)_{12}, (M_p^{-1})_{21}, (I))$ in C^1 which, in turn, determines gluing data for the sheeves $\hat{\mathcal{F}}_1$ and $\hat{\mathcal{F}}_2$. If $\hat{\mathcal{F}}$ is the sheaf obtained by gluing via this data, then $\hat{\mathcal{F}} \cong \check{G}$; hence, a KMD matrix sequence is the matrix representation for a gluing map for G. By Proposition 4.1 (2), two matrix sequences are KMD-equivalent exactly when the corresponding 1-cocycles are cohomologous. Thus, as in the local class, the equivalence classes of matrix sequences arising in the traditional theorem may be identified with the cohomology classes in $\mathcal{S} \subset H^1$.

5. Applications. Returning to the local case, we state some applications of the results of section 3; details will appear elsewhere. Notation throughout this section will be as in section 3. All V-modules will be torsion-free of finite rank.

Using the correspondence $G \longmapsto \check{G}$ defined in section 3, one may prove the following classification theorem for extensions.

Theorem 5.1. Let G and H be V-modules. There is a one-to-one correspondence between $\mathrm{Ext}_V(G, H)$ and equivalence classes of matrices of the form

$$\begin{bmatrix} I_k & 0 & 0 & 0 \\ 0 & I_{k'} & 0 & 0 \\ \Delta & 0 & I_\ell & 0 \\ C & \Gamma & 0 & I_{\ell'} \end{bmatrix}$$

where $\begin{bmatrix} I_k & 0 \\ \Delta & I_\ell \end{bmatrix}$ is the matrix for a fixed gluing map for H, and $\begin{bmatrix} I_{k'} & 0 \\ \Gamma & I_{\ell'} \end{bmatrix}$ is the matrix for a fixed gluing map for G.

This theorem provides an explicit method for computing the Baer sum of two extensions in terms of their matrix representatives. In conjunction with the concept of a splitting field for a module, this is especially useful. The following definition is due to E.L. Lady.

Definition 5.2. Let K be a field such that $Q \subset K \subset Q^*$; let $R = K \cap V^*$. Then K is a splitting field for a V-module G if $R \otimes G$ is the direct sum of a free R-module and a divisible R-module.

It is not difficult to show that every V-module G of rank n and p-rank k has a Kuroš matrix representative of the form $\begin{bmatrix} I_k & 0 \\ \Delta & I_{n-k} \end{bmatrix}$ where Δ is an $(n - k) \times k$ matrix with entries in V^*. Such a matrix representative will be called a V^*-matrix representative for G. Let $Q(\Delta)$ denote the field obtained by adjoining to Q the entries in the matrix Δ.

Proposition 5.3. 1) Every module G has a unique minimal splitting field K_G.
2) If $\begin{bmatrix} I_k & 0 \\ \Delta & I_{n-k} \end{bmatrix}$ is a V^*-matrix representative for G, then $K_G = Q(\Delta)$.
3) If G is quasi-isomorphic to H, then $K_G = K_H$.

Let K and R be as in Definition 5.2; let $\mathcal{A}$ be the full subcategory of V-modules with objects all torsion-free, finite rank V-modules which are split by K. Using Theorem 5.1, we obtain the following results.

Theorem 5.4. 1) The set of extensions $0 \longrightarrow H \longrightarrow E \longrightarrow G \longrightarrow 0$ where E is in $\mathcal{A}$ forms a submodule of $\mathrm{Ext}_V(G, H)$, denoted $\mathrm{RExt}_V(G, H)$.
2) If $0 \longrightarrow A \longrightarrow B \longrightarrow C \longrightarrow 0$ is an exact sequence of modules in $\mathcal{A}$, then the sequences

$$0 \longrightarrow \mathrm{Hom}(G, A) \longrightarrow \mathrm{Hom}(G, B) \longrightarrow \mathrm{Hom}(G, C)$$
$$\longrightarrow \mathrm{RExt}(G, A) \longrightarrow \mathrm{RExt}(G, B) \longrightarrow \mathrm{RExt}(G, C) \longrightarrow 0, \text{ and}$$
$$0 \longrightarrow \mathrm{Hom}(C, G) \longrightarrow \mathrm{Hom}(B, G) \longrightarrow \mathrm{Hom}(A, G)$$
$$\longrightarrow \mathrm{RExt}(C, G) \longrightarrow \mathrm{RExt}(B, G) \longrightarrow \mathrm{RExt}(A, G) \longrightarrow 0$$

are exact for all G in $\mathcal{A}$.
3) Let K be a finite extension of Q. A reduced module H in $\mathcal{A}$ satisfies $\mathrm{RExt}(G, H) = 0$, for all G in $\mathcal{A}$, if and only if H is

isomorphic to a direct sum of copies of R.

The final theorems of this section illustrate that the rank and p-rank of a module G, together with the transcendence degree (over Q) of its minimal splitting field K_G determine certain properties of G.

Definition 5.5. If G is a V-module of rank n and p-rank k, then the corank of G = n - k.

Theorem 5.6. Let G be a V-module with tr. $\deg._Q K_G$ = (p-rank G)(corank G). Then

1) G is strongly indecomposable.
2) Every submodule H of G with p-rank H < p-rank G is free; every torsion-free quotient K of G with rank K $\leq$ corank G is divisible.
3) Let H be a module with tr. $\deg._Q K_H$ = (p-rank H)(corank H). If Hom(G, H) $\neq 0$ and Hom(H, G) $\neq 0$, then G is quasi-isomorphic to H.

Theorem 5.7. Let G be a module of p-rank k and corank ℓ having no free or divisible summands. If

$$\text{tr. deg.}_Q K_G \geq \begin{cases} (k-2)(\ell-2)+2, & \text{if } \ell \geq 3 \\ \frac{k}{2}+1, & \text{if } \ell = 2 \text{ and } k \text{ is even} \end{cases},$$

then G has the Krull-Schmidt property.

Remark. If k is odd and $\ell = 2$, it is not difficult to show that G always has the Krull-Schmidt property, regardless of the transcendence degree of K_G.

REFERENCES

1. D.M. Arnold and E.L. Lady, Endomorphism rings and direct sums of torsion-free abelian groups, Trans. Amer. Math. Soc. 211(1975), 225-237.

2. L. Fuchs, Infinite Abelian Groups. v. 2, Academic Press, New York, 1973.

3. J. Frenkel, Cohomologie non-abélienne et éspaces fibrés, Bull. Soc. Math. France, 85(1957), 135-218.

4. R. Godement, Topologie Algébrique et Théorie des Faisceaux, Hermann, Paris, 1964.

5. A. Grothendieck, A General Theory of Fibre Spaces with Structure Sheaf, University of Kansas Mathematics Department, 1955.

6. ______________ and J. Dieudonné, Eléments de Géometrie Algébrique I, Springer, Berlin, Heidelburg, and New York, 1971.

7. I. Kaplansky, Infinite Abelian Groups, 2nd Edition, University of Michigan Press, Ann Arbor, 1971.

8. E.L. Lady, Splitting fields for torsion-free modules over discrete valuation rings, I, preprint.

9. I.G. MacDonald, Algebraic Geometry : Introduction to Schemes, Benjamin, New York, 1968.

GENERA AND DIRECT SUM DECOMPOSITIONS OF TORSION FREE MODULES

David M. Arnold[1]

0. Introduction. Direct sum decompositions of finite rank torsion free abelian groups are notoriously complicated. The first examples of pathological decompositions of these groups were provided by B. Jónsson (see Fuchs [4]). He also introduced the notion of quasi-isomorphism of groups and proved a Krull-Schmidt theorem for quasi-decompositions. Unfortunately, this theorem gives little information about group decompositions, primarily due to the abundance of examples of groups that are indecomposable but not strongly indecomposable (e.g. Fuchs [4], §90).

Numerous other examples of pathological decompositions have subsequently been published, giving rise to the suspicion that the subject was too complicated to be worthy of further study. From this point of view the results of L. Lady [9] are startling: If K is the Grothendieck group (modulo split exact sequences) of the category of finite rank torsion free abelian groups and if T is the torsion subgroup of K, then K/T is a free abelian group. In other words, the complexity of direct sum decompositions is intimately related to T. Furthermore, he introduced the notion of near-isomorphism of groups (written $A \sim B$), a relation weaker than isomorphism but stronger than quasi-isomorphism, and proved that $T = \{[A] - [B] : A \sim B\}$. He also proved that if $A^n \sim B^n$ then $A \sim B$; if $A \oplus C \sim B \oplus C$ then $A \sim B$; and if the endomorphism ring of A is commutative and if $A \sim B$, then A is indecomposable iff B is indecomposable. The latter result suggests that, in contrast to quasi-isomorphism, near-isomorphism preserves indecomposability. Finally, near-isomorphism coincides with isomorphism if the groups are divisible by almost all primes. These results are analogous to earlier results of Jacobinski [7] for lattices over orders if near-isomorphism is replaced by genus class. Another parallel between the two subjects appears in Lady [8],

1 This research was partially supported by NSF Grant MPS71-2778 A04.

where the full force of the Jordan-Zassenhaus theorem for Z-orders is used to prove that a finite rank torsion free abelian group has, up to isomorphism, only finitely many summands.

This paper establishes a connection between the theory of lattices over orders and torsion free modules of finite rank over a Dedekind subring of an algebraic number field. A simultaneous generalization of genus class for lattices over orders and near isomorphism for finite rank torsion free abelian groups is given. Many of the classical results for lattices over orders are shown to be true in this setting, giving rise to some previously unknown theorems involving near-isomorphism and direct sum decompositions. In addition, most of the results of Lady [9] are rederived.

Genus class is an essential tool for studying lattices over orders. It seems likely that near-isomorphism is destined to play the same role in the theory of torsion free abelian groups of finite rank.

This paper is semi-expository in the sense that many of the basic ideas are borrowed from the theory of lattices over orders (e.g. as found in Swan-Evans [17]) or from Lady [9]. On the other hand, the generalizations do not appear to be immediate.

Let R be a Dedekind domain with quotient field K, an algebraic number field; let Λ be a Z-reduced R-torsion free R-algebra such that $K \otimes_R \Lambda$ is a finite dimensional K-algebra; and let L, M, N be Z-reduced R-torsion free right Λ-modules of finite R-rank (not necessarily finitely R-generated).

Define M and N to be in the <u>same R-genus</u>, written $M \vee N$, if for any non-zero ideal I of R there is a Λ-exact sequence $0 \to M \to N \to T \to 0$ with $I + \mathrm{Ann}(T) = R$, where <u>$\mathrm{Ann}(T)$</u> $= \{r \in R \mid rT = 0\}$.

As a special case, it is well known (e.g. see Reiner [13]) that if Λ is an <u>R-order</u> (Λ is finitely R-generated and $K \otimes_R \Lambda$ is semi-simple), and if M, N are <u>Λ-lattices</u> (M and N are finitely R-generated), then $M \vee N$ iff M_P and N_P are Λ_P-isomorphic for all maximal ideals P of R, where $X_P = R_P \otimes_R X$ and R_P is the localization of R at P.

If $R = \Lambda = Z$ then $M \vee N$ iff M and N are nearly isomorphic.

As a consequence of Section 1, problems of genus and direct sum decompositions are reduced to the case that $K \otimes_R \Lambda$ is semi-simple and L, M, N are finitely Λ-generated projective right Λ-modules.

The finitely generated Λ-projective case is considered in Section 2. The more general results are developed in Section 3, while further applications to finite rank torsion free abelian groups are given in Section 4.

Theorem 1. (a) There is an integer, n, with $M^n \simeq N^n$ iff $M \vee N$.

(b) If $(M \oplus L) \vee (N \oplus L)$, then $M \vee N$.

(c) If $M \vee N^k$, then $M \simeq N^{k-1} \oplus L$ with $L \vee N$.

(d) If $M \vee N$ and if $f \in \mathrm{Hom}_\Lambda(M, L^n)$, $g \in \mathrm{Hom}_\Lambda(L^n, M)$ with gf monic, then $M \oplus M' \simeq N \oplus L$ for some $M' \vee L$.

Call M a genus-summand of N if for each non-zero ideal I of R there is $f \in \mathrm{Hom}_\Lambda(M,N)$ and $g \in \mathrm{Hom}_\Lambda(N,M)$ such that gf is monic and $\mathrm{Ann}(M/gf(M)) + I = R$. If Λ is an R-order and if M, N are Λ-lattices, then M is a genus summand of N iff M_P is a Λ_P-summand of N_P for all maximal ideals, P, of R.

Theorem 2. If M is a genus summand of N, then $N \simeq X \oplus M'$ for some $X \vee M$. Consequently, if $M \vee N$ then M is Λ-indecomposable iff N is Λ-indecomposable.

Define $P_\Lambda(M)$ to be the category of right Λ-modules N such that N is a summand of M^n for some n. Let $K_0(M)$ be $K_0(P_\Lambda(M), \oplus)$, i.e. the abelian group with isomorphism classes of modules in $P_\Lambda(M)$ as generators and with relations $(N_1) + (N_2) - (N_1 \oplus N_2)$.

Theorem 3. (a) $K_0(M)$ is finitely generated and the torsion subgroup of $K_0(M)$ is $\{[N_1] - [N_2] \mid N_i \in P_\Lambda(M)$ and $N_1 \vee N_2\}$.

(b) There are only finitely many isomorphism classes of Λ-indecomposable modules in $P_\Lambda(M)$.

(c) Given M, there is $n \in Z$ such that every genus class of Λ-modules in $P_\Lambda(M)$ contains $\leq n$ isomorphism classes.

Theorem 4. Let $\Lambda' = E_\Lambda(M)/N$, where $E_\Lambda(M)$ is the Λ-endomorphism ring of M and N is the nil radical of $E_\Lambda(M)$. Suppose that $K \otimes_R \Lambda' = K_1 \oplus .. \oplus K_n$, where each K_i is a simple K-algebra, and that no K_i is a totally definite quaternion algebra. Then $M \oplus M' \simeq M \oplus M$ implies that $M' \simeq M$

Corollary 5. Suppose that $\Lambda' = E_\Lambda(M)/N$ satisfies the hypotheses of Theorem 4. If L is a genus summand of M^n for some $n \in Z$ and if $L \oplus M \simeq L \oplus N$, then $M \simeq N$.

Our development parallels that of Swan-Evans [17] with the following exceptions: (i) Since R/P and $\Lambda/P\Lambda$ are finite for all maximal ideals P of R we can avoid using completions; (ii) The stable range results of Warfield [20] allow simplification of some of the arguments; (iii) No explicit reference to the Jordan-Zassenhaus theorem for R is needed since Lady [8] proves that any torsion free abelian group of finite rank has, up to isomorphism, only finitely many summands and any Λ-projective module is, as an abelian group, torsion free of finite rank; (iv) Since M and N need not be finitely R-generated it is not necessarily true that localization commutes with $\mathrm{Hom}(M,N)$.

The proofs of Theorem 3.b and 3.c are left to the reader. (See Swan-Evans [17] and make the necessary modification as suggested by the arguments in Section 2.)

The author acknowledges his gratitude to Professor R. B. Warfield, Jr. for various conversations and encouragement regarding the contents of this paper, in particular, for an early version of Proposition 3.6.c, the suggestion that Proposition 3.5 might be true, and a preliminary draft of [20].

1. Preliminaries.

Notation. In this section, R is an integral domain, Λ is an R-torsion free R-algebra, and L, M, N are R-torsion free right Λ-modules.

The relation v is an equivalence relation on the category of R-torsion free right Λ-modules, noting that if $f : M \to N$ is monic and if $0 \neq r \in \mathrm{Ann}(N/f(M))$, then $g = f^{-1}r : N \to M$, $fg = r : N \to N$, and $gf = r : M \to M$. The genus class of M, denoted by G_M, is the equivalence class of M.

Lemma 1.1. If $M \vee N$ and if $N \vee L$ then M is isomorphic to a Λ-summand of $N \oplus L$.

Proof. Choose monomorphisms $f_1 : N \to M$, $f_2 : L \to M$ with $\mathrm{Ann}(M/f_1(N)) + \mathrm{Ann}(M/f_2(L)) = R$, say $1 = r_1 + r_2$ with $r_iM \subseteq \text{Image } f_i$. Then $g_1 = f_1^{-1}r_1 : M \to N$ and $g_2 = f_2^{-1}r_2 : M \to L$. Define $\phi: N \oplus L \to M$ by $\phi = (f_1, f_2)$ and $\theta: M \to N \oplus L$ by $\theta = (g_1, g_2)$. Thus, $\phi\theta = f_1g_1 + f_2g_2 = r_1 + r_2 = 1$, and $M \oplus \mathrm{Ker}\,\phi \simeq N \oplus L$.

As a consequence of Lemma 1.1, if $M \in P_\Lambda(L)$ and if $M \vee N$, then $N \in P_\Lambda(L)$.

Theorem 1.2. Let $\Phi = E_\Lambda(L)/N$, where N is the nil radical of $E_\Lambda(L)$, and assume that N is nilpotent. Let $F : P_\Lambda(L) \to P_\Phi(\Phi)$ be the functor defined by $F(M) = \mathrm{Hom}_\Lambda(L,M)/\mathrm{Hom}_\Lambda(L,M)N$. Then (a) F induces a bijection from isomorphism classes in $P_\Lambda(L)$ to isomorphism classes in $P_\Phi(\Phi)$, and (b) F induces a bijection from R-genus classes in $P_\Lambda(L)$ to R-genus classes in $P_\Phi(\Phi)$.

Proof. Let $E = E_\Lambda(L)$. Then F is the composite of functors $H : P_\Lambda(L) \to P_E(E)$, defined by $H(M) = \mathrm{Hom}_\Lambda(L,M)$, and $G : P_E(E) \to P_\Phi(\Phi)$, defined by $G(B) = B/BN$. The functor H is a category equivalence with inverse $T : P_E(E) \to P_\Lambda(L)$, given by $T(B) = B \otimes_E L$ (e.g. see Arnold-Lady [1]).

(a) It is well known (e.g. Swan [19], p. 89, ff.) that G induces a bijection from isomorphism classes in $P_E(E)$ to isomorphism classes in $P_\Phi(\Phi)$. This proves (a).

(b) Assume that M and N are in $P_\Lambda(L)$ with $M \vee N$. To verify that $F(M) \vee F(N)$, let I be a non-zero ideal of R. Then there is a Λ-monomorphism $F : M \to N$ with $\mathrm{Ann}(N/f(M)) + I = R$. If $0 \neq r \in \mathrm{Ann}(N/f(M))$, then there is $g : N \to M$ with $gf = r : M \to M$ and $fg = r : N \to N$. Furthermore,

$F(g)F(f) = F(gf) = F(r) = r : F(M) \to F(M)$ and $F(f)F(g) = r : F(N) \to F(N)$. Note that $E_\Lambda(L)$, hence Φ, is R-torsion free, so that $F(M)$ is R-torsion free and $F(f) : F(M) \to F(N)$ is monic (since $F(g)F(f) = r \neq 0$). Since $F(f)F(g) = r$, $r \in \mathrm{Ann}(F(N)/\mathrm{Image}\ F(f)) = J$, and it follows that $I + J = R$, i.e. $F(M) \vee F(N)$.

For the converse, first observe that by the preceding argument, if $H(M) \vee H(N)$ then $TH(M) \vee TH(N)$, hence $M \vee N$. In view of (a), it is now sufficient to prove that if B, C are in $P_E(E)$ with $(B/BN) \vee (C/CN)$ then $B \vee C$.

Let I be a non-zero ideal of R and let $0 \to B/BN \to C/CN \to T \to 0$ be a Φ-exact sequence with $\mathrm{Ann}(T) + I = R$, say $1 = s + r$ with $s \in I$, $r \in \mathrm{Ann}(T)$, $f' : B/BN \to C/CN$ and $g' : C/CN \to B/BN$ with $f'g' = r$ and $g'f' = r$. Since B and C are E-projective (being in $P_E(E)$), there is $f : B \to C$ and $g : C \to B$ with $h = fg - r : C \to CN$ and $gf - r : B \to BN$. It follows that $h^k = (fg - r)^k = 0$ for some k. Since r commutes with fg and C is R-torsion free, the binomial theorem guarantees that g is monic and that $r^kC \subseteq f(B) \subseteq C$. Similarly, f is monic. Therefore, $B \vee C$ since $1 = r + s$ and $r^k \in \mathrm{Ann}(C/f(B))$ imply that $\mathrm{Ann}(C/f(B)) + I = R$.

Corollary 1.3. Assume the notation of Theorem 1.2.

(a) If $X_0 = X_1 \oplus X_2$ in $P_\Phi(\Phi)$, then there is $M_0 = M_1 \oplus M_2$ in $P_\Lambda(L)$ with $F(M_i) \simeq X_i$ for $0 \leq i \leq 2$.

(b) If $F(M) = X$ and $F(N) = Y$, then X is a genus summand of Y iff M is a genus summand of N.

2. Finitely generated projective modules

Notation. In this section, R is an Dedekind domain with quotient field K, an algebraic number field; Λ is a Z-reduced R-torsion free R-algebra; $K \otimes_R \Lambda$ is a semi-simple finite dimensional K-algebra; and L, M, N are finitely generated projective right Λ-modules.

Since K is a finite dimensional extension of Q, the field of rational numbers, $QR = K$. Thus, if I is a non-zero ideal of R, then there is a non-zero integer n with $nR \subseteq I \subseteq R$, and R/nR, hence R/I, is finite since the additive group of R is a torsion free abelian group of finite rank. If P is a maximal ideal of R, then $\Lambda/P\Lambda$ is a finite R/P-module since the additive group of Λ is torsion free of finite rank. Note that the additive group of M is a reduced torsion free abelian group of finite rank. Thus there is a non-zero integer n with $nX \neq X$ for all non-zero subgroups X of M (e.g. Lady [9]).

Proposition 2.1. The following are equivalent:
(a) $M \vee N$; (b) M_P and N_P are Λ_P-isomorphic for all maximal ideals P of R; (c) M/PM and N/PN are $\Lambda/P\Lambda$-isomorphic for all maximal ideals P of R.

Proof. ((a) ⇒ (b)) Given a maximal ideal, P, there is a Λ-exact sequence $0 \to M \to N \to T \to 0$ with $\mathrm{Ann}(T) + P = R$. Then $0 \to M_P \to N_P \to T_P \to 0$ is Λ_P-exact and $T_P = 0$, since $(\mathrm{Ann}(T))_P = R_P$.

((b) ⇒ (c)) This is a consequence of the observation that $M/PM \simeq M_P/PM_P$ and $N/PN \simeq N_P$ as $\Lambda/P\Lambda \simeq \Lambda_P/P\Lambda_P$-modules.

((c) ⇒ (a)) Let I be a non-zero ideal of R. It is sufficient to replace I with I_0, where I_0 is a non-zero ideal of R with $I_0{}^k \subseteq nI$ for some $k \in Z$, where n is an integer with $nX \neq X$ for all non-zero subgroups X of $M \oplus N$. (See remarks preceding Proposition 2.1.)

In particular, $I = P_1P_2\ldots P_k$ may be chosen to be a product of distinct maximal ideals of R with $IX \neq X$ for all non-zero R-submodules X of $M \oplus N$. Thus, $M/IM \simeq \Sigma \oplus M/P_iM \simeq \Sigma \oplus N/P_iN \simeq N/IN$ as Λ-modules.

Since M and N are Λ-projective, there are $f : M \to N$ and $g : N \to M$ with $fg - 1 : N \to IN$ and $gf - 1 : M \to IM$. Consequently, $I(\mathrm{Ker}\ fg) = \mathrm{Ker}\ fg = 0$ by the assumptions on I; i.e. fg is monic. Similarly, gf is monic and so f is monic. Since gf is monic, there is $0 \neq r \in R$ with $rN \subseteq fg(N) \subseteq f(M) \subseteq N$ (N has finite R-rank, e.g. see Warfield [20]). Let $T = N/f(M)$ and note that $IT = T$ since $N = fg(N) + IN = f(M) + IN$. By induction on the number of distince maximal ideals dividing $\mathrm{Ann}(T)$, it follows that $\mathrm{Ann}(T) + I = R$, so that $M \vee N$.

We summarize some of the basic properties of orders (e.g. see Swan-Evans [17] or Reiner [13]). If R is Dedekind, then Λ is an R-order in the semi-simple K-algebra $K \otimes_R \Lambda$ if Λ is finitely R-generated, and a maximal R-order in $K \otimes_R \Lambda$ if there is no R-order in $K \otimes_R \Lambda$ properly containing Λ.

Every R-order, Λ, is contained in a maximal R-order, Γ, and $r\Gamma \subseteq \Lambda \subseteq \Gamma$ for some $0 \neq r \in R$. Maximal R-orders are right hereditary. If Λ is a maximal R-order and if S is a multiplicatively closed subset of R, then Λ_S is a maximal R_S-order. Conversely, if Λ_P is a maximal R_P-order for all maximal ideals, P, of R then Λ is a maximal R-order.

If R is semi-local and Dedekind, Λ is a maximal R-order, and M, N are finitely generated projective Λ-modules, then the following are equivalent: $M \vee N$; M and N are Λ-isomorphic; $K \otimes_R M$ and $K \otimes_R N$ are $K \otimes_R \Lambda$-isomorphic.

Lemma 2.2. (Beaumont-Pierce [3], Pierce [11]). There is $0 \neq r \in R$ with $r\Gamma \subseteq \Lambda \subseteq \Gamma$, where $\Gamma = \Gamma_1 \oplus \ldots \oplus \Gamma_n$ is a ring direct sum and each Γ_i is a maximal order over a Dedekind domain R_i. In particular, Γ is right hereditary.

Proof. Write $K' = K \otimes_R \Lambda = K_1 \oplus \ldots \oplus K_n$ as a direct sum of simple algebras. Let Λ_i be the image of Λ under the projection of K' to K_i so that $\Lambda \subseteq \Lambda_1 \oplus \ldots \oplus \Lambda_n$; $\Lambda_i \subseteq K_i$; and $s(\Lambda_1 \oplus \ldots \oplus \Lambda_n) \subseteq \Lambda$ for some $0 \neq s \in R$.

Now assume that $K \otimes_R \Lambda$ is simple. By Pierce [11], there is $0 \neq t \in R$ with $t\Lambda \subseteq Sa_1 \oplus \ldots \oplus Sa_k$, where $R \subseteq S = \Lambda \cap F$; F is a subfield of the center of $K \otimes_R \Lambda$; and $a_1, a_2, \ldots, a_k$ is an F-basis of $K \otimes_R \Lambda$ contained in Λ. Since S is Noetherian, Λ is finitely S-generated. Let D be the integral closure of S in its quotient field, $F = QS$. Then D/S is finite.

Now $\Lambda' = D\Lambda$ is a D-order. Embed Λ' in a maximal D-order, Γ. Then $v\Gamma \subseteq \Lambda \subseteq \Gamma$ for some $0 \neq v \in R$, and the proof is complete.

Proposition 2.3. There is a (non-empty) finite set $\pi = \{P_1, P_2, \ldots P_n\}$ of maximal ideals of R, depending only on Λ, such that the following are equivalent: (a) $M \vee N$; (b) M_P and N_P are Λ_P-isomorphic for all $P \in \pi$;

(c) M/PM and N/PN are $\Lambda/P\Lambda$-isomorphic for all $P \in \pi$.

Proof. As in the proof of Proposition 2.1, (b)$\Leftrightarrow$(c) as long as $\pi \supseteq \pi_1 = \{P \mid P \supseteq I'\}$, where I' is a non-zero ideal of R with $I'X \neq X$ for all non-zero R-submodules X of $M \oplus N$.

((b) $\Rightarrow$ (a)) Let $\pi = \pi_1 \cup \{P \mid P \supseteq rR\}$, where $0 \neq r \in R$ is given by Lemma 2.2. Suppose that P is a maximal ideal of R with $P \notin \pi$. By Lemma 2.2, $\Lambda_P = \Gamma_P = \Lambda_1 \oplus \ldots \oplus \Lambda_n$, where $\Lambda_i = (\Gamma_i)_P$ is a $D_i = (R_i)_P$-order; $M_P = M_1 \oplus \ldots \oplus M_n$ and $N_P = N_1 \oplus \ldots \oplus N_n$. Thus each Λ_i is a maximal D_i-order, where D_i is semi-local and Dedekind. Since $M \vee N$ and $QR = K$, $K \otimes_R M_i \simeq K \otimes_R N_i \simeq F_i \otimes_{D_i} M_i \simeq F_i \otimes_{D_i} N_i$, where $QD_i = F_i$ is the quotient field of D_i. By the remarks preceding Lemma 2.2, M_i and N_i are Λ_i-isomorphic. Thus M_P and N_P are Λ_P-isomorphic for all maximal ideals, P, of R. Now apply Proposition 2.1 to see that $M \vee N$.

((a) $\Rightarrow$ (b)) This is clear.

Corollary 2.4. Let I be an ideal of R with $I \subseteq P_1 P_2 \ldots P_n$, where $\pi = \{P_1, P_2, \ldots, P_n\}$ is given by Prop. 2.3, and let N be a finitely generated projective Λ-module. Then there are, up to isomorphism, only finitely many Λ-submodules N_j of N such that $\mathrm{Ann}(N/N_j) + I = R$. ($N_j$ need not be Λ-projective.)

Proof. Note that $(N_j)_P \simeq (N)_P$ for all $P \in \pi$. As in the proof of Proposition 2.3, $\Gamma_P \simeq \Lambda_P$ for all primes $P \notin \pi$. Thus $(N_j\Gamma)_P \simeq (N\Gamma)_P$ for all maximal ideals, P, of R. By Proposition 2.1, $(N\Gamma) \vee (N_j\Gamma)$ since $N_j\Gamma \subseteq N\Gamma$, $N\Gamma$ is finitely generated Γ-projective, and Γ is right hereditary. Therefore, $N_j\Gamma$ is finitely generated Γ-projective. By Lemma 1.1, $N_j\Gamma$ is a summand of $N\Gamma$, so there are only finitely many $N_j\Gamma$. ($N\Gamma \oplus N\Gamma$ is a torsion free abelian group of finite rank, so apply results of Lady [8].) But $rN_j\Gamma \subseteq N_j \subseteq N_j\Gamma$, where $r\Gamma \subseteq \Lambda \subseteq \Gamma$, and $(N_j\Gamma)/(rN_j\Gamma)$ is finite, so there are only finitely many N_j, up to isomorphism.

Proposition 2.5. Suppose that $M \vee N$. If I is a non-zero ideal of R, then there is a Λ-exact sequence $0 \to M \to N \to U \to 0$ such that $\mathrm{Ann}(U) + I = R$;

$U = U_1 \oplus \ldots \oplus U_n$; each U_i is a simple Λ-module; and $\mathrm{Ann}(U_i) + \mathrm{Ann}(U_j) = R$ if $i \neq j$.

Proof. It is sufficient to assume that $I \subseteq P_1 P_2 \ldots P_n$, where $\pi = \{P_1, P_2, \ldots, P_n\}$ is given by Proposition 2.3.(Replace I by $IP_1 P_2 \ldots P_n$.) By Corollary 2.4, there are only finitely many $N_j \subseteq N$ with $\mathrm{Ann}(N/N_j) + I = R$.

The remainder of the proof is as in Swan-Evans [17], p. 108 ff. Briefly, by adjusting I it is sufficient to assume the existence of $0 \to M \xrightarrow{f} N \to T \to 0$ such that $\mathrm{Ann}(T) + I = R$, and if $f(M) \subseteq N_i \subseteq N_{i+1} \subseteq N$, then there are monomorphisms $0 \to N_i \to N_{i+1} \to S \to 0$ with S simple and $\mathrm{Ann}(S)$ any of an infinite number of primes of R. A composition series of T lifts to a chain $f(M) = N_0 \subseteq N_1 \subseteq \ldots \subseteq N_k = M$ with N_{i+1}/N_i simple. Modifying the inclusions $N_i \to N_{i+1}$ gives the desired result.

We say that M satisfies Eichler's condition if $K \otimes_R E_\Lambda(M) = K_1 \oplus \ldots \oplus K_n$, where each K_i is a simple K-algebra but not a totally definite quaternion algebra (where K_i is a totally definite quaternion algebra if K_i is a division algebra with center F_i such that F_i-dimension $K_i = 4$; any embedding of F_i into the field of complex numbers has image contained in Re, the field of real numbers; and $\mathrm{Re} \otimes_{F_i} K_i$ is the real quaternion algebra, i.e. the Hamiltonian quaternions). Note that, in particular, K_i must be a non-commutative division algebra.

Note that if $r\Lambda \subseteq M \subseteq \Lambda$, then M satisfies Eichler's condition iff $K \otimes_R \Lambda = K_1 \oplus \ldots \oplus K_n$, where each K_i is a simple K-algebra but not a totally definite quaternion algebra, since $K \otimes_R E_\Lambda(M) = K \otimes_R E_\Lambda(\Lambda) = K \otimes_R \Lambda$. If M and Λ are finitely R-generated then $K \otimes_R E_\Lambda(M) \cong E_K(K \otimes_R M)$, so that our definition coincides, in this case, with the definition given by Swan-Evans [17].

Lemma 2.6. Suppose that M (not necessarily Λ-projective) is a submodule of Λ satisfying Eichler's condition such that $\mathrm{Ann}(\Lambda/M) + P_1 P_2 \ldots P_n = R$, where $\pi = \{P_1, P_2, \ldots P_n\}$ is given by Prop. 2.3. Then there is a finite set, π_0, of prime ideals of R, such that if U is a simple Λ-module with

$\mathrm{Ann}(U) \not\subset \pi_0$, if I is an ideal of R with $\mathrm{Ann}(U) + I = R$, and if $f,g : M \to U$ are Λ-epimorphisms, then there is a Λ-automorphism, α, of M such that $\alpha : \mathrm{Ker}\, g \to \mathrm{Ker}\, f$ is an isomorphism and $(1 - \alpha)(M) \subseteq IM$.

Proof. By Lemma 2.2, there is $0 \neq r \in R$ with $r\Gamma \subseteq \Lambda \subseteq \Gamma = \Gamma_1 \oplus \ldots \oplus \Gamma_n$, where each Γ_i is a maximal order over a Dedekind domain, R_i, and $K_i = K \otimes_R \Gamma_i$ is simple. Since $QR = K$ and $QR_i = F_i$, the quotient field of R_i, it follows that $K_i \simeq F_i \otimes_{R_i} \Gamma_i \simeq E_{K_i}(K_i) \simeq E_{K_i}(F_i \otimes_{R_i} M\Gamma_i)$ is not a totally definite quaternion algebra. Thus there is a set of primes, π_i, of R_i such that the lemma is true for $M\Gamma_i$, a finitely generated projective Γ_i-module. (The proof is decidely non-trivial. See Roggenkamp [15], p. 44 ff., or Swan-Evans [17], p. 177 ff.)

Define $\pi_0 = \{P \mid P$ is a prime ideal of R such that either $P \in \pi$ (as given by Prop. 2.3) or else $PR_i \subseteq P_i$ for some $P_i \in \pi_i\}$.

Now assume the hypotheses of the Lemma. Then U is a simple Γ-module (noting that $\mathrm{Ann}(U) + rR = R$). Lift f and g to Γ-epimorphisms $f' : M\Gamma \to U$, $g' : M\Gamma \to U$. By the preceding remarks, there is an automorphism β of $M\Gamma$ such that $\beta : \mathrm{Ker}\, g' \to \mathrm{Ker}\, f'$ is an isomorphism and $(1 - \beta)(M\Gamma) \subseteq rIM\Gamma \subseteq IM$, since $r\Gamma \subseteq \Lambda$ and $\mathrm{Ann}(U) + rI = R$. Let α be the restriction of β to M. Then $(1 - \alpha)(M) \subseteq IM$. Moreover, for $m \in M$, $\beta^{-1}(1 - \alpha)(m) = \beta^{-1}(m) - m \in \beta^{-1}(rIM\Gamma) \subseteq rI\beta^{-1}(M\Gamma) \subseteq rIM\Gamma \subseteq IM$. Thus, α is an automorphism of M, since $\beta^{-1}(M) \subseteq M$, with $(1 - \alpha)(M) \subseteq IM$ and $\alpha : \mathrm{Ker}\, g \to \mathrm{Ker}\, f$ is an isomorphism.

Proposition 2.7. Suppose that Λ satisfies Eichler's condition. If $\Lambda \oplus M \simeq \Lambda \oplus \Lambda$, then $M \simeq \Lambda$.

Proof. By Corollary 2.4, there are only finitely many M satisfying the hypotheses of Lemma 2.6 so that we can choose a finite set of prime ideals, π_0, of R such that Lemma 2.6 is true for any such M.

In particular, if $\Lambda \oplus M \simeq \Lambda \oplus \Lambda$, then $M \vee \Lambda$. (Apply the proof of Corollary 3.2.b.) Thus there is $0 \to M \xrightarrow{g} \Lambda \to T \to 0$ with $\mathrm{Ann}(T) + I = R$, where $I = P_1 P_2 \ldots P_m$, $\pi_0 = \{P_1, P_2, \ldots, P_m\}$, $T = T_1 \oplus \ldots \oplus T_n$; T_i is simple, and $\mathrm{Ann}(T_i) + \mathrm{Ann}(T_j) = R$ if $i \neq j$ (Proposition 2.5).

We seek $0 \to \Lambda \xrightarrow{f} \Lambda \to D \to 0$ and $0 \to \Lambda \xrightarrow{s} \Lambda \to E \to 0$ such that $D \simeq T \oplus E$. The exact sequence $0 \to \Lambda \oplus M \xrightarrow{1\oplus g} \Lambda \oplus \Lambda \to T \to 0$ induces an exact sequence $0 \to \Lambda \oplus \Lambda \xrightarrow{h} \Lambda \oplus \Lambda \to T \to 0$. (Replace $\Lambda \oplus M$ by $\Lambda \oplus \Lambda$.) Then $h_S : \Lambda_S \oplus \Lambda_S \to \Lambda_S \oplus \Lambda_S$ is an isomorphism, where $S = R \backslash \cup \{P | P \in \pi_0\}$, since $\mathrm{Ann}(T) + I = R$ guarantees that $\mathrm{Ann}(T)_S = R_S$ and $T_S = 0$.

As in Warfield [20], $\Lambda_S/J(\Lambda_S)$ is artinian, where $J(\Lambda_S)$ is the Jacobson radical of Λ_S, 1 is in the stable range of Λ_S, and $h_S = (1,\lambda)$: $\Lambda_S \oplus \Lambda_S \to \Lambda_S \oplus \Lambda_S$ for some $\lambda = f/s \in \Lambda_S$, where $f \in \Lambda$, $s \in S$, and λ is a unit for Λ_S. But $(s,f) = sh_S = sh : \Lambda \oplus \Lambda \to \Lambda \oplus \Lambda$, so that $C = (\Lambda \oplus \Lambda)/$ Image $sh \simeq E \oplus D$, where $E = \Lambda/s\Lambda$ and $D = \Lambda/f\Lambda$. Since λ is a unit of Λ_S, f is monic. Thus, D and E are finite R-modules.

On the other hand, there is an exact sequence (*) $0 \to T \xrightarrow{s} C \to E \oplus E \to 0$ since $T = \mathrm{Cokernel}\ h$, $C = \mathrm{Cokernel}\ (sh)$, and $E \oplus E = \mathrm{Cokernel}\ (s)$. Consequently, $s \notin \mathrm{Ann}(T_i)$ for any i, $sR + \mathrm{Ann}\ T = R$, $sT = T$, and multiplication by s is an automorphism of T. Also inclusion induces an epimorphism of $C = (\Lambda \oplus \Lambda)/\mathrm{Image}\ sh$ onto $T = (\Lambda \oplus \Lambda)/\mathrm{Image}\ h$, since $sh = hs$ implies that Image $sh \subseteq$ Image h. Thus (*) is a split exact sequence, i.e. $C \simeq T \oplus E \oplus E$. But $C \simeq E \oplus D$ so that $D \simeq T \oplus E$ by the classical Krull-Schmidt theorem, noting that D, T, E are finite Λ-modules (e.g. Swan [19]).

We now have an exact sequence $0 \to M \xrightarrow{sg} \Lambda \to U \to 0$ and an exact sequence $0 \to T \xrightarrow{s} U \to E \to 0$, where $T = \Lambda/g(M)$ and $E = \Lambda/s\Lambda$. As above, $U \simeq T \oplus E$ (since $sT = T$), so $U \simeq D$, as Λ-modules.

To summarize, there are exact sequences $0 \to M \xrightarrow{e} \Lambda \xrightarrow{\pi} D \to 0$ and $0 \to \Lambda \xrightarrow{f} \Lambda \xrightarrow{\sigma} D \to 0$. Let $D = D_0 \supset D_1 \supset \ldots D_k = 0$ be a composition series of the Λ-module, D, with $D_i/D_{i+1} = S_i$ simple Λ-modules. Let $M_i = \pi^{-1}(D_i)$ and $N_i = \sigma^{-1}(d_i)$, so that $M_0 = \Lambda \supset M_1 \supset \ldots \supset M_k \simeq M$, $N_0 = \Lambda \supset N_1 \supset \ldots \supset N_k \simeq \Lambda$, and $S_i \simeq M_i/M_{i+1} \simeq N_i/N_{i+1}$. By the choice of T, $\mathrm{Ann}(T) = \pi_i \mathrm{Ann}(T_i)$, $\mathrm{Ann}(T_i) \notin \pi_0$. Also, $\mathrm{Ann}(E) \supseteq sR$ and $s \in R\backslash\cup\{P|P \in \pi_0\}$. Thus, $\mathrm{Ann}(D) = \mathrm{Ann}(T \oplus E)$ implies that $\mathrm{Ann}(D) + J = R$, where $J = \Pi\{P|P \in \pi_0\}$. Consequently, each M_i and N_i satisfies the hypotheses of Lemma 2.6 since $\pi_0 \supseteq \pi$ and by the choice of π_0.

Now $0 \to M_1 \to \Lambda \to S_0 \to 0$ and $0 \to N_1 \to \Lambda \to S_0 \to 0$ are exact sequences with S_0 simple and $\mathrm{Ann}(S_0) \notin \pi_0$. By Lemma 2.6, $M_1 \simeq N_1$. Also $0 \to M_2 \to M_1 \to S_1 \to 0$ and $0 \to N_2 \to N_1 \to S_1 \to 0$ are exact, $\mathrm{Ann}(S_1) \notin \pi_0$, and Lemma 2.6 implies that $M_2 \simeq N_2$. By induction, $M \simeq M_k \simeq N_k \simeq \Lambda$, as desired.

3. Torsion free modules of finite rank

Notation. In this section, R is a Dedekind domain with quotient field, K, an algebraic number field; Λ is a Z-reduced R-torsion free R-algebra; $K \otimes_R \Lambda$ is a finite dimensional K-algebra; and L, M, N are Z-reduced R-torsion free right Λ-modules of finite R-rank.

We first demonstrate how Theorem 1.2 can be used to apply the results of §2 to this setting.

Let $E = E_\Lambda(L)$, and let N be the nil radical of E. Then E is a Z-reduced, R-torsion free subalgebra of $K' = K \otimes_R E$, and K' is a finite dimensional K-algebra. Thus K' is artinian, so that the Jacobson radical of K' = nil radical of $K' = N'$ is nilpotent. Embed E in K' (via $x \to 1 \otimes x$), so that $N = N' \cap E$ is nilpotent. In particular, $K \otimes (E/N)$ is a semi-simple finite dimensional K-algebra. By Beaumont-Pierce [2], E/N is isomorphic to a subgroup of E, hence is Z-reduced.

In summary, E/N is a Z-reduced R-torsion free R-algebra and $K \otimes_R (E/N)$ is a semi-simple finite dimensional K-algebra, precisely the hypotheses of §2.

Corollary 3.1. Given L, there is a non-zero ideal, I, of R such that if M, N are in $P_\Lambda(L)$, then $M \vee N$ iff there is a Λ-exact sequence $0 \to M \to N \to T \to 0$ with $\mathrm{Ann}(T) + I = R$.

Proof. ($\Leftarrow$) Let $I = P_1 P_2 \dots P_n$, where $\pi = \{P_1, P_2, \dots, P_n\}$ is given by Proposition 2.3 depending only on $E_\Lambda(L)/N$. As in the proof of Theorem 1.2, there is an $E_\Lambda(L)/N$ exact sequence $0 \to F(M) \to F(N) \to T' \to 0$, where

$\text{Ann } T' + I = R$ (i.e. $\text{Ann}(T) \subseteq \text{Ann}(T')$). As a consequence of Proposition 2.3, $F(M) \vee F(N)$, so that $M \vee N$ follows from Theorem 1.2.

Remark. Corollary 3.1 represents a significant improvement over the definition of $M \vee N$ in the sense that only a single ideal must be tested to show that $M \vee N$. On the other hand, for a given L, this ideal is difficult to discover (e.g. see proof of Proposition 2.3 and 2.2).

Corollary 3.2. (a) If $M^k \vee N^k$, then $M \vee N$.

(b) If $(M \oplus L) \vee (N \oplus L)$, then $M \vee N$.

(c) If $M \vee N \vee L$, then $M \oplus M' \simeq N \oplus L$ for some $M' \vee M$.

Proof. (a) By letting $L = M \oplus N$ and applying Theorem 1.2 and Corollary 1.3, we may assume that $K \otimes_R \Lambda$ is semi-simple and that M, N are finitely Λ-generated projective right Λ-modules. As a consequence of Proposition 2.1, $(M/PM)^k$ is $\Lambda/P\Lambda$-isomorphic to $(N/PN)^k$ for all maximal ideals P of R. Since $\Lambda/P\Lambda$ is finite, the classical Krull-Schmidt theorem (Swan [19]), guarantees that M/PM and N/PN are $\Lambda/P\Lambda$-isomorphic for all P. Thus, by Proposition 2.1, $M \vee N$.

The proof of (b) is similar and (c) follows from Lemma 1.1 and (b).

Proposition 3.3. If M is a genus summand of N, then $N = X \oplus Y$ for some $X \vee M$.

Proof. First assume that $K \otimes_R \Lambda$ is semi-simple and that M, N are finitely generated Λ-projective. Choose, via Lemma 2.2, a non-zero $r \in R$ with $r\Gamma \subseteq \Lambda \subseteq \Gamma$, where Γ is right hereditary. We may adjust r so that $r \in J = \pi\{P_i | P_i \in x\}$, where π is given by Proposition 2.3. Then there are Λ-homorphisms $f : M \to N$ and $g : N \to M$ with $gf = s : M \to M$, where $1 = s + tr$ for some $s, t \in R$. (Since M is a genus summand of N, let $I = Rr$.) Then $sM = gf(M) \subseteq X \subseteq M$, where $X = g(N)$. Since X is a submodule of a finitely generated projective Λ-module M, $g : N \to X$ is epic, and Γ is right hereditary there is a Λ-homomorphism $h : X \to N$ with $gh = rt : X \to X$. (Use the proof given by Swan-Evans [17], Lemma 6.1, p. 104.) Then $f + h : X \to N$

and $g(f + h) = s + rt = 1$, i.e. $N \simeq X \oplus \operatorname{Ker} g$. Apply Proposition 2.3 to see that $X \vee M$. (i.e. $sM \subseteq X \subseteq M$, $1 = s + tr$, and $r \in J$ imply that $X_P \simeq M_P$ for all $P \in \pi$.)

For the general case, apply Theorem 1.2 and Corollary 1.2 with $L = M \oplus N$.

Corollary 3.4. (a) If $L \vee (M \oplus N)$, then $L = L_1 \oplus L_2$ with $L_1 \vee M$, $L_2 \vee N$.

(b) If $M \vee N$, then M is Λ-indecomposable iff N is Λ-indecomposable.

(c) If $M \vee N^k$, then $M \simeq N^{k-1} \oplus L$ with $L \vee N$.

Proof. (a) Apply Proposition 3.3 and Corollary 3.2b.

(b) Clear.

(c) Repeated application of (a) and Corollary 3.2c.

Call M a quasi-summand of N if there is $f \in \operatorname{Hom}_\Lambda(M,N)$, $g \in \operatorname{Hom}_\Lambda(N,M)$ such that $gf : M \to M$ is monic. In particular, there is $0 \neq r \in R$ with $rM \subseteq gf(M) \subseteq g(N) \subseteq M$.

Proposition 3.5. If $M \vee N$ and if M is a quasi-summand of L^n for some n, then M is a summand of $N \oplus L$.

Proof. As usual, it is sufficient to assume that $K \otimes_R \Lambda$ is semi-simple and L, M, N are finitely generated Λ-projective.

Choose $f : M \to L^k$, $g : L^k \to M$ with gf monic and $\operatorname{Ann}(M/gf(M)) \neq 0$. By Proposition 2.5, there is $f_1 : N \to M$ with $\operatorname{Ann}(T) + \operatorname{Ann}(M/gf(M)) = R$, where $T = M/f_1(N) = T_1 \oplus \ldots \oplus T_m$, each T_i is a simple Λ-module, and $\operatorname{Ann}(T_i) + \operatorname{Ann}(T_j) = R$ if $i \neq j$. Then $g(L^k) + f_1(N) = M$, i.e. the composite map $L^k \to M \to T$ is epic (since $\operatorname{Ann}(T) + \operatorname{Ann}(T) + \operatorname{Ann}(M/g(L^k)) = R$).

For $1 \leq i \leq m$, let $P_i = \operatorname{Ann}(T_i)$ a maximal ideal of R. By the Chinese remainder theorem there is $r_i \in R$ with $r_i \in P_j \backslash P_i$ for all $j \neq i$. Since T_i is simple, there is $g_i : L \to M$ such that the composite $L \to M \to T_i$ is non-zero, hence epic. Let $h = r_1 g_1 + r_2 g_2 + \ldots + r_m g_m$. If $t = t_1 + \ldots + t_m \in T$ then $t_i = r_i g_i(x_i)$ for some $x_i \in L$ (since $r_i \notin P_i = \operatorname{Ann}(T_i)$). Thus $h(x_1 + \ldots + x_n) = r_1 g_1(x_1) + \ldots + r_m g_m(M) = t \in T = M/f_1(N)$, i.e. $M = h(L) + f_1(N)$. Consequently, $h \oplus f_1 : L \oplus N \to M$ is epic. Since M is projective, $L \oplus N \simeq M \oplus M'$, as desired. This completes the proof of Proposition 3.5.

Let $[M]$ be the element of $K_0(L)$ associated with the isomorphism class of M. Then $[M] = [N]$ in $K_0(L)$ iff $M \oplus L^k \simeq N \oplus L^k$ for some $k \in Z$. (e.g. see Swan-Evans [17].)

Proposition 3.6. (a) The torsion subgroup of $K_0(L)$ is $\{[M] - [N] | M \vee N\}$ and is finite.

(b) $K_0(L)$ is finitely generated.

(c) $M \vee N$ iff $M^k \simeq N^k$ for some k.

Proof. ((a), (b)) If $m([M]) - [N]) = 0$ then $[M^m] - [N^m] = 0$, i.e. $M^m \oplus L^s \simeq N^m \oplus L^s$ for some $s \in Z$. By Corollary 3.2, $M \vee N$ so that the torsion subgroup of $K_0(L)$ is contained in $\{[M] - [N] | M \vee N\}$.

Via Theorem 1.2, $K_0(L) \simeq K_0(E_\Lambda(L)/N)$, so we may assume that $L = \Lambda$ and $K \otimes_R \Lambda$ is semi-simple. Let $\pi = \{P_1, P_2, \ldots, P_n\}$ be the set of maximal ideals given by Prop. 2.3. Define $\phi : K_0(\Lambda) \to \pi_{P \in \pi} K_0(\Lambda/P\Lambda)$ by $\phi(M) = ([M/PM])_{P \in \pi}$. Then ϕ is a well defined homomorphism. Since $\Lambda/P\Lambda$ is artinian, $K_0(\Lambda/P\Lambda)$ is finitely generated free. (See Swan [19].) Thus the image of ϕ is a finitely generated free abelian group. Moreover, the torsion subgroup of $K_0(\Lambda)$ is contained in $\text{Ker}\ \phi = \{[M] - [N] | M \vee N\}$ by Prop. 2.3 and the Krull-Schmidt theorem for finitely generated projective $\Lambda/P\Lambda$-modules.

It remains to prove that $\text{Ker}\ \phi$ is finite. By Proposition 3.5, if $M \vee N$ then $M \oplus M' \simeq N \oplus L$, with $M' \vee L$ by Corollary 3.2.b. Then $[M] - [N] = [L] - [M']$ in $K_0(L)$. By Lemma 1.1, M' is a summand of $L \oplus L$, and as noted earlier, Lady [8] proved that there are only finitely many M', up to isomorphism. Thus $\text{Ker}\ \phi$ is finite.

(c) By Lemma 1.1, if $M \vee N$ then $N \in P_\Lambda(M)$. Let H_M be the subgroup of $K_0(M)$ generated by $[M]$, and let $G = \{[N] + H_M | N \vee M\}$. Then H is a subgroup of $K_0(M)/H_M$ since $N \vee M$ implies that $N \oplus N' \simeq M \oplus M$ with $N' \vee M$, $[N] + H_M + [N'] + H_M = 0 + H_M$. If $N_1 \vee M \vee N_2$, then $M \oplus M' \simeq N_1 \oplus N_2$, so $M' \vee M$ and $[N_1] + H_M + [N_2] + H_M = [M'] + H_M$.

As above, G is finite so $k([N] + H_M) = [N^k] + H_M = 0 + H_M$ for some

$0 \neq k \in Z$. Thus $N^k \oplus M^s \simeq M^{k+s}$ for some $s \in Z$. If $s = 0$, then the proof is complete. So suppose that $s \geq 1$. By Warfield [20], 2 is in the stable range of M, i.e. $N^k \oplus M^{s-1} \simeq M^{k+s-2} \oplus M'$ with $M \oplus M' \simeq M \oplus M$. Thus $N^k \oplus M^{s-1} \simeq M^{k+s-2} \oplus M'$ with $M \oplus M' \simeq M \oplus M$ and $N^k \oplus M^{s-1} \simeq M^{k+s-1}$ if $k + s \geq 2$. If $k \geq 2$ then eventually, $N^k \simeq M^k$. Finally, assume that $k = s = 1$, i.e. $N \oplus M \simeq M \oplus M$. Then $M \oplus M' \simeq N \oplus N$ for some $M' \vee N$, so $N \oplus N \oplus N \simeq N \oplus M \oplus M' \simeq M \oplus M \oplus M' \simeq M \oplus N \oplus N \simeq M \oplus M \oplus N$. Since 2 is in the stable range of N, $M \oplus M \simeq N \oplus N'$ with $N \oplus N' \simeq N \oplus N$, i.e. $M \oplus M \simeq N \oplus N$. The proof is now complete.

Corollary 3.7. The following are equivalent:

(a) $M \oplus M \simeq M \oplus N$ implies that $M \simeq N$.

(b) If X is a genus summand of M^n for some n, then $X \oplus M \simeq X \oplus N$ implies that $M \simeq N$.

(c) If $\Lambda' \oplus M' \simeq \Lambda' \oplus \Lambda'$, where $\Lambda' = E_\Lambda(M)/\mathcal{N}$ and $\mathcal{N}$ is the nil radical of $E_\Lambda(M)$, then $\Lambda' \simeq M'$.

Proof. (c) $\Leftrightarrow$ (a) follows from Theorem 1.2 with $L = M$.

(b) $\Rightarrow$ (a) is clear.

((a) $\Rightarrow$ (b)) As a consequence of Proposition 3.3, $X' \oplus N \simeq M^n$ for some $n \in Z$ and $X' \vee X$. But $M^{2n} \simeq X' \oplus X' \oplus N \oplus N$, and X is isomorphic to a summand of $X' \oplus X'$ by Corollary 3.2.c. Thus, it is sufficient to assume that X is a summand of M^n, say $X \oplus X' = M^n$.

Now $X \oplus X' \oplus M \simeq M^{n+1} \simeq X \oplus X' \oplus N \simeq M^n \oplus N$, since $X \oplus M \simeq X \oplus N$. Since 2 is in the stable range of M, $M^{n-1} \oplus N \simeq M^{n-1} \oplus M'$, where $M \oplus M' \simeq M \oplus M$. By induction, $M \oplus M \simeq M \oplus N$. By (a), $M \simeq N$, as required.

Corollary 3.8. Suppose that $rM \subset N \subset M$ for some $0 \neq r \in R$. If $E_\Lambda(N)/\mathcal{N}$ satisfies Eichler's condition, where $\mathcal{N}$ is the nil radical of $E_\Lambda(N)$, then $X \oplus M \simeq X \oplus L$, where X is a genus summand of M^n for some n, implies that $M \simeq L$.

Proof. Note that $K \otimes_R (E_\Lambda(N)/\mathcal{N}) \simeq K \otimes_R (E_\Lambda(M)/\mathcal{N}')$, where $\mathcal{N}'$ is the nil radical of $E_\Lambda(M)$. Now apply Proposition 2.7 and Corollary 3.7.

<u>Corollary 3.9</u>. Suppose that $\Lambda' = E_\Lambda(M)/N$ satisfies Eichler's condition, where N is the nil radical of $E_\Lambda(M)$, and that $r\Gamma \subseteq \Lambda \subseteq \Gamma$ for some $0 \neq r \in R$, where Γ is a torsion free R-algebra. Then $M \vee N$ and $M\Gamma \simeq N\Gamma$ iff $M \oplus M\Gamma \simeq N \oplus M\Gamma$.

In particular, if $G_{M\Gamma}$ is the Γ-isomorphism class of $M\Gamma$, then $M \vee N$ iff $M \oplus M\Gamma \simeq N \oplus M\Gamma$.

<u>Proof</u>. ($\Leftarrow$) By Corollary 3.2.b, $M \vee N$. But $M\Gamma \oplus M\Gamma \simeq N\Gamma \oplus M\Gamma$. Also $M\Gamma$ is R-torsion free, $M\Gamma/M$ is bounded, and $E_\Gamma(M\Gamma)/N' = \Gamma'$ satisfies Eichler's condition since $K \otimes_R \Lambda' \simeq K \otimes_R \Gamma'$. Now apply Corollary 3.8 to see that $M\Gamma \simeq N\Gamma$.

($\Rightarrow$) Since $M\Gamma/M$ is bounded, $M \oplus X \simeq N \oplus M\Gamma$ for some Λ-module, X, with $X \vee M\Gamma$ (Proposition 3.5 and Corollary 3.2.b). By Corollary 3.2.c, X is a Λ-summand of $M\Gamma \oplus M\Gamma$ and it follows that $X = X\Gamma$. As noted above, $E_\Gamma(M\Gamma)/N'$ satisfies Eichler's condition and $M\Gamma \oplus X \simeq N\Gamma \oplus M\Gamma \simeq M\Gamma \oplus M\Gamma$. Thus, $X \simeq M\Gamma$ and $M \oplus M\Gamma \simeq N \oplus M\Gamma$, as desired.

<u>Remark</u>. Jacobinski [7] calls M in the <u>restricted genus of N</u> iff $M \vee N$ and $M\Gamma \simeq N\Gamma$, where M is a Λ-lattice, Λ is an R-order, and Γ is a maximal R-order.

4. <u>Applications to torsion free abelian groups of finite rank</u>

<u>Notation</u>. In this section L, M, N are reduced torsion free abelian groups of finite rank, i.e. $R = \Lambda = Z$ and $Q = K$.

Following Fuchs [4], call M and N <u>quasi-isomorphic</u> if M is isomorphic to a subgroup of finite index in N. We write <u>$E'(M)$</u> for $E_Z(M)/N$, where N is the nil radical of $E_Z(M)$.

<u>Corollary 4.1</u>. Suppose that M and N are quasi-isomorphic. Then $E'(M)$ satisfies Eichler's condition iff $E'(N)$ does.

<u>Proof</u>. Observe that $Q \otimes_Z E'(M) \simeq Q \otimes_Z E'(N)$.

The group, M, is <u>completely decomposable</u> if $M = M_1 \oplus \ldots \oplus M_n$, where rank $(M_i) = 1$, for $1 \leq i \leq n$, and <u>almost completely decomposable</u> if M is quasi-isomorphic to a completely decomposable group.

Corollary 4.2. If M is almost completely decomposable then $E'(M)$ satisfies Eichler's condition.

Proof. In view of Corollary 4.1, it is sufficient to assume that $M = M_1 \oplus \ldots \oplus M_n$ with rank $(M_i) = 1$ for $1 \leq i \leq n$.

First of all, N = nil radical of $E_Z(M) = H = \sum \oplus \{\mathrm{Hom}_Z(M_i, M_j) \mid M_i \not\cong M_j$. Briefly, if $f_i : M_i \to M_j$ for some $M_i \not\cong M_j$, then $f_i E_Z(M)$ is nilpotent: since if $m = m_1 + m_2 + \ldots + m_n \in M$, then $(m)f_i g f_i h = (m_i)f_i g f_i h = (m_j'')g f_i h = (m_i')f_i h = 0$, where $(m_i)f_i = m_j''$ and $(m_j'')g = m_1' + m_2' + \ldots + M_n'$, since if $m_i' \neq 0$, then g induces a non-zero map $M_j \to M_i$ so that $M_i \simeq M_j$, a contradiction. Thus $H \subseteq N$. Conversely, H is a two sided ideal of $E_Z(M)$ and nil radical $(E_Z(M)/H) = 0$, since $E_Z(M)/H$ is a ring direct sum of matrix rings so that $N \subseteq H$.

Consequently, $E'(M) = \Gamma_1 \oplus \ldots \oplus \Gamma_m$ where $\Gamma_i = M_{n_i}(E(M_i))$ (the ring of $n_i \times n_i$ matrices over $E(M_i)$, a subring of Q) and $\{M_1, \ldots, M_m\}$ is a set of representatives of isomorphism classes of $\{M_1, M_2, \ldots, M_n\}$. Now $Q \otimes_Z \Gamma_i = M_{n_i}(Q)$, which is simple but not a totally definite quaternion algebra. (e.g. $M_{n_i}(Q)$ is never a non-commutative division algebra.)

Corollary 4.3. (Lady [10]). Suppose that M is almost completely decomposable. The following are equivalent:

(a) $M \oplus L \simeq N \oplus L$ for some L;

(b) $M \vee N$;

(c) $M \oplus L \simeq N \oplus L$ for some completely decomposable L;

(d) $M^k \simeq N^k$ for some $k \in Z$.

Proof. In view of the results of §3, it is sufficient to prove that (b) ⟹ (c). By Corollary 4.2, $E'(M)$ satisfies Eichler's condition.

Let $mM \subseteq L \subseteq M$, where L is completely decomposable. By Proposition 3.5, $M \oplus M' \simeq N \oplus L$ for some $M' \vee L$ (by Corollary 3.2.b). Thus it is sufficient to prove that if L is completely decomposable and $M' \vee L$, then $M' \simeq L$. By Corollary 3.2, M' is a summand of $L \oplus L$. But it is well known that summands of completely decomposable groups are completely decomposable and that if M' and L are completely decomposable with M' quasi-isomorphic to L, then $M' \simeq L$ (e.g. see Fuchs [4]).

Remark. Fuchs-Loonstra [5] have proved that for any $k \geq 2$ there are almost completely decomposable torsion free groups M and N of rank two such that $M^k \not\simeq N^k$ but $M^n \not\simeq N^n$ if $1 \leq n \leq k - 1$. In particular, $M \vee N$ but M and N are not isomorphic.

Combining a result of Arnold-Lady [1] and Corollary 3.8 we have:

Corollary 4.4. Suppose that $M \oplus L \simeq N \oplus L$. Then $M \simeq N$ if either (a) L and M have no quasi-summands in common, or (b) $E'(M)$ satisfies Eichler's condition and L is a genus summand of M^n for some n.

Example 4.5. Swan [19], proves that if Λ is the integral group ring of the generalized quaternion group of order 32, then $\Lambda' \oplus \Lambda \simeq \Lambda \oplus \Lambda$ for some $\Lambda' \not\simeq \Lambda$. By Zassenhaus [21], $\Lambda = E_Z(M)$ for some torsion free abelian group of rank 32 (since Λ is a free abelian group of rank 32). Thus there is $M \oplus M' \simeq M \oplus M$ with $M' \not\simeq M$. On the other hand $M \vee M'$.

If M is a torsion free abelian group of finite rank then $mM \subseteq M_1 \oplus M_2 \oplus \ldots M_n \subseteq M$ for some $0 \neq m \in Z$ where each M_i is strongly indecomposable (i.e. $m_i M_i \subseteq B \oplus C \subseteq M_i$ for $0 \neq m_k \in Z$ implies that $B = 0$ or $C = 0$).

Example 4.6. Following are examples of groups M such that $E'(M)$ satisfies Eichler's condition:

(i) $E'(M)$ is commutative ($Q \otimes_Z E'(M)$ is commutative).

(ii) M is quasi-isomorphic to B^n for some $n > 1$, where B is strongly indecomposable ($Q \otimes_Z E'(M) = M_n(D)$ where $D = Q \otimes E'(B)$ is a division algebra, and $M_n(D)$ is not a division algebra since $n > 1$).

We conclude with a series of open problems.

(1) Describe (in group theoretic terms?) all M such that $E'(M)$ satisfies

Eichler's condition.

In view of Corollary 4.1, it is sufficient to assume that $M = M_1 \oplus \ldots \oplus M_n$, where each M_i is strongly indecomposable. One must then be prepared to describe the nil radical of $Q \otimes_Z E_Z(M)$ in terms of $Q \otimes_Z E_Z(M_i)$. (See Corollary 4.2 for the case that all M_i have rank 1.) Finally, can the strongly indecomposable groups, M, such that $Q \otimes_Z E'(M)$ is not a totally definite quaternion algebra be described in group theoretic terms?

(2) Find necessary and sufficient condition such that $M \oplus M \simeq M \oplus N$ implies that $M \simeq N$.

In view of Corollary 3.7, this is equivalent to: $E'(M) \oplus L \simeq E'(M) \oplus E'(M)$ implies $L \simeq E'(M)$. This question is unsolved even for Λ' an R-order. In particular, is "Λ' satisfies Eichler's condition" necessary?

(3) Find all M such that G_M is the isomorphism class of M.

As observed by Lady [9], if $E'(M)$ is right principal or M is divisible by almost all primes, then $M \vee N$ iff $M \simeq N$. As noted in the proof of Corollary 4.3, completely decomposable groups also have this property.

(4) Find all M such that $M \oplus L \simeq N \oplus L$ iff $M \vee N$.

By Corollary 4.3, this includes all almost completely decomposable groups. Also note Corollary 3.9, i.e. embed $E'(M)$ in $\Gamma = \Gamma_1 \oplus \ldots \oplus \Gamma_n$ such that Γ_i is a maximal R_i-order. If $G_{M\Gamma_i}$ is the Γ_i-isomorphism class of $M\Gamma_i$ for all i, then $M \vee N$ iff $M \oplus M\Gamma \simeq N \oplus M\Gamma$.

REFERENCES

[1] D. M. Arnold and E. L. Lady, Endomorphism rings and direct sums of torsion free abelian groups, Trans. A.M.S. 211(1975), 225-237.

[2] R. A. Beaumont and R. S. Pierce, Subrings of algebraic number fields, Acta. Sci. Math. Szeged 22(1961), 202-216.

[3] ________________, Torsion free rings, Ill. Jour. Math. 5(1961), 61-98.

[4] L. Fuchs, Infinite Abelian Groups, Vol. II, Academic Press, New York, 1973.

[5] L. Fuchs and F. Loonstra, On direct decompositions of torsion free abelian groups of finite rank, Rend. Sem. Mat. Univ. Padova 44(1970), 75-83.

[6] K. R. Goodearl, Power cancellation of groups and modules, Pacific J. Math., to appear.

[7] H. Jacobinski, Genera and decompositions of lattices over orders, Acta. Math. 121(1968), 1-29.

[8] E. L. Lady, Summands of finite rank torsion free abelian groups, J. Alg. 32(1974), 51-52.

[9] __________________, Nearly isomorphic torsion free abelian groups, J. Alg. 35(1975), 235-238.

[10] __________________, Almost completely decomposable torsion free abelian groups, Proc. A.M.S. 45(1974), 41-47.

[11] R. S. Pierce, Subrings of simple algebras, Michigan Math. J. 7(1960), 241-243.

[12] J. D. Reid, On the ring of quasi-endomorphisms of a torsion free group, Topics in Abelian-Groups, Chicago, 1963, 51-68.

[13] I. Reiner, Maximal Orders, Academic Press, New York, 1975.

[14] __________________, A survey of integral representation theory, Bull. A.M.S. 76(1970), 159-227.

[15] K. W. Roggenkamp, Lattices over Orders II, Springer-Verlag Lecture notes No. 142, Springer, Berlin, 1970.

[16] A. V. Roiter, On integral representations belonging to a genus, Izv. Akad. Nauk. USSR 30(1966), 1315-1324.

[17] R. G. Swan, and E. G. Evans, K-Theory of Finite Groups and Orders, Springer-Verlag Lecture Notes No. 149, Springer, Berlin, 1970.

[18] R. G. Swan, Projective modules over group rings and maximal orders, Ann. of Math. (2) 76(1962), 55-61.

[19] __________________, Algebraic K-Theory, Springer-Verlag Lecture Notes No. 76, Springer, Berlin, 1968.

[20] R. B. Warfield, Jr., Cancellation for modules and the stable range of endomorphism rings, to appear.

[21] H. Zassenhaus, Orders as endomorphism rings of modules of the same rank, J. London Math. Soc. 42(1967), 180-182.

QUASI-PURE-INJECTIVITY AND QUASI-PURE PROJECTIVITY

J. D. Reid

1. Introduction. In his well known book [5], L. Fuchs suggests the problem of "developing properties of" quasi-pure-injective and quasi-pure-projective abelian groups. This paper is a survey of results on this problem obtained recently by the author, in conjunction with various others as indicated in the references. The aim has been to develop enough properties of the groups in question to characterize them. Our discussion should give the reader a fairly complete picture of the status of the characterization problem. We touch on generalizations only to the extent that they bear directly on this problem.

Recall that an abelian group G is quasi-pure-injective (qpi) if every pure exact sequence $0 \to H \to G \to L \to 0$ induces an exact sequence $0 \to \mathrm{Hom}(L, G) \to \mathrm{Hom}(G, G) \to \mathrm{Hom}(H, G) \to 0$. Dually, G is quasi-pure-projective (qpp) if every pure exact sequence $0 \to H \to G \to L \to 0$ induces an exact sequence $0 \to \mathrm{Hom}(G, H) \to \mathrm{Hom}(G, G) \to \mathrm{Hom}(G, L) \to 0$. (Here, and throughout this paper, Hom and $\otimes$ indicate the functors with base ring the ring of integers Z.) Our object is to present complete characterizations of the qpi and qpp torsion free groups of finite rank, as well as of the qpi and qpp torsion groups.

Section 2 treats the qpi torsion free groups of finite rank. Section 3 introduces a generalization of the projectivity condition which is needed to show that the characterization of the qpp torsion free groups of finite rank given there is a complete one. Section 4 treats the torsion groups and we close in Section 5 with a short discussion of two open problems that the author finds interesting. Our notation is standard, though we might mention that if x is an element of the torsion free group G, we write $|x|$ for the type of the pure rank-1 subgroup of G generated by x; and if G is a homogeneous torsion free group we denote by $|G|$ the common type of the elements of G. (In Section 5, however, $|S|$ is used to denote the cardinality of the sets). The endomorphism ring of any group G is denoted by $E(G)$.

Proofs are generally omitted, but each result has associated with it a reference indicating where a proof may be found as well as the authors to whom the result is due.

2. Quasi-pure-injective torsion free groups. In this section we deal exclusively with reduced torsion free groups of finite rank. As is well known, this is an extremely intractable class of groups, rich in pathology. It is especially surprising then that the torsion free qpi and qpp groups of finite rank admit an extraordinarily explicit and detailed structure theory. We believe that the results we are about to describe are of interest for another reason as well, in that they exemplify the fact that the general theory of torsion free groups of finite rank has progressed to the point that it can handle non-trivial problems. The elements of that theory needed here are contained in Arnold [1], Reid [9] and Warfield [14], in addition to Fuchs [6].

The following decomposition theorem splits the discussion into that of homogeneous groups and strongly indecomposable non-homogeneous groups.

Theorem 1. [2] Let G be a reduced torsion free group of finite rank. Then G is qpi if and only if $G = H_1 \oplus \ldots \oplus H_n \oplus G_1 \oplus \ldots \oplus G_m$, where each H_i is qpi and is the direct sum of finite number of copies of a homogeneous strongly indecomposable qpi group; each G_j is non-homogeneous strongly indecomposable qpi; and if X and Y are pure rank-1 subgroups of different summands in this decomposition of G, then $\sup(|X|, |Y|) = |Q|$, where Q is the additive group of rational numbers.

Two points are perhaps worthy of note here. First, Theorem 1 gives an actual decomposition of G, not merely a quasi-decomposition. Secondly, the terms H_i and G_j in this decomposition are uniquely determined, just as in the primary decomposition of a torsion group or the decomposition of a completely decomposable module into its homogeneous components. This uniqueness follows easily from the condition on the types of the pure rank-1 subgroups of the various summands.

To describe the homogeneous qpi groups we recall first that a torsion free group is said to be strongly homogeneous if its group of automorphisms acts transitively on the set of pure rank-1 subgroups. For convenience in stating the following theorem on homogeneous strongly indecomposable qpi groups we will say that, if R is a subring of an algebraic number field (= integral domain of characteristic 0 and finite Z-rank), then a rational prime p has degree 1 in R if R/pR is the prime field of characteristic p. We call a subring R of an algebraic number field a strongly homogeneous ring if every element of R is a rational integral multiple of a unit in R. Note that the additive group R^+ of a strongly homogeneous ring is a strongly homogeneous group but R^+ may well be strongly homogeneous without R being a strongly homogeneous ring.

Theorem 2. [2] Let G be a reduced torsion free abelian group of finite rank. Then the following are equivalent: (a) G is homogeneous strongly indecomposable and qpi; (b) G is strongly homogeneous and every pure subgroup of G is strongly indecomposable; (c) $G \approx R \otimes A$ where R is a strongly homogeneous ring in which some rational prime has degree 1 and A is a rank-1 group with $|E(A)| = |R \cap Q|$; (d) $G \approx R \otimes A$, where R is a strongly homogeneous ring whose additive group is strongly indecomposable qpi and A is a rank-1 group with $|E(A)| = |R \cap Q|$. If these conditions hold, the ring R is $E(G)$.

The statement of Theorem 2 in [2] differs from that here in that there the notion of a pointwise locally rational ring was introduced. The defining property of such a ring R is that R be a subring of an algebraic number field and that for each $r \in R$ there is a rational prime p with $pR \neq R$, and an element $a \in R \cap Q$ such that $r \equiv a (\mathrm{mod}\ pR)$. It turns out, however [10], that a strongly homogeneous ring is pointwise locally rational if and only if some prime p has degree 1 in R. We have therefore used this perhaps more natural, certainly more standard, condition here.

We might remark also, concerning the rings R occurring in Theorem 2, that they are, under the regular representation, the full endomorphism rings of their additive groups. In general, it is unknown when this phenomenon occurs. The following result, which covers the case at hand as well as, for example, the ring of p-adic integers, may be of interest. (See [8] for a more general statement, and further results on this problem. See also [12].)

Proposition 1. Let R be a torsion free ring (with identity) and let S be the pure subgroup of R generated by the identity, 1, of R. The left regular representation is an isomorphism of R onto the endomorphism ring of its additive group R^+ if and only if $\mathrm{Hom}(R/S, R) = 0$.

Proof. Clearly $\mathrm{Hom}(R/S, R)$ is isomorphic to the left ideal L of $E(R^+)$ consisting of those endomorphisms that annihilate 1. Thus if $E(R^+)$ consists of left multiplications r_ℓ by elements $r \in R$ then $\mathrm{Hom}(R/S, R) = 0$. On the other hand, if this condition holds and $f \in E(R^+)$, then the endomorphism $g = f(1)_\ell - f$ annihilates 1, hence is zero, so $f = f(1)_\ell$.

Getting back to the problem at hand, Theorem 1 and Theorem 2 will handle the case of homogeneous qpi groups once we know the conditions under which a direct sum of copies of a strongly indecomposable homogeneous qpi group is again qpi. It is perhaps somewhat surprising that this is not always the case.

Theorem 3. [2] Let H be a reduced homogeneous strongly indecomposable qpi torsion free group of finite rank and let G be the direct sum of m copies of H, $m > 1$. Then G is qpi if and only if $R/R \cap Q$ is divisible, where $R = E(H)$ as in Theorem 2.

Notice that the rings R arising in Theorem 2 must have $R/R \cap Q$ divisible by at least one prime, the prime of degree 1 in R, but not necessarily by all primes as required by Theorem 3. Both cases actually occur, as shown by examples given in [2].

We complete the characterization of torsion free qpi groups of finite rank with:

Theorem 4. [2] Let G be a reduced torsion free abelian group of finite rank. Then G is non-homogeneous strongly indecomposable qpi if and only if G is a torsion free R-module, where (i) R is a strongly homogeneous ring in which some rational prime has degree 1; (ii) if X and Y are distinct R-pure R-submodules of G with R-rank 1, then $\mathrm{Hom}_R(X, Y) = 0$; (iii) $|E(A)| = |R \cap Q|$ for all pure rank-1 subgroups, A, of G; and (iv) the typeset of G is infinite. Furthermore, $R \approx E(G)$.

Examples of the groups described in Theorem 4 are given in [2]. Our statement of this theorem differs from that in [2] in the same way that our statement of Theorem 2 differs from [2], as mentioned above.

3. Quasi-pure-projective torsion free groups. The property of being a qpi group is not a quasi-isomorphism invariant. It does not seem obvious, but the property of being a (torsion free) qpp group is invariant under quasi-isomorphism. To see this, as well as to derive one of our basic results (Theorem 6) we generalize the qpp notion as follows. A torsion free abelian group G is called (for lack of a better name) almost quasi-pure projective (aqpp) if there exists a non-zero integer n, depending only on G, such that for every pure exact sequence $0 \to H \to G \to G/H \to 0$, in the induced sequence $0 \to \mathrm{Hom}(G, H) \to \mathrm{Hom}(G, G) \to \mathrm{Hom}(G, G/H)$ we have $n\,\mathrm{Hom}(G, G/H) \subseteq \mathrm{Image}\,\mathrm{Hom}(G, G)$. A torsion free group G is called (for lack of a name) qpqp if for every pure subgroup H of G there exists a non-zero integer $n = n_H$, depending on H, such that the above condition holds. Then we have:

Proposition 2. [11] If A and B are quasi-isomorphic torsion free groups of finite rank, and A is qpqp, then B is qpqp.

There are qpqp groups which are not qpp. However, one can show [11] that a strongly indecomposable qpqp group has rank-1 endomorphism ring, and then establish:

Theorem 5. [11] Every strongly indecomposable qpqp torsion free group of finite rank if qpp.

Thus for strongly indecomposable groups the generalizations given above are only apparent, though this is not obvious. Moreover we have:

Theorem 6. [11] Let G be a torsion free aqpp group of finite rank. Then G is either strongly indecomposable or is a direct sum of isomorphic rank-1 groups.

We may therefore change the "almost" in aqpp to "actually". Using the theorem above and applying the results of Section 2 by use of Warfield's duality theory [14], which interchanges qpp and qpi when it applies, we can characterize the qpp groups as follows.

<u>Theorem 7</u>. [2] Let G be a reduced torsion free abelian group of finite rank. Then G is qpp if and only if G is homogeneous completely decomposable or G is strongly indecomposable, rank G = Z/pZ-dimension of G/pG for all primes p with $pG \neq G$, and if X and Y are distinct pure subgroups of G with rank G/X = rank G/Y = 1, then Hom(G/X, G/Y) = 0. In the latter case, E(G) is isomorphic to a subring of Q.

4. <u>Torsion groups</u>. In contrast to the case of torsion free groups we will see that the qpp and qpi torsion groups are exactly the obvious ones. It is immediate that it suffices to consider the case of p-groups, so we limit our attention to those in this section.

<u>Theorem 8</u>. [4] The p-group G is qpi if and only if G is a direct sum of a divisible group and a torsion complete group.

This theorem follows easily from Leptin's characterization of torsion complete groups (e.g. Section 69 of [6]) and the exact sequence for Pext. Standard results on basic subgroups yield:

<u>Theorem 9</u>. [4] A non-reduced p-group G is qpp if and only if G is a direct sum of a divisible group and a bounded group.

The case of reduced qpp groups is more interesting. The countable case is easy, since it is not hard to show that a reduced qpp group cannot have elements of infinite height, hence is a direct sum of cyclic groups. Unfortunately there seemed to be no particular virtue in splitting off the countable case, and for the general case we had to use the generalized continuum hypothesis. Thus:

<u>Theorem 10</u>. [4] (with GCH) A reduced p-group is qpp if and only if it is a direct sum of cyclic groups.

The general idea of the proof is as follows. Given a p-group G that is qpp, one uses GCH to show that G and B have the same cardinality. This allows one to reduce to the case that G and B have the same final rank. Using a theorem of Hill [7] (see also [13]) one can then find two disjoint basic subgroups B_1, B_2 of G and endomorphisms $\theta_i : G \to B_i$ with $\operatorname{Ker} \theta_i \subset B_i$. This last condition is the critical one, for it and $B_1 \cap B_2 = 0$ imply that the product map $\theta_1 \times \theta_2 : G \to B_1 \times B_2$ is monic. Since finite products are coproducts, this is an embedding of G into a direct sum of cyclic groups. Hence G, as a subgroup of such a direct sum, is itself a direct sum of cyclic groups. We found this proof somewhat amusing.

We remark that we have been informed [3] by K. Benabdallah that he and colleagues have obtained (independently) these results on qpp and qpi torsion groups. Interestingly enough, these authors also required GCH for the reduced qpp case according to Professor Benabdallah. Not yet having seen their work, we do not know how the two approaches compare otherwise.

5. Two questions. One cannot help but wonder whether there is a proof of Theorem 10 which does not rely on GCH. As mentioned above, this hypothesis is needed to show that, for a reduced p-group G that is qpp, one has $|G| = |B|$, where B is basic in G. (We use $|S|$ to indicate the cardinality of the set S in this section.) In order to indicate what is involved, we include a sketch of this proof.

Assume then that G is qpp and that $|B| < |G|$ for some basic B in G. Then $|D| = |G|$ where $D = G/B$. If we write $E(G, B) = \{\sigma \in E(G) \mid \sigma B \subseteq B\}$, then the fact that G is qpp, and B is pure in G, yield

(*) $$E(D) \approx E(G, B)/\operatorname{Hom}(G, B).$$

Since G is reduced, restriction gives a monomorphism of $E(G, B)$ into $E(B)$. Thus $2^{|G|} = 2^{|D|} = |E(D)| \le |E(G, B)| \le |E(B)| = 2^{|B|}$, an inequality which, in view of $|B| < |G|$, violates GCH.

Our second question concerns cardinalities of endomorphism rings also, but this time for torsion free qpp groups. We do not know of any strongly indecomposable qpp torsion free abelian groups of infinite rank. (Note however that the p-adic integers are strongly indecomposable qpi and infinite rank.)

Suppose G is any qpp torsion free group of infinite rank, and let F be a full free subgroup of G (i.e. G/F is torsion). Let V be a rational vector space such that $(|V| =)\ \dim_Q V = \text{rank } G\ (= |G|)$. Since F and G have the same rank and F is free, there is an epimorphism of F onto V which, by the injectivity of V, can be lifted to an epimorphism $\eta : G \to V$. Let $K = \text{kernel } \eta$. Since V is torsion free, K is pure in G and an argument similar to that yielding (*) above allows us to conclude that $|E(G)| = 2^{|G|}$.

Now the large (strongly) indecomposable groups that have been constructed tend to have small (relative to the size of the group) endomorphism rings. The p-adic integers, being their own endomorphism rings, satisfy $|G| = |E(G)|$, and we admit that $Z(p^\infty)$ satisfies $|E(G)| = 2^{|G|}$, but we know of no strongly indecomposable torsion free group G of infinite rank satisfying $|E(G)| = 2^{|G|}$. Are there any? In particular, are there any strongly indecomposable qpp groups of infinite rank? Are there integral domains of arbitrary cardinality that are equal (under the regular representation) to their endomorphism rings?

REFERENCES

1. D. M. Arnold, Strongly homogeneous torsion free abelian groups of finite rank, Proc. Amer. Math. Soc. 56(1976), 67-72.

2. D. M. Arnold, B. O'Brien and J. D. Reid, Quasi-pure injective and projective torsion free abelian groups of finite rank, Proc. London Math. Soc., to appear.

3. K. Benabdallah, oral communication.

4. William Berlinghoff and J. D. Reid, Quasi-pure projective and injective torsion groups, Proc. Amer. Math. Soc., to appear.

5. L. Fuchs, Infinite Abelian Groups, Vol I, Academic Press, New York, 1970.

6. ________, ______________________, Vol II, Academic Press, New York, 1973.

7. Paul Hill, The covering theorem for upper basic subgroups, Michigan Math. J. 18(1971), 187-192.

8. G. P. Niedzwicki and J. D. Ried, E-rings (tent.), in prep.

9. J. D. Reid, On the ring of quasi-endomorphisms of a torsion free group, Topics in Abelian Groups, 51-68, Chicago, 1963.

10. J. D. Reid, On subrings of algebraic number fields, in prep.

11. J. D. Reid, C. I. Vinsonhaler and W. J. Wickless, Quasi-pure projectivity and two generalizations, to appear.

12. P. Schultz, The endomorphism ring of the additive group of a ring, J. Australian Math. Soc. 15(1973), 60-69.

13. Dalton Tarwater and Elbert Walker, Decompositions of direct sums of cyclic p-groups, Rocky Mountain J. Math. 2(1972), 275-282.

14. R. B. Warfield, Jr., Homomorphisms and duality of torsion-free groups, Math. Z. 107(1968), 189-200.

SUR LES GROUPES QUASI-P-NETS INJECTIFS ET PROJECTIFS

K. Benabdallah, R. Bradley et A. Laroche[1]

0. Introduction. Récemment plusieurs équipes de chercheurs ce sont intéressées au problème 17 de L. Fuchs dans lequel il demande de caractériser les groupes quasi-purs injectifs et projectifs. D. Arnold, C.I. Vinsonhaler et W.J. Wickles [1] ont résolu le problème pour les groupes sans torsion de rang 2. La solution a été étendue aux groupes sans torsion de rang fini par D. Arnold, B. O'Brien et J.D. Reid, [2]. Les auteurs ont résolu le cas torsion[2] dans [3] et [5].

Pour nous permettre d'utiliser des techniques semblables à celles utilisées dans le cas torsion, nous avons introduit les notions de groupes quasi-p-pur injectifs et projectifs. Pour ces groupes, la théorie des sous-groupes de p-base s'avère très utile et nous obtenons de nombreux résultats dans le cas injectif ainsi que des caractérisation complètes dans le cas projectif dans [6] et [4]. Dans ce travail, nous étudions les groupes quasi p-nets injectifs et les groupes quasi-p-nets projectifs. Le symbol p désigne tout le long de cet article un nombre entier premier fixe.

1. Définition et résultats généraux. Soit $H \leq G$, on désigne par i_H et ν_H, respectivement, l'inclusion de H dans G et l'épimorphisme canonique de G sur G/H. On dit qu'un groupe G est quasi-injectif par rapport à un sous-groupe H de G, si pour tout homomorphisme $f\colon H \to G$, il existe un endomorphisme $\overline{f}$ de G tel que $\overline{f} \circ i_H = f$. Dualement, on dit que G est quasi-projectif par rapport à H si pour tout homomorphisme $f\colon G \to G/H$, il existe un endomorphisme $\overline{f}$ de G tel que $\nu_H \circ \overline{f} = f$. Pour tout groupe G, on dénote par $I(G)$ l'ensemble de tous les sous groupes de G par rapport auxquels G est quasi-injectif. De même, on dénote par $P(G)$ l'ensemble de tous les sous-groupes ayant la propriété duale.

---

1. Cette recherche a été partiellement subventionnée par le Conseil National de Recherche du Canada fond no A5591.

2. Ce cas a été aussi résolu par une équipe dirigée par J.D. Reid.

Proposition 1.1. Soit G un groupe. Alors

(1) $\mathcal{I}(G)$ contient tous les facteurs directs de G;

(2) $H \in \mathcal{I}(G)$ et A facteur direct de $H \Rightarrow A \in \mathcal{I}(G)$;

(3) $A \in \mathcal{I}(G)$, $A \leq H$ et H sous-groupe totalement invariant de $G \Rightarrow A \in \mathcal{I}(H)$;

(4) $A \in \mathcal{I}(G)$, $A \leq H$ et H facteur direct de $G \Rightarrow A \in \mathcal{I}(H)$;

(5) $H \in \mathcal{I}(G)$ et H isomorphe à $G \Rightarrow H$ est facteur direct de G;

(6) pour G torsion, $A \in \mathcal{I}(G) \Leftrightarrow A_p \in \mathcal{I}(G_p)$ pour tout nombre premier p.

$\mathcal{P}(G)$ satisfait des propriétés semblables a celles contenues dans la proposition précédente. Nous laissons au lecteur le soin de les formuler. Soit p un nombre premier, un sous-groupe H de G est dit p-net si $H \cap pG = pH$, et p-pur si $H \cap p^nG = p^nH$ pour tout entier positif n. Un groupe G est dit quasi-p-net injectif (q.p.n.i.) si $\mathcal{I}(G)$ contient tous les sous-groupes p-nets de G. Il est dit quasi-p-net projectif (q.p.n.p.) si $\mathcal{P}(G)$ contient tous les sous-groupes p-nets de G. De même nous définissons les groupes quasi-p-pur injectifs (q.p.p.i.) et quasi-p-pur projectifs ($q.p.^3$). Il est clair que les groupes q.p.n.i. et q.p.n.p. sont aussi respectivement q.p.p.i. et $q.p.^3$. Pour des résultats sur ces derniers voir [6] et [4]. Notamment, nous avons besoin ici de la caractérisation suivante des groupes $q.p.^3$, établie dans [4].

Théorème[1] 1.2. Un groupe G est $q.p.^3$ si et seulement si il est de l'une des cinq formes suivantes.

(1) $G = D \oplus B \oplus L$

(2) $G = D \oplus B \oplus S$

(3) $G = A \oplus L$

(4) $G = A \oplus (\bigoplus_{q \neq p} C_q)$

(5) $G = D \oplus B \oplus (\bigoplus_{q \neq p} C_q)$

où D est un p-groupe divisible, B un p-groupe borné, L un groupe libre, S un $Q^{(p)}$-module libre, A un p-groupe somme directe de groupes cycliques, et C_q un q-groupe quasi-projectif pour un nombre premier q.

Nous terminons cette section avec une liste de résultats dont nous nous servons par la suite quelque fois implicitement.

---

1. L'hypothèse généralisée du contenu est utilisée pour montrer que les groupes $q.p.^3$ sont des formes indiquées.

Théorème 1.3. Tout p-groupe divisible est q.p.n.p. et q.p.n.i.

Théorème 1.4. Tout facteur direct d'un groupe q.p.n.p. (q.p.n.i.) est q.p.n.p. (q.p.n.i.)

Théorème 1.5. Un groupe torsion G est q.p.n.p. (q.p.n.i.) si et seulement si G_p est q.p.n.p. (q.p.n.i.) et G_q est quasi-projectif (quasi-injectif) pour tout q distinct de p.

2. Les groupes quasi-p-nets-injectifs torsions. Par le théorème 1.5, l'étude des groupes quasi-p-nets-injectifs torsions se réduit à celle des p-groupes.

Lemme 2.1. Si $G = \langle x\rangle \oplus \langle y\rangle \oplus D$ où $O(x) = p^n$, $O(y) = p^m$, $n < m$ et $D \simeq Z(p^\infty)$, alors G n'est pas q.p.n.i.

Preuve. Considérons $H = \langle x-z\rangle$ où $z \in D$ et $O(z) = p^m$. H est alors net dans G. On définit l'homomorphisme $f\colon H \to G$ par $f(x-z) = y$. Si f s'étend à un endomorphisme φ de G, $\varphi(p^n(x-z)) = \varphi(-p^n z) = p^n y$ et $h(-p^n z) = \infty > h(p^n y) = n$. Ceci est une contradiction. Donc G n'est pas q.p.n.i. □

La proposition suivante nous permet de caractériser les p-groupes quasi-p-nets-injectifs non réduits.

Proposition 2.2. Soit $H = \bigoplus_{i\in I} \langle x_i\rangle$ où $O(x_i) = p^n$ pour tout $i \in I$. Soient D un p-groupe divisible et $G = H \oplus D$. Alors G est q.p.n.i.

Preuve. Soit A un sous-groupe p-net de G et soit $f\colon A \to G$ un homomorphisme quelconque. Comme $A \cap D = p^n A$, on a $f(A \cap D) \subseteq p^n G = D$ et f induit un homomorphisme $f'\colon (A + D)/D \to G/D$ défini par $f'(a + D) = f(a) + D$.

Comme H est quasi-injectif, il existe un homomorphisme $\bar{\varphi}\colon G/D \to G/D$ tel que $\bar{\varphi}$ coïncide avec f' sur $(A + D)/D$. On définit alors $\varphi\colon G \to G$ par $\varphi = \alpha.\bar{\varphi}.\nu_D$ où $\alpha\colon G/D \to H$ est l'isomorphisme canonique. Si $a \in A$, $\varphi(a) = \alpha.\bar{\varphi}(a + D) = \alpha.f'(a + D) = \alpha(f(a) + D)$.

Donc $f(a) - \varphi(a) \in D$ pour tout $a \in A$ et on a un homomorphisme $(f-\varphi).i_A\colon A \to D$. Mais D est injectif, et donc il existe un homomorphisme $\psi\colon G \to D$ tel que $\psi.i_A = (f-\varphi).i_A$. Soit $\bar{f} = \psi + \varphi\colon G \to G$. Si $a \in A$,

$\overline{f}(a) = (\psi + \varphi)(a) = \psi(a) + \varphi(a) = f(a) - \varphi(a) + \varphi(a) = f(a)$. Donc G est q.p.n.i. □

On s'intéresse maintenant au cas réduit.

<u>Lemme 2.3</u>. Si $G = \langle x\rangle \oplus \langle y\rangle \oplus \langle z\rangle$ où $O(x) = p^n$, $O(y) = p^m$, $O(z) = p^r$ et $n < m < r$, alors G n'est pas q.p.n.i.

<u>Preuve</u>. Soit $H = \langle x-pz\rangle$. Alors H est p-net dans G. Soit $f: H \to G$ l'homomorphisme défini par $f(x-pz) = y$. Si f s'étend à $\overline{f}: G \to G$, $\overline{f}(p^n(x-pz)) = \overline{f}(-p^{n+1}z) = p^n y$ et $h_p(-p^{n+1}z) = n + 1 > h(p^n y) = n$, ce qui est une contradiction. Donc G n'est pas q.p.n.i. □

<u>Lemme 2.4</u>. Si $G = H \oplus K$ où $H = \bigoplus_{i\in I} \langle x_i\rangle$, $O(x_i) = p^n$ $\forall i \in I$, et $K = \bigoplus_{j\in J} \langle y_j\rangle$, $O(y_j) = p^m$ $\forall_j \in J$, $n < m$, et si A est un sous-groupe p-net dans G, alors on peut écrire $A = A_1 \oplus A_2 \oplus A_3$, $G = A_1 \oplus A_2 \oplus G'$ où $A_3 \leq G'$, et $A_3 \cong \bigoplus_{k\in K} Z(p^{n_k})$ où $n < n_k < m$ $\forall k \in K$.

<u>Preuve</u>. Considérons $A = \bigoplus_{\lambda\in\Lambda} \langle a_\lambda\rangle$. Alors si $\Lambda_1 = \{\lambda \in \Lambda | O(a_\lambda) = p^m\}$, $\Lambda_2 = \{\lambda \in \Lambda | O(a_\lambda) = p^n\}$, $\Lambda_3 = \Lambda - (\Lambda_1 \cup \Lambda_2)$, posons $A_i = \bigoplus_{\lambda\in\Lambda_i} \langle a_\lambda\rangle$. Alors $A = A_1 \oplus A_2 \oplus A_3$ où A_3 est tel que décrit dans l'énoncé.

Alors $(A_1 \oplus A_2)[p] = (p^{m-1}A_1) \oplus (p^{n-1}A_2)$ et donc $A_1 \oplus A_2$ est pur dans G car tout élément du socle de $A_1 \oplus A_2$ est de même hauteur dans $A_1 \oplus A_2$ et dans G. Donc $G = A_1 \oplus A_2 \oplus G'$, et on peut supposer que $A = A_1 \oplus A_2 \oplus A_3'$ où $A_3' \leq G'$. □

<u>Théorème 2.5</u>. Si $G = H \oplus K$ où $H = \bigoplus_{t\in T} \langle x_t\rangle$, $O(x_t) = p^n$ $\forall t \in T$, $K = \bigoplus_{j\in J} \langle y_j\rangle$, $O(y_j) = p^m$ $\forall_j \in J$, $n < m$, alors G est q.p.n.i.

<u>Preuve</u>. Soit A un sous-groupe p-net de G. Le lemme précédent nous permet de nous réduire au cas où $A = \bigoplus_{\lambda\in\Lambda} \langle a_\lambda\rangle$, $p^n < O(a_\lambda) < p^m$ $\forall\lambda \in \Lambda$. Soit $f: A \to G$ un homomorphisme quelconque. On a que $A \cap K = \bigoplus_{i\in I} \langle c_i\rangle$. On peut vérifier sans trop de difficulté que si la hauteur de c_i est r_i dans G elle est aussi r_i dans A pour $i \in I$ pour lesquels $r_i \leq n$. Par contre, si $r_i > n$ alors $f(c_i) \in p^n G \subset K$ car $c_i = p^n a_i$ où $a_i \in A$. Mais $O(c_i) = p^{m-r_i}$, donc $O(f(x_i)) \leq p^{m-r_i}$ d'où $h(f(x_i)) \geq r_i$. On en déduit que dans tous les cas $f(c_i) = p^{r_i} g_i$, $g_i \in G$ $\forall i \in I$. Posons maintenant $c_i = p^{r_i} k_i$, $k_i \in K$. Alors

$K' = \bigoplus_{i\in I} \langle k_i\rangle$ est un sous-groupe pur de K; en fait K' est un facteur direct de K. La correspondance $k_i \to g_i$ où $f(c_i) = p^{r_i} g_i$ définit un homomorphisme de K' dans G qui s'étend facilement à un homomorphisme f' de K dans G qui coïncide avec f sur $A \cap K$.

Si $a_\lambda = x_\lambda + y_\lambda$ où $x_\lambda \in H$, $y_\lambda \in K$, $p^n a_\lambda = p^n y_\lambda \in A \cap K$ d'où $p^n f(a_\lambda) = p^n f'(y_\lambda)$ et $f(a_\lambda) - f'(y_\lambda) \in G[p^n]$. Considérons $B = \langle\{x_\lambda\}\lambda\in\Lambda\rangle \subseteq H$. On définit $\varphi\colon B \to G[p^n]$ par $\varphi(x_\lambda) = f(a_\lambda) - f'(y_\lambda)$. Si $\alpha_1 x_1 + \ldots + \alpha_s x_s = 0$, $\alpha_1 a_1 + \ldots + \alpha_s a_s = \alpha_1 y_1 + \ldots + \alpha_s y_s \in A \cap K$, d'où $f(\alpha_1 a_1 + \ldots + \alpha_s a_s) = f'(\alpha_1 y_1 + \ldots + \alpha_s y_s)$ et donc φ est bien défini.

Comme $G[p^n]$ est quasi-injectif, il existe un endomorphisme $\varphi'\colon H \to G[p^n]$ qui coïncide avec φ sur B. Soit $\overline{f} = \varphi' \oplus f'$. Alors $\overline{f}$ étend f et G est q.p.n.i. □

On est maintenant en mesure de donner la caractérisation complète des groupes q.p.n.i. torsion.

<u>Théorème 2.6</u>. Un groupe torsion G est q.p.n.i. si et seulement si G_p est la somme directe de deux groupes quasi-injectifs et G_q est quasi-injectif pour tout premier q distinct de p.

<u>Preuve</u>. Si G_p est non réduit, $G_p = D \oplus R$, $D \neq 0$, D divisible, R réduit et si R n'est pas quasi-injectif, on peut trouver un facteur direct de G_p de la forme du lemme 2.1. Celui-ci devant être q.p.n.i., on obtient une contradiction.

Si G_p est réduit et si G_p n'est pas somme directe de deux groupes quasi-projectifs, alors il existe $\langle x\rangle \oplus \langle y\rangle \oplus \langle z\rangle$ facteur direct de G_p tel que $O(x) < O(y) < O(z)$ ce qui contredit le lemme 2.3. La proposition 2.2 et le théorème 2.5 nous donne l'implication inverse. □

<u>3. Les groupes quasi-p-nets-injectifs mixtes</u>. On traite d'abord le cas p-divisible. Un groupe p-divisible G tel que $G_p = 0$ est un $Q^{(p)}$-module. De plus, un sous-groupe H de G est p-net si et seulement si H est un sous-$Q^{(p)}$-module de G. Donc, pour caractériser les groupes p-divisibles q.p.n.i. G tels que $G_p = 0$, il suffit de caractériser les $Q^{(p)}$-module quasi-injectifs.

<u>Théorème 3.1</u>. Un $Q^{(p)}$-module G est quasi-injectif si et seulement si G est soit divisible, soit torsion avec partie q-primaire somme directe de groupes cocycliques isomorphes, et ce pour tout q premier (différent de p).

<u>Preuve</u>. Si G est torsion, tout sous-groupe de G est un sous-$Q^{(p)}$-module de G et donc G doit être quasi-injectif comme groupe et donc de la forme décrite plus haut. Si G est sans torsion, alors G doit être divisible. En effet, si $x \in G$ est tel que $h_q(x) = 0$ pour un q premier, on considère $H = Q^{(p)}qx$ et on définit $f: H \to G$ par $f(qx) = x$. Il existe un $Q^{(p)}$-endomorphisme φ de G tel que $\varphi(qx) = x$, et $x = q\varphi(x)$. Ce qui est une contradiction.

Maintenant, on montre que G ne peut être mixte à moins qu'il ne soit divisible. Supposons donc que G est mixte. Il suffit de voir que $K = A \oplus B$ où $\langle x\rangle = B \simeq Z(q^n)$ et $A_q = 0$, $A \neq A_t$ n'est pas quasi-injectif. Soit $a \in A \setminus A_t$ et soit $H = Q^{(p)}qa$. On définit $f: H \to K$ par $f(qa) = x$. Si K était quasi-injectif, on aurait $\varphi: K \to K$ tel que: $\varphi(qa) = x$. Mais, comme ci-dessus, ceci est une contradiction. La réciproque est immédiate, car tous les sous-groupes d'un q-groupe sont p-nets, dès que $q \neq p$. □

On démontre aisément le résultat suivant:

<u>Lemme 3.2</u>. Si G est q.p.n.i., alors pour chaque nombre premier $q \neq p$, G_q est facteur direct de G et G/G_q est q-divisible.

<u>Théorème 3.3</u>. Un groupe q.p.n.i. réduit, est soit torsion soit sans torsion.

<u>Preuve</u>. Soit G un groupe q.p.n.i. réduit, alors G_p est q.p.n.i. et par le théorème 2.6 on voit que G_p est un facteur direct de G. Ecrivons $G = A \oplus G_p$. Si A n'est pas torsion, alors il n'est pas p-divisible en vue du théorème 3.1. Donc il existe $x \in A$, $O(x) = \infty$ et $h_p(x) = 0$. On montre alors que $G_p = 0$. Supposons $G_p \neq 0$ et soit $\langle y\rangle$ un facteur direct non nul de G_p. Posons $H = \langle y+px\rangle$. Alors H est p-net dans G. Soit $f: H \to G$ définie par $f(y+px) = x$, et soit $\bar{f}$ son extension à G. Alors si $O(y) = p^n$, $p^n(y+px) = p^{n+1}x$, et

$$\bar{f}(p^{n+1}x) = p^n\,\bar{f}(y+px) = p^n x = p^{n+1}\bar{f}(x),$$

et comme A est un facteur direct de G tel que $A_p = 0$, on obtient une contradiction. Donc $G_p = 0$. Mais A s'écrit, pour chaque nombre premier q, comme

$A = A_q \oplus B$, car A_q est quasi-injectif. Il reste a montrer que $A_q = 0$. Comme A n'est pas torsion, il existe dans B un élément sans torsion z de p-hauteur 0. Alors $\langle z\rangle$ est p-net et la correspondance $z \to x$ où $x \in A_q$ définit un homomorphisme de $\langle z\rangle$ dans A. Celui-ci se relevant à un endomorphisme de A, et B étant q-divisible par le lemme 3.2, on conclut que $h_q(x) = \infty$. Donc A_q est q-divisible et par suite, divisible. Comme G est réduit, on voit que $A_q = 0$. Nous avons montré donc que si G contient un élément non nul d'ordre infini, G doit être sans torsion. □

Corollaire 3.4. Si G est un groupe q.p.n.i. mixte, alors pour tout nombre prémier $q \neq p$, G_q est un groupe divisible.

On examine maintenant le cas où G n'est pas réduit. Le lemme suivant est très utile.

Lemme 3.5. Si $G = D \oplus R$ où R est q.p.n.i. et si pour tout sous-groupe p-net H de G et pour tout homomorphisme $f: H \to G$, $f(H \cap D) \subseteq D$, alors G est q.p.n.i.

Preuve. Soient H un sous-groupe p-net de G et $f: H \to G$ un homomorphisme quelconque. Alors $(H+D)/D$ est p-net dans $G/D \simeq R$, et $f(H \cap D) \subseteq D$ implique qu'on peut définir $\bar{f}: (H+D)/D \to G/D$ par $\bar{f}(h+D) = f(h) + D$. Comme R est q.p.n.i., il existe $\bar{\varphi}: G/D \to G/D$ tel que $\bar{\varphi}|(H+D)/D = \bar{f}$.

On définit $\varphi_1: G \to G$ par $\varphi_1 = i.\alpha.\bar{\varphi}.\nu_D$ où $\alpha: G/D \to R$, $i: R \to G$ sont les homomorphismes évidents. Si $h \in H$ et si $f(h) = d + r$, $d \in D$, $r \in R$, $\varphi_1(h) = \alpha(f(h) + D) = r$ d'où $(f-\varphi_1).i_H: H \to D$. On prolonge $(f-\varphi_1).i_H$ à $\varphi_2: G \to D$ par injectivité de D. On note aussi $\varphi_2: G \to G$ l'homomorphisme $j.\varphi_2$, où $j: D \to G$ est l'inclusion. Alors $\varphi_1 + \varphi_2: G \to G$ est tel que

$$(\varphi_1 + \varphi_2)(h) = \varphi_1(h) + f(h) - \varphi_1(h) = f(h)$$

pour $h \in H$. Donc G est q.p.n.i. □

Théorème 3.6. Si R est réduit sans torsion et si $G = D \oplus R$ où D est divisible, alors G est q.p.n.i. si et seulement si R est q.p.n.i.

Preuve. Supposons que R soit q.p.n.i. Soient H p-net dans G et $f: H \to G$ un homomorphisme quelconque. Alors, si $x \in H \cap D$, $x = pd = ph$, $d \in D$, $h \in H$ d'où

$d-h \in G[p] \subseteq D$ et $h \in H \cap D$. Donc $H \cap D \subseteq H_p^1$ et $f(H \cap D) \subseteq G_p^1$. Mais, par le théorème 3.1, R est p-réduit car réduit d'où $G_p^1 = D$. Par le lemme précédent, G est q.p.n.i. □

<u>Théorème 3.7</u>. Si $G = D \oplus R$, où $R \simeq \oplus Z(p^n)$ et D est divisible, alors G est q.p.n.i..

<u>Preuve</u>. Soit H p-net dans G et soit $f: H \to G$ un homomorphisme quelconque. Alors $H \cap D = p^n H$ car si $h \in H \cap D$, $h = pd = ph_1$, $h_1 = \alpha_1 + r_1$, $d_1 \in D$, $r_1 \in R$, $pr_1 = pd - pd_1 = 0$, d'où $r_1 = p^{n-1}r_1'$ et $h_1 = pg_1 = ph_2$ etc... Donc $f(H \cap D) \subseteq p^n G = D$ et G est q.p.n.i. par le lemme 3.5. □

<u>Lemme 3.8</u>. Si $G = D \oplus A \oplus B$ où $D \simeq Q$ ou $Z(p^\infty)$, $A = \langle x \rangle \simeq Z(p^n)$, $B = \langle y \rangle \simeq Z(p^m)$ $n < m$, alors G n'est pas q.p.n.i.

<u>Preuve</u>. Soit $z \in D$ d'ordre $> p^m$. Alors $H = \langle x+z \rangle$ est p-net dans G. Soit $f: \langle x+z \rangle \to G$ défini par $f(x+z) = y$. Si G est q.p.n.i., il existe $\varphi: G \to G$ tel que $\varphi(p^n z) = p^n y$. Mais $h_p(p^n z) = \infty$ et $h_p(p^n y) = n$ ce qui est une contradiction. □

<u>Théorème 3.9</u>. Un groupe mixte G est q.p.n.i. si et seulement si G est d'une des deux formes suivantes

(1) (1) $G = D \oplus R$ (2) $G = D \oplus (\oplus Z(p^n))$,

où D est divisible et R q.p.n.i. réduit sans torsion.

<u>Preuve</u>. Supposons que G soit q.p.n.i. mixte. On écrit $G = D \oplus R$ où D est divisible et R est réduit. $D = A \oplus (\oplus B_q)$ où $A \simeq \oplus Q$, $B_q \simeq \oplus Z(q^\infty)$. R étant q.p.n.i. réduit, R n'est pas mixte. Si R est sans torsion, G est de la forme (1). Si R est torsion, $A \neq 0$ car G est mixte. Alors $R = R_p$ par corollaire 3.4. De plus $R \simeq \oplus Z(p^n)$ par le lemme 3.8. Donc G est de la forme (2). Si G est de la forme (1) ou (2), alors G est q.p.n.i. par les théorèmes 3.6 et 3.7 respectivement. □

On réduit également le cas sans torsion au cas sans torsion p-réduit.

<u>Théorème 3.10</u>. Un groupe sans torsion G est q.p.n.i. si et seulement si $G = D \oplus R$ où R est q.p.n.i. p-réduit et D est divisible.

Preuve. Supposons que G soit q.p.n.i. Soit $D = \{x \in G \mid h_p(x) = \infty\}$. Alors D est q.p.n.i. et donc divisible par le théorème 3.1. Donc $G = D \oplus R$ où R est p-réduit. La réciproque, découle du théorème 3.6. □

Nous concluons cette section en observant qu'il reste donc à caractériser les groupes sans torsion p-réduit q.p.n.i.. Ceux-ci étant exactement les groupes q.p.p.i., nous reférons le lecteur à [6] où plusieurs résultats les concernant sont établis.

4. Les groupes quasi-p-nets-projectifs. On sait par le théorème 1.2 qu'un groupe q.p.n.p. ne peut s'écrire que sous cinq formes. Nous nous attacherons à vérifier dans ce qui suit si un groupe s'écrivant sous une de ces formes est q.p.n.p.

Lemme 4.1. Si $G = R \oplus D$ où $D \simeq Z(p^\infty)$ et $R \simeq Z(p^n)$, $n \geq 1$, alors G n'est pas q.p.n.p.

Preuve. On écrit $R = \langle x \rangle$ et $D = \langle y_o, y_1, \dots, y_m, \dots \rangle$ avec $py_o = 0$, $py_m = y_{m-1}$, pour $m \geq 1$. Soit $H = \langle x+y_n \rangle$. Alors H est p-net dans G. De plus $G/H = (D+H)/H$ est divisible. Définissons $f: \langle y_o \rangle \to G/H$ par $f(y_o) = y_1 + H$. Cet homomorphisme est bien défini. On prolonge f à $\overline{f}: G \to G/H$ par injectivité de G/H. On suppose que G est q.p.n.p. Il existe donc un endomorphisme φ de G tel que $\varphi(y_o) - y_1 \in H$. Donc $\varphi(y_o) - y_1 = a(x+y_n)$, $a \in Z$. Comme $\varphi(y_o) - y_1 \in D$, $ax = 0$, d'où $a = p^n b$, $b \in Z$. Donc $\varphi(y_o) - y_1 = by_o$, $\varphi(py_o) - y_o = bpy_o$, $y_o = 0$, ce qui est une contradiction. □

Théorème 4.2. Si G est un p-groupe non réduit, alors G est q.p.n.p. si et seulement si G est divisible.

Preuve. Si G n'est divisible, G possède un facteur direct cyclique fini $\langle x \rangle$. On écrit donc $G = D \oplus \langle x \rangle \oplus R$ où D est divisible non nul, $\langle x \rangle \simeq Z(p^n)$. Il existe donc un facteur direct H de G tel que $H \simeq Z(p^n) \oplus Z(p^\infty)$. Par le théorème 1.4 ceci contredit le lemme 4.1. □

Lemme 4.3. Si $G = \langle x \rangle \oplus \langle y \rangle \oplus K$ où $O(x) = p^s$, $O(y) = p^t$ et K est soit sans torsion non nul, soit cyclique d'ordre p^u est $s<t<u$, alors G n'est pas q.p.n.p.

Preuve. Soit $z \in K$ tel que $O(z) = \infty$ ou p^u selon le cas. Alors $H = \langle x+pz \rangle$ est p-net dans G. Si $p^n z \in H$, $p^{n-1}x = 0$, d'où $n-1 \geq s$ et $O(z+H) = p^{s+1}$. Mais y est d'ordre $p^t \geq p^{s+1}$. On peut donc définir $f: G \to G/H$ par $f(y) = z+H$, $f(x) = f(K) = 0$. On suppose que G est q.p.n.p. Il existe donc un endomorphisme φ de G tel que $\varphi(y) - z \in H$ d'où $\varphi(y) - z = d(x+pz)$, $d \in Z$. Donc $\varphi(p^t y) = (dp+1)p^t z$ d'où $p^t z = 0$, ce qui est une contradiction. □

Théorème 4.4. Si $G = A \oplus B$ où A et B sont quasi-projectifs tels que $G_t = A_p \oplus B$, alors G est q.p.n.p.

Preuve. On écrit $A = \bigoplus_{i\in I} A_i$, $B = \bigoplus_{j\in J} B_j$, $B_j \simeq Z(p^n)$, $A_i = \langle a_i \rangle$ où $O(a_i) = p^m$ ou ∞ où $n < m$. Soit H un sous-groupe p-net de G et soit $f: G \to G/H$ un homomorphisme quelconque. Il suffit de voir que pour $A_i = \langle a \rangle$, $B_j = \langle b \rangle$ il existe $\varphi_1: \langle a \rangle \to G$, $\varphi_2: \langle b \rangle \to G$ tel que $\nu_H \cdot \varphi_1 = f|_{A_i}$ $\nu_H \cdot \varphi_2 = f|_{B_j}$. Si $f(a) = g + H$, on définit $\varphi_1(a) = g$. ϕ_1 est bien défini par maximalité de l'ordre de a et $\nu_H \cdot \varphi_1(a) = f(a)$.

Pour trouver φ_2, on choisit $g_1 \in G$ tel que $f(b) = g_1 + H$. Si $O(g_1) \leq p^n$, on définit $\varphi_2(b) = g_1$ et on obtient le résultat. Supposons donc que $O(g_1) > p^n$. On a que $O(g_1+H) = p^s \leq p^n$ car $f(b) = g_1 + H$. Donc $p^s g_1 \in H$. Comme $O(g_1) > p^n$, $g_1 = a_o + b_o$, $a_o \in A$, $b_o \in B$, $a_o \neq 0$. Comme H est p-net, $p^s g_1 = p^s a_o + p^s b_o = ph_1$, $h_1 \in H$, $h_1 = a_1 + b_1$, $a_1 \in A$, $b_1 \in B$.

Donc $p^s a_o = pa_1$, $p^s b_o = pb_1$. Comme A et B sont quasi-projectifs, $a_1 = p^{s-1}a_2$, $b_1 = p^{s-1}b_2$. Donc $a_1 + b_1 = p^{s-1}(a_2+b_2) = p^{s-1}g_2$, $g_2 \in G$, d'où $h_1 = ph_2$, $h_2 = a_3 + b_3$ etc... On répète le processus et on obtient que $p^s g_1 = p^s h_{s+1}$, $h_{s+1} \in H$. Donc $p^s(g_1 - h_{s+1}) = 0$. On définit alors $\varphi_2(b) = g_1 - h_{s+1}$ et on obtient le résultat. □

Théorème 4.5. Si $G = K \oplus E$ où K est somme directe de copies de $Q^{(p)}$ et E est un p-groupe quasi-projectif, alors G est q.p.n.p.

Preuve. Soit H un sous-groupe p-net dans G et soit $f: G \to G/H$ un homomorphisme quelconque. Par le même raisonnement que celui utilisé dans la preuve précédente, il existe $\varphi_1: E \to G$ tel que $\nu_H \cdot \varphi_1 = f.i_E$. Il suffit donc de voir qu'il existe $\varphi_2: K \to G$ tel que $\nu_H \cdot \varphi_2 = f.i_K$.

Il est clair que $(G/H)_p^1$, le sous-groupe des éléments de p-hauteur infinie de G/H, est égal à (K+H)/H. Soit θ: $(K+H)/H \to K/(K\cap H)$ l'isomorphisme canonique. Alors $\theta.f.i_K$: $K \to K/(K\cap H)$ où $K/(K\cap H)$ est un $Q^{(p)}$-module. Comme K est un $Q^{(p)}$-module projectif, il existe φ: $K \to K$ tel que $\nu_{H\cap K}.\varphi = \theta.f.i_K$. Soit φ_2: $K \to G$ défini par $\varphi_2 = i_K.\varphi$. Alors, si $k \in K$ et $f(k) = k_o + H$, $k_o \in K$, on a que $\varphi(k) + H \cap K = k_o + H \cap K$, d'où $\varphi_2(k) - k_o \in H$ et $\varphi_2(k) + H = f(k)$. □

<u>Théorème 4.6</u>. Si $G = D \oplus K$ où D est un p-groupe divisible et K est soit libre, soit somme directe de copies de $Q^{(p)}$, alors G est q.p.n.p.

<u>Preuve</u>. Par le théorème 1.2, il suffit de vérifier que tout sous-groupe p-net de G est aussi p-pur dans G. Soit H p-net dans G et soient $g \in G$, $h \in H$ tels que $p^n g = h$. Alors $p^n g = ph_1$, $h_1 \in H$. Donc $p(p^{n-1}g-h_1) = 0$ et $p^{n-1}g-h_1 \in D$, d'où $p^{n-1}g - h_1 = p^{n-1}d$ et $h_1 = p^{n-1}g_1$ où $g_1 = (g-d)$. On répète le processus pour trouver $h = p^n h_n$. □

Nous sommes maintenant en mesure de caractériser complètement des groupes quasi-p-nets-projectifs.

<u>Théorème 4.7</u>. Un groupe G est q.p.n.p. si et seulement si il est d'une des six formes suivantes:

(1) $G = D \oplus (\underset{q\, p}{\oplus} C_q)$	(3) $G = E \oplus L$	(5) $G = D \oplus L$
(2) $G = E \oplus F \oplus (\underset{q \neq p}{\oplus} C_q)$	(4) $G = E \oplus S$	(6) $G = D \oplus S$,

où D est un p-groupe divisible, E et F sont des p-groupes quasi-projectifs, C_q est un q-groupe quasi-projectif, q est un nombre premier, L est libre, et S est somme directe de copies de $Q^{(p)}$.

<u>Preuve</u>. Supposons d'abord que G est q.p.n.p. Par le théorème 1.2, G ne peut s'écrire que de cinq façons. Si $G = D \oplus B \oplus (\underset{q \neq p}{\oplus} C_q)$ où B est un p-groupe borné, alors, si $D = 0$, alors $B = E \oplus F$ par le lemme 4.3 et G est de la forme (2), si $D \neq 0$, alors $B = 0$ par le théorème 4.2 et G est de la forme (1).

Si $G = A \oplus (\underset{q \neq p}{\oplus} C_q)$ où A est un p-groupe somme directe de groupes cycliques, alors, par le lemme 4.3, G est de la forme (2). Si $G = A \oplus L$, G est de la forme (3) par le lemme 4.3. Si $G = D \oplus B \oplus S$, alors, si $D \neq 0$ on a $B = 0$ par le théorème 4.2 d'où G est de la forme (6), si $B \neq 0$, alors $D = 0$ par le théorème 4.2 et $B = E$ par le lemme 4.3, d'où G est de la forme (4).

Enfin, si $G = D \oplus B \oplus L$, si $D \neq 0$ alors $B = 0$ et G est de la forme (5), si $D = 0$ alors G est de la forme (3) par le lemme 4.3. Inversement, si G est d'une des six formes précédentes, G est q.p.n.p. par les théorèmes 1.3, 4.4, 4.5 et 4.6. □

On remarque que le fait que les groupes de la forme (1) à (6) soient q.p.n.p. est indépendant de l'hypothèse généralisée du continu.

Bibliographie.

[1] Arnold, D.M., Vinsonhaler, C.I. and Wickless, W.J., Quasi-pure projective and injective torsion-free abelian groups of rank 2, Rocky Mountain J. Math., 6, 1976, p. 61-70.

[2] Arnold, D.M., O'Brien, B. and Reid, J.D., Torsion-free abelian q.p.i. and q.p.p. groups, à paraître.

[3] Benabdallah, K. et Bradley, R., Sur les groupes quasi-purs-projectifs torsions, Can. J. Math., à paraître.

[4] Benabdallah, K. et Bradley, R., Sur les groupes quasi-p-purs-projectifs, Acta Math. Acad. Sci. Hungar., vol. 33, 1979, à paraître.

[5] Benabdallah, K. et Laroche, A., Sur le problème 17 de L. Fuchs, Annales des Sciences Mathématiques du Québec, vol 1, no 1, 1976, p. 63-65.

[6] Benabdallah, K. et Laroche, A., Quasi-p-pure injective groups, Can. J. Math. à paraître.

[7] Fuchs, L., Infinite Abelian Groups, vol I, II, Academic Press, New York, 1970, 1973.

WHITEHEAD'S PROBLEM

R. J. Nunke

Whitehead's problem is to prove the following statement.

(1) If $\mathrm{Ext}(A, Z) = 0$, then A is free.

The reference to Ext in this statement can be removed by taking a free resolution for A. It then becomes

(2) If F is a free abelian group with subgroup R and every homomorphism of R into Z extends to a homomorphism of F into Z, then R is a direct summand of F.

This is the form in which the problem was stated by A. Ehrenfeucht [3] in 1955. Ehrenfeucht says that the problem was presented by J. H. C. Whitehead in Warsaw in May 1952, hence the name.

Another way to remove Ext is to take an injective resolution for Z, for example embedding Z in the rational numbers Q. We then get

(3) If every homomorphism of A into Q/Z can be lifted to a homomorphism of A into Q, then A is free.

This is essentially the form in which the problem was considered by K. Stein [11] in 1951. Both Stein and Ehrenfeucht solved the problem with the additional hypothesis that A be countable.

Many people: J. Rotman, myself, S. Chase, P. Griffith have studied this problem obtaining meager results. In view of the paucity of results it is perhaps better to rephrase Whitehead's problem thus

(4) Determine all abelian groups A such that $\mathrm{Ext}(A, Z) = 0$.

The wisdom of this rephrasing is justified by the recent results of Shelah [9, 10] that (1) is independent of ZFC.

A solution to Whitehead's problem has implications outside the theory of abelian groups. Stein applied (3) to the second Cousin problem. He only required (3) for A countable and had a positive solution. There are other applications suggested by (3). If Q is replaced by the real numbers, then Q/Z becomes the circle group and the homomorphisms of A into Q/Z are characters of A. This suggests the duality discrete and compact abelian topological groups. Two results of Dixmier [2] call for a solution of Whitehead's problem.

(5) A locally compact abelian group is arcwise connected if and only if it has the form

$$R^n \oplus \hat{D}$$

where R is the additive group of real numbers, D is a discrete group with $\mathrm{Ext}(C, Z) = 0$ and $\hat{D}$ is the dual of D.

(6) A locally compact abelian group is locally arcwise connected if and only if it has the form

$$R^n \oplus \hat{D} \oplus E$$

where R and D are as in (4) and E is discrete.

From (5) and (1) follows

(7) A compact arcwise connected abelian group is a solenoid.

There are also applications to algebraic topology. The functor Ext enters the study of topology via the Universal Coefficient theorem.

(8) $H^n(X; G) = \mathrm{Hom}(H_n(X); G) \oplus \mathrm{Ext}(H_{n-1}(X); G)$ where $n \geq 1$, $H^n(X; G)$ is the

n-th singular cohomology group of X with coefficients in the abelian group G and $H_n(X)$ is the n-th integral singular homology group of X.

In the Summer of 1953 Edwin Spanier asked me if the vanishing of all integral cohomology groups of a space from some dimension n on implied the vanishing of all cohomology groups with any coefficient group from some dimension on. From the Universal Coefficient Theorem we see that $H^k(X; Z) = H^{k+1}(X, Z)$ implies $Hom(H_k(X), Z) = Ext(H_k(X), Z) = 0$. This suggests trying to show that

(9) If $Hom(A, Z) = 0 = Ext(A, Z)$, then $A = 0$.

Once (8) has been proved another application of the Universal Coefficient Theorem gives the following result.

(10) If $H^k(X; Z) = 0$ for $k \geq n > 0$, then $H^k(X; G) = 0$ for all G and $k \geq n+1$.

Notice the shift from n to $n+1$ between the two clauses in (10). The obstruction to improving $n+1$ to n in the conclusion of (10) lies in (8). From the hypothesis of (10) we get only $Ext(H_{n-1}(X), Z) = 0$. If this implied that $H_{n-1}(X)$ were free, we could conclude that $Ext(H_{n-1}(X), G) = 0$ which would give the desired result. In other words if Whitehead's problem were solved we could put n for $n+1$ in (10).

A final connection between Whitehead's Problem and Topology is contained in the observation that while it is easy to construct a connected space whose integral homology sequence is a prescribed sequence $Z, A_1, A_2, A_3, \ldots$ of abelian groups, it is much more difficult to find a space whose integral cohomology groups is this sequence. In fact it cannot be shown that this can always be done. The source of the difficulty is in the Universal Coefficient Theorem. For $n \geq 1$ we must factor each A_n as

$$A_n = Hom(H_n, Z) \oplus Ext(H_{n-1}, Z)$$

with $H_0 = Z$. J. Rotman and I investigated this situation when the homology groups are countable in [7].

As a simple question of the sort just discussed we may consider

(11) Can the additive group Q of rational numbers be an integral cohomology group of a space?

In view of (7) and the fact that Q is indecomposable and that Hom(A, Z) is not divisible,(10) is equivalent to

(12) Is Q = Ext(A, Z) for some Z?

Shelah's results show that the answer to (11) is no if V = L.

Abelian groups A with Ext(A, Z) = 0 are commonly referred to as Whitehead groups. The term has been further abbreviated by J. Rotman [8] to W-groups. I have assembled most of the known results in the diagram which appears on the next page. Each word or phrase in the diagram refers to a class or property of abelian groups. Lines slanting downward connect more inclusive to less inclusive classes.

In order to interpret the diagram a glossary is needed which I will provide now.

An abelian group is:

ω_1-free iff every countable subgroup is free.

strongly ω_1-free iff every countable subgroup is contained in countable free subgroup such that the factor group is ω_1-free.

ω-separable iff every pure subgroup of finite rank is a free direct summand.

ω_1-separable iff every countable subgroup is contained in a countable free direct summand.

ω_1-coseparable iff it is ω_1-free and every subgroup of countable index contains a direct summand of countable index.

In addition, for any property P, a group is hereditarily P if every subgroup is P. A property P is hereditary if P is equivalent to hereditarily P.

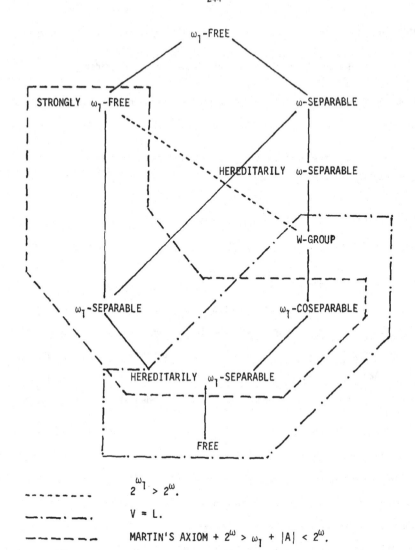

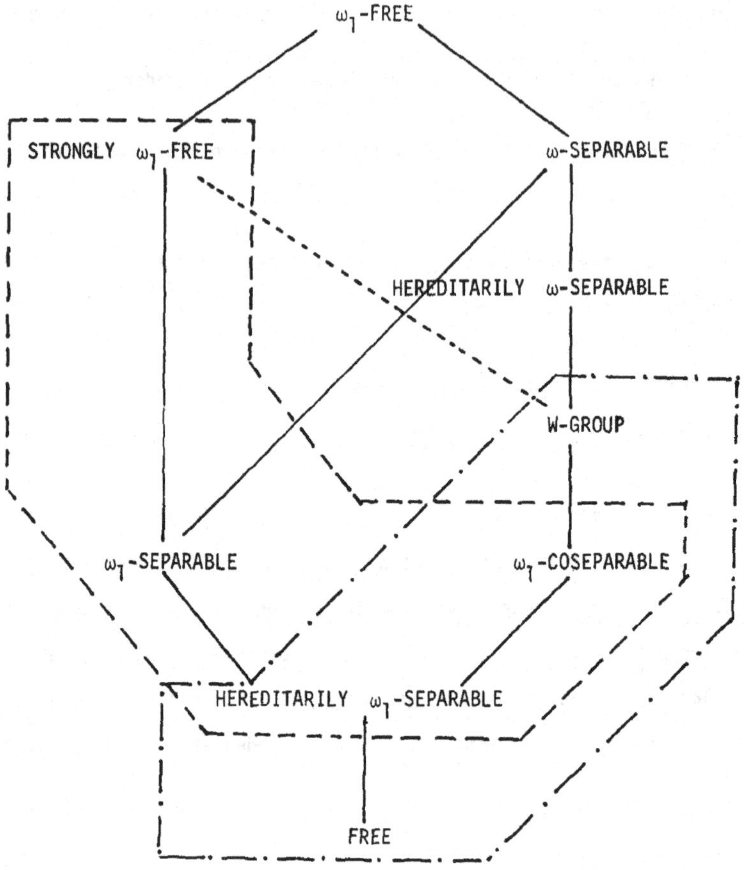

- - - - - - - - $2^{\omega_1} > 2^{\omega}$.

—·—·— $V = L$.

— — — — — MARTIN'S AXIOM + $2^{\omega} > \omega_1$ + $|A| < 2^{\omega}$.

Since Ext is right exact being a W-group is a hereditary property. Hence if all W-groups have a property P then they are all hereditarily P. If A is cyclic of finite order, $A \cong \text{Ext}(A, Z)$. Thus all W-groups are torsion-free.

A word is necessary here about the dotted and dashed lines in the diagram. Without them we have the location of the class of W-groups as far as it is known using just the set theoretic axioms of ZFC. In the presence of the axiom of constructability or of Martin's Axiom collapsing takes place. This is indicated by the dashed ovals. In the presence of the Continuum Hypothesis the inclusion indicated by the dashed line is provable.

I will now prove the implications in the upper right hand corner of the diagram. The first step is to prove (9).

Suppose that $\text{Hom}(A, Z) = 0 = \text{Ext}(A, Z)$. Let p be a prime and apply Hom(A,) to the exact sequence

$$Z \rightarrowtail Z \twoheadrightarrow Z/pZ$$

to get the exact sequence

$$\text{Hom}(A, Z) \to \text{Hom}(A, Z/pZ) \to \text{Ext}(A, Z).$$

The hypotheses then imply $\text{Hom}(A, Z/pZ) = 0$ which can happen only if A is p-divisible. Since p is arbitrary, A is divisible and also torsion free as we have already seen. If $A \neq 0$, it has Q as a direct summand giving $\text{Ext}(Q, Z) = 0$ because being a W-group is hereditary. But $\text{Ext}(Q, Z) \neq 0$ so $A = 0$ as desired.

If A is torsion-free of rank one and $\text{Hom}(A, Z) \neq 0$, then $A \cong Z$. Thus we have the corollary

(13) If A is a rank one W-group, then $A \cong Z$.

The next step is to show

(14) W-groups are ω-separable.

Let B be a pure subgroup of finite rank of the W-group A. Suppose first that B has rank one. In the exact sequence

$$B \rightarrowtail A \rightarrow A/B$$

A/B is torsion-free by purity and B is cyclic by (12). Thus

$$\mathrm{Hom}(A, B) \rightarrow \mathrm{Hom}(B, B) \twoheadrightarrow \mathrm{Ext}(A/B, B)$$

is exact, the right hand map being epic because the next term in the sequence is $\mathrm{Ext}(A, B) \cong \mathrm{Ext}(A, Z) = 0$. Also $\mathrm{Hom}(B, B) \cong Z$ making $\mathrm{Ext}(A/B, B)$ cyclic. Since A/B is torsion-free this group is also divisible, hence 0. Thus the left hand map in the sequence is epic showing that B is a direct summand of A.

The proof is now completed by induction on the rank of B. Let C be a pure rank one subgroup of B. The above result shows that C is a direct summand of A. Then A/C is a W-group with pure subgroup B/C with rank one less than the rank of B.

Since subgroups of W-groups are W-groups, the class of hereditarily ω-separable groups can now be put into the diagram. It is well known that ω-separable groups are ω_1-free so this class can be put in place. The statement that W-groups are ω_1-free is, of course, the result proved by Stein and by Ehrenfeucht.

The implications beginning at free and proceeding up the left hand part of the diagram are well known. The implication

(15) If $2^{\omega_1} > 2^{\omega}$, then W-groups are strongly ω_1-free was proved by S. Chase [1].

The concept of a ω_1-coseparable group is due to P. Griffith [5]. He showed the following.

(16) If F is free of countable rank, then

$$\mathrm{Ext}(A, F) = 0 \text{ if and only if } A \text{ is } \omega_1\text{-coseparable}$$

(17) A group is hereditarily ω_1-separable if and only if it is both ω_1-separable and ω_1-coseparable.

H is not apparent from the diagram whether all these classes of groups are distinct. In fact an affirmative answer to the original Whitehead problem implies that all the classes enclosed in the lower right hand oval are the same. It is known that some of the classes are distinct. A construction due to Paul Hill gives an ω_1-separable group that is not free. The direct product of countably many copies of Z is ω-separable but not hereditarily ω-separable and an example due to S. Shelah gives an ω_1-free group that is not ω-separable.

The diagram shows the status of Whitehead's problem before Shelah's discoveries. Shelah's results go in two different directions. On the one hand he showed that assuming the axiom of constructibility $(V = L)$ all W-groups are free. Thus the lower right hand portion of the diagram collapses. On the other hand under the assumption of Martin's axiom and the denial of the continuum hypothesis $(2^{\omega} > \omega_1)$, we have

(18) If $|A|<2^{\omega}$ and A is strongly ω_1-free, then A is ω_1-coseparable.

This produces the collapse indicated on the left of the diagram. In order to see this we require the following result due to A. Mekler.

(19) If every strongly ω_1-free group of cardinal less than the cardinal λ is ω_1-coseparable, then every strongly ω_1-free group of cardinal less than λ is ω_1-separable.

To see this suppose $|A|<\lambda$, A is strongly ω_1-free and B is a countable subgroup of A. Then there is a countable subgroup C of A containing B such that A/C is ω_1-free. It is easy to see that A/C is in fact strongly ω_1-free and $|A/C|\leq|A|<\lambda$. Now A/C is ω_1-coseparable by the hypotheses of (19) and C is free of countable rank, hence C is a direct summand of A by (16). This shows A is ω_1-separable. Now, using (17), it is easy to verify that all the classes in the oval to the left are the same.

Some relationships in the diagram are worthy of further comment. Chase's result showing that W-groups are strongly ω_1-free requires $2^{\omega_1} > 2^{\omega}$. Shelah's result in the other direction requires Martin's Axiom and $2^{\omega} > \omega_1$ which contradicts $2^{\omega_1} > 2^{\omega}$. H is not known whether the hypothesis $2^{\omega_1} > 2^{\omega}$ is needed for Chase's result. Notice also the part of the diagram lying above the class of W-groups. The relations here were discovered in the first attempts to solve Whitehead's problem. The most recent discoveries leave this part of the diagram unchanged. We still don't know whether hereditarily ω-separable groups are W-groups.

In this exposition I have referred to Shelah's results many times. Some have said that they settle Whitehead's problem because the show that the statement 'A group is a W-group if and only if it is free' is consistent with and independent of ZFC. I have tried to show that if the problem is interpreted to mean locating the class of W-groups among the classes of groups lying between the free groups and the ω_1-free groups, there is still much to be learned. Neither $V = L$ nor Martin's Axiom have much appeal in their own right so we are left to determine what set theoretic properties are required to settle Whitehead's problem. Since the continuum hypothesis is often invoked in studying this subject it is interesting to note Shelah's recent claim that the existence of a non-free W-group is consistent with the generalized continuum hypothesis.

I am omitting an outline of Shelah's results because P. Eklov has given an excellent and easily accessible exposition of them in [4].

In closing I would like to return to the question

(12) Whether Q has the form Ext(A, Z) and give a partial negative answer. I will begin with a positive result.

(20) If Ext(A, Z) is countable and A is torsion-free, then there is an exact sequence

$$F \rightarrowtail E \twoheadrightarrow A$$

with E free of countable rank and E a W-group.

Suppose P is the direct product of countably many copies of Z and F is the direct sum. Then

$$F \rightarrowtail P \twoheadrightarrow P/F$$

is exact and as is well known ([6] for example) P/F is cotorsion. Suppose the elements of $\mathrm{Ext}(A, Z)$ have been enumerated $X_1, X_2, \ldots,$ and $\mathrm{Ext}(A, P)$ is identified with the direct product of countably many copies of $\mathrm{Ext}(A, Z)$. There is then an $X \in \mathrm{Ext}(A, P)$ whose i-th coordinate is X_i. Now

$$\mathrm{Ext}(A, F) \rightarrow \mathrm{Ext}(A, P) \rightarrow \mathrm{Ext}(A, P/F) = 0$$

is exact. The zero appears because A is torsion-free and P/F is cotorsion. Thus the left hand map is epic. The sequence required for (20) is the element of $\mathrm{Ext}(A, F)$ mapping onto X.

To see that E is a W-group consider the exact sequence

$$\mathrm{Hom}(F, Z) \rightarrow \mathrm{Ext}(A, Z) \twoheadrightarrow \mathrm{Ext}(E, Z).$$

It is easily verified that the i-th coordinate projection $\pi_i \in \mathrm{Hom}(F, Z)$ maps to X_i. Thus the left hand map is epic and $\mathrm{Ext}(E, Z) = 0$.

(21) If all W-groups are ω_1-separable, then for A torsion-free $\mathrm{Ext}(A, Z) = 0$ or is uncountable. In particular $\mathrm{Ext}(A, Z) \neq Q$.

For suppose A is torsion-free and $\mathrm{Ext}(A, Z)$ is countable. Look at the sequence given by (20). Since E is a W-group, it is ω_1-free. Then the countable subgroup F is contained in a countable free direct summand C of E. Thus $A = (C/F) \oplus B$ where C/F is countable torsion-free and B is a direct summand of E. Thus B is a W-groups so $\mathrm{Ext}(A, Z) = \mathrm{Ext}(C/F, Z)$ is countable contradicting the fact that C/F is countable and torsion-free. To derive the result on Q we note that, since Q is divisible, A must be torsion-free so the first part of (20) applies.

REFERENCES

1. S. Chase, On group extensions and a problem of Whitehead, Topics in Abelian Groups (1962) 173-193.

2. J. Dixmier, Quelques proprietes des groupes abelians localement compacts, Bull. Sci. Math. (2) 81(1957) 3-48.

3. A Ehrenfeucht, On a problem of Whitehead concerning abelian groups, Bull. Acad. Polon. Sci. III 3 (1955) 127-128.

4. P. Eklov, Whitehead's problem is undecideable, Amer. Math. Monthly 83 (1976) 775-788.

5. P. Griffith, Separability of torsion-free groups and a problem of J.H.C. Whitehead, Ill. Journ. Math. 12(1968) 654-657-659.

6. R. Nunke, Slender groups, Acta Szeged, 23(1962) 67-73.

7. R. Nunke and J. Rotman, Singular cohomology groups, Journ. London Math. Soc. 37(1962) 301-306.

8. J. Rotman, On a problem of Baer and a problem of Whitehead in abelian groups, Acta. Math. Sci. Hung. 12(1961) 245-254.

9. S. Shelah, Infinite abelian groups - Whitehead problem and some constructions, Is. Journ. Math. 18(1974) 243-256.

10. ___________, A compactness theorem for singular cardinals, free algebras, Whitehead problem and transversals, Is. Jour. Math. 21(1975) 319-349.

11. K. Stein, Analytische Functionen mehrer komplexer Veranderlichen zu vorgegebenen Periodizitätsmoduln und das zweite Cousinsche Problem, Math. Ann. 123(1951) 201-222.

METHODS OF LOGIC IN ABELIAN GROUP THEORY

Paul C. Eklof[1]

0. Introduction. In this paper we survey contributions of two branches of logic, namely model theory and set theory, to the study of abelian groups. Model theory is the study of the properties of algebraic structures expressible in formal mathematical languages. In the case of abelian groups, the ordinary first-order predicate calculus (or $L_{\omega\omega}$) is not very expressive, but infinitary languages have proved to be quite useful. In particular, the "back-and-forth criterion" (Theorem 1.3) for infinitary logic is an analog of familiar methods in abelian group theory; this analogy can be exploited to generalize in an unusual direction some well-known results of group theory (cf. Theorems 1.5 and 1.6). Moreover, model theory together with set theory provides tools for analyzing the structure of "almost free" groups. Central to this analysis is the notion from set theory of a stationary set, which we use to define an invariant of κ-free groups (Theorem 2.5). In addition, group theoretic results are affected by set theoretic axioms. (This was made most dramatically clear in Shelah's results on Whitehead's problem, which are not discussed here since they are dealt with in another paper in this volume - although their influence is certainly felt here.) One of the results of the second half of this paper is that additional information is obtained when the Axiom of Constructibility is assumed (2.10 and 2.12). Moreover, certain large cardinals motivated by model theoretic ideas, the compact cardinals, enter unavoidably into the study of κ-free groups (2.6 and 2.15). At the end of the paper we list a number of open problems about "almost free" groups.

We let ZF denote the axioms of Zermelo-Frankel Set Theory. (These comprise the familiar, commonly accepted, assumptions about the properties of sets.) Also, ZFC denotes ZF plus the Axiom of Choice. We shall always identify a cardinal with the first ordinal of that cardinality, and identify an ordinal with the set of its predecessors. Thus, we identify $\aleph_1$ the first uncountable cardinal, with $\omega_1 = \{\nu | \nu$ is a countable ordinal$\} = \{\nu | \nu < \omega_1\}$. An infinite cardinal κ is

1 Research partially supported by NSF grants MPS74-06409-A01 and MCS76-12014

called <u>regular</u> if its cofinality is κ, or - equivalently - if κ is not the union of fewer than κ sets each of cardinality $< \kappa$; otherwise κ is <u>singular</u>. Let κ^+ denote the successor cardinal of κ , and $|A|$ denote the cardinality of A .

1. <u>Model theory</u>. We begin by defining a class of languages suitable for studying abelian groups. The basic building blocks are the <u>atomic formulas</u>, which are expressions of the form

$$\sum_{i=1}^{r} n_i x_i = 0$$

where the n_i are integers and the x_i are variable symbols. We construct more complex expressions using the logical symbols: $\neg$ ("not"), $\bigwedge$ ("and") and $\exists$ ("there exists"). If λ and κ are infinite cardinals, define $L_{\lambda\kappa}$ to be the closure of the class of atomic formulas under the following rules:

(i) if $\varphi \in L_{\lambda\kappa}$, then $\neg\varphi \in L_{\lambda\kappa}$;

(ii) if $\{\psi_i | i \in I\} \subseteq L_{\lambda\kappa}$ and $|I| < \lambda$, then $\bigwedge\{\psi_i | i \in I\} \in L_{\lambda\kappa}$

(iii) if $\{x_j | j \in J\}$ is a set of variable symbols and $|J| < \kappa$ and $\varphi \in L_{\lambda\kappa}$, then $\exists\{x_j | j \in J\}\varphi \in L_{\lambda\kappa}$.

The members of $L_{\lambda\kappa}$ are called <u>formulas</u>. A <u>sentence</u> of $L_{\lambda\kappa}$ is a formula in which every variable is bounded by a quantifier. Given a sentence φ of $L_{\lambda\kappa}$ and an abelian group A it makes sense to ask whether or not φ is <u>true</u> in A or, synonomously, A is a <u>model</u> of φ ; if so, we write $A \models \varphi$. If T is a set of sentences of $L_{\lambda\kappa}$ we say A is a <u>model of T</u> , denoted $A \models T$, if $A \models \varphi$ for all $\varphi \in T$.

For example, the formulas of $L_{\omega\omega}$ are finite expressions since we allow only conjunction and quantification over <u>finite</u> sets. The formulas of $L_{\omega\omega}$ are called <u>first-order</u> (or <u>elementary</u>) statements. Consider, for example the following formulas of $L_{\omega\omega}$. (Note that for convenience we use $\forall x\varphi$ as an abbreviation for $\neg \exists x \neg \varphi$ and $\bigvee\{\psi_i | i \in I\}$ as an abbreviation for $\neg \bigwedge\{\neg \psi_i : i \in I\}$).

φ_1: $\exists y(2y - x = 0)$

φ_2: $\forall x \exists y(2y - x = 0)$

φ_p: $\forall x((x=0) \vee (\neg(px = 0)))$

Then φ_1 is not a sentence; its truth in a group A depends on an evaluation of the <u>free variable</u> x. We write $A \models \varphi_1[a]$ if φ_1 is true in A when we let $x = a \in A$. On the other hand, φ_2 is a sentence which is true in a group A if and only if A is 2-divisible; and A is a model of φ_p if and only if A has no p-torsion. The property that A is torsion-free is thus expressed by the infinite set $\{\varphi_p : p \text{ a prime}\}$ of sentences of $L_{\omega\omega}$ or the single sentence $\bigwedge_p \varphi_p$ of $L_{\omega_1\omega}$. On the other hand, the property of being a torsion group is naturally expressed as the following sentence of $L_{\omega_1\omega}$:

$$\forall x \bigvee \{nx = 0 \mid n \neq 0\}.$$

In fact we can <u>prove</u> that this property can not be expressed by even an infinite set of sentences of $L_{\omega\omega}$:

<u>1.1 Theorem.</u> There are groups A and B such that A is a torsion group and B is not, yet A and B are models of precisely the same sentences of $L_{\omega\omega}$.

The key to the proof of Theorem 1.1 is the:

<u>1.2 Compactness Theorem for</u> $L_{\omega\omega}$. (Gödel-Mal'cev). If T is a set of sentences of $L_{\omega\omega}$ such that every subset of T of cardinality $< \omega$ has a model, then T has a model.

For expositions of the proof of Theorems 1.1 and 1.2, see [2] and [9]. Theorem 1.1 suggests that $L_{\omega\omega}$ is rather too weak to express many interesting properties of abelian groups, and, in fact, from the point of view of $L_{\omega\omega}$ we have, in principle, a complete understanding of infinite abelian groups! Thus W. Szmielew [23] proved that there is an algorithm for deciding whether a given sentence of $L_{\omega\omega}$ holds in a given group A, solely on the basis of certain simple algebraic invariants of A such as the Ulm invariants, the rank of A/p^nA etc. (See [10] for another approach to Szmielew's results).

The Compactness Theorem for $L_{\omega\omega}$ is a major tool for studying the model theory of $L_{\omega\omega}$ which is lacking for most of the languages $L_{\lambda\kappa}$. Let us call a cardinal $\kappa > \omega$

weakly (resp. strongly) compact if the Compactness Theorem (i.e. 1.2 with ω replaced everywhere by κ) holds for sets of sentences T of $L_{\kappa\kappa}$ of cardinality κ (resp. of arbitrary cardinality). (In this definition, $L_{\kappa\kappa}$ is defined more broadly than our language of abelian groups - that is, the formulas may refer to any given collection of relations and functions; even in abelian group theory such general languages are useful as we will see in the proof of 2.6) Such cardinals cannot be proved to exist on the basis of ZFC ; in fact, it follows from Gödel's Incompleteness Theorems that the assumption that compact cardinals exist can't even be proven consistent with ZFC . (Hence it is consistent - if ZFC is consistent - to assume that compact cardinals do not exist. But to date no one has proved from ZFC that compact cardinals do not exist - despite intensive study of the concept.) It can be proved that if compact cardinals exist, then they must be very large. For example, if κ is weakly compact, then it is weakly inaccessible, i.e. it is a regular limit cardinal; moreover it is a fixed point of weakly inaccessibles, i.e. it is the κth weakly inaccessible cardinal; moreover it is the κth fixed point of weakly inaccessibles, and so forth. If κ is strongly compact, then it is a measurable cardinal; hence κ is strongly inaccessible (i.e. $\lambda < \kappa$ implies $2^{\lambda} < \kappa$) and κ is the κth weakly compact cardinal.

Now we consider the more expressive languages $L_{\lambda\kappa}$ with $\lambda > \omega$. For any $\kappa \geq \omega$ let us define $L_{\infty\kappa} = \bigcup_{\lambda} L_{\lambda\kappa}$. For λ a cardinal or the symbol ∞ , we shall write $A \equiv_{\lambda\kappa} B$ if exactly the same sentences of $L_{\lambda\kappa}$ are true in A and B . There is a very useful algebraic criterion for this relationship which - even without the motivation from logic - is interesting in its own right. Let us define a partial isomorphism from A to B to be an isomorphism from a subgroup of A to a subgroup of B .

1.3 Back and Forth Theorem. (C. Karp et al.) For any groups A and B , $A \equiv_{\infty\kappa} B$ if and only if there is a non-empty set $\mathcal{I}$ of partial isomorphisms from A to B with the κ-extendability property that for any $f \in \mathcal{I}$ and any subset X of A (resp. B) of cardinality $< \kappa$, there is an $f' \in \mathcal{I}$ with $f \subseteq f'$ and X contained in the domain (resp. range) of f' . (See [1] for an excellent exposition of this subject.)

The basic idea of extending partial isomorphisms is, of course, a familiar one in abelian group theory, for example, in Ulm's Theorem and its generalizations. The original form of Ulm's Theorem is "explained" by Theorem 1.5 below and:

1.4 Corollary. If A and B are countable, $A \equiv_{\infty\omega} B$ if and only if $A \cong B$.

The generalization of Ulm's Theorem to totally projective groups was motivated by the aim of finding a larger class of groups such that ω-extendability implies isomorphism. The following result goes in a different direction:

1.5 Theorem. (Barwise-Eklof [3]) If λ is a regular cardinal $> \omega$, and A and B are reduced torsion groups, then $A \equiv_{\lambda\omega} B$ if and only if the Ulm-Kaplansky invariants satisfy:

$$\min\{f_\sigma(A),\lambda\} = \min\{f_\sigma(B),\lambda\}$$

for all $\sigma < \lambda$.

We shall give another, new, example of an extension of a familiar theorem in an unusual direction, which at the same time illustrates the method of proof of 1.5.

A subset G of a group A is said to be definable by a formula of $L_{\lambda\omega}$ if there is a formula φ of $L_{\lambda\omega}$ with one free variable x such that

$$G = \{a \in A \mid A \models \varphi[a]\}$$

i.e. G is exactly the set of elements which make φ true when a is substituted for x.

Kaplansky characterized the fully invariant subgroups of fully transitive groups (see [18]). Any countable p-group is fully transitive but uncountable p-groups may not be fully transitive. However we have:

1.6 Theorem. Let A be a reduced p-group, where $p \neq 2$. Let κ be a regular cardinal $> \omega$. Then a subgroup G of A is definable by a formula of $L_{\kappa\omega}$ if and only if G is of the form

$$A(\{\alpha_n\}) = \{a \in A : U(a) \geq (\alpha_0,\alpha_1,\ldots)\}$$

for a unique increasing sequence $\{\alpha_n | n \in \omega\}$ of ordinals $< \kappa$ or ∞'s which is a U-sequence (relative to A) as defined in Kaplansky [18], pp 58f. [Here U(a) is the usual height sequence, $(h(a), h(pa), h(p^2a), \ldots)$.]

Proof. (sketch) Kaplansky [18] proved the following results:

(1) If S is a subgroup of A such that whenever $b \in S$, $a \in A$ and $U(b) \le U(a)$ it follows that $a \in S$, then S is of the form $A(\{\alpha_n\})$ for a unique U-sequence $\{\alpha_n\}$.

(2) (using $p \neq 2$). If S is a subgroup of A such that whenever $b \in S$, $a \in A$ and $U(b) = U(a)$ it follows that $a \in S$, then S also has the property that whenever $b \in S$, $a \in A$ and $U(b) \le U(a)$ it follows that $a \in S$.

In Barwise-Eklof [3] (see also [1]), the following is proved, using ideas from the Kaplansky-Mackey proof of Ulm's Theorem and a more sophisticated version of the Back and Forth Theorem.

(3) If $a, b \in A$ of the same order such that $\min\{\kappa, h(p^n a)\} = \min\{\kappa, h(p^n a)\}$ for all $n \in \omega$, then for any formula ψ of $L_{\kappa\omega}$ with one free variable x, $A \models \psi[a]$ if and only if $A \models \psi[b]$.

Now suppose S is a subgroup of A defined by a formula of φ of $L_{\kappa\omega}$. Let $b \in S$, $a \in A$ such that $U(b) = U(a)$. By (3) we conclude that $A \models \varphi[a]$ since $A \models \varphi[b]$; hence $a \in S$. From this and (2) we see that S satisfies the hypothesis of (1). By (1) S is of the form $A(\{\alpha_n\})$ for some unique U-sequence $\{\alpha_n\}$. Suppose that $\kappa < \sigma_n < \infty$ for some n, and let m be minimal with that property. Then there must be a gap between α_{m-1} and α_m (i.e. $\alpha_{m-1} + 1 < \alpha_m$), so by definition of a U-sequence there is a U-sequence $\{\beta_n\}$ with $\beta_n = \alpha_n$ for all $n < m$; $\kappa < \beta_m < \alpha_m$; and $\beta_n = \infty$ iff $\alpha_n = \infty$. By Lemma 24 of [18] there exist elements $a, b \in A$ such that $U(a) = \{\alpha_n\}$ and $U(b) = \{\beta_n\}$. Since $S = A(\{\alpha_n\})$, $a \in S$ but $b \notin S$. However by (3), a and b satisfy the same formulas of $L_{\kappa\omega}$, which contradicts the assumption that S is defined by a formula of $L_{\kappa\omega}$, and thus proves the theorem in one direction.

For the converse, if S is of the form $A(\{\alpha_n\})$ where each α_n is either $< \kappa$ or equals ∞, it is relatively straightforward to see that S is definable by a formula of $L_{\kappa\omega}$. The point is that the property $h(x) \geq \alpha$ is naturally expressible by a formula of $L_{\lambda\omega}$ if $\lambda \geq |\alpha| + \omega$; for example,

$$h(x) \geq \omega \text{ by: } \bigwedge\{\exists y(p^n y = x) \mid n \in \omega\}$$

$$h(x) \geq \omega + 1 \text{ by: } \exists z(pz = x \wedge \bigwedge\{\exists y(p^n y = z) \mid n \in \omega\})$$

This completes the proof of 1.6.

1.6 Remarks (1) We could also, of course, have imitated Kaplansky and stated Theorem 1.6 for the case when $p = 2$ but A has at most 2 Ulm invariants equal to one and if it has exactly two, they correspond to successive ordinals. The counter-example in [18] shows that some hypothesis on A is necessary in 1.6. We could also generalize to non-reduced groups.

(2) A more natural way to filter $L_{\infty\omega}$ for the purposes discussed above is according to the "quantifier rank" of a formula of $L_{\infty\omega}$ rather than by the size of the conjunctions. This leads to results analogous to 1.5 and 1.6 with an arbitrary limit ordinal in place of the regular cardinal κ . (See [1] or [3] for an exact statement of the analog of 1.5).

(3) The relation $A \equiv_{\infty\omega} B$ is "absolute" i.e. it continues to hold even if the universe of sets is changed. By contrast, $A \equiv_{\infty\omega} B$ if and only if $A \cong B$ in some (Boolean-valued) extension of the universe of sets (see [12]) .

2. Set Theory. We want to consider now another kind of contribution which logic can make to the study of abelian groups, namely an analysis of the pathologies of infinite groups and the difficulties of classifying them. Here another branch of logic, set theory, makes contributions in two ways: notions from set theory prove to be useful tools; and results may depend upon foundational assumptions.

Recall that a group is called κ-free if every subgroup of cardinality $< \kappa$ is free. (We shall assume throughout this section that κ is uncountable). The problem we study is the existence of groups which are κ-free but not κ^+-free. The property of being κ-free is expressed by the following sentence of $L_{\infty\kappa}$.

$$\bigwedge_{\lambda<\kappa} \forall\{x_\nu : \nu<\lambda\}\exists\{y_\mu : \mu<\lambda\}[\bigwedge_{\mu<\lambda} \bigvee\{y_\mu = \sum_\nu n_\nu x_\nu \mid (n_\nu) \in I\}$$

$$\wedge \bigwedge_{\nu<\lambda} \bigvee\{x_\nu = \sum_\mu n_\mu y_\mu \mid (n_\mu) \in I\} \wedge \bigwedge\{\neg (\sum_\mu m_\mu y_\mu = 0) \mid (m_\mu) \in J\}]$$

where I is the set of all sequences of integers (n_i) which are zero in all but finitely many places and J is I minus the sequence which is zero everywhere.

A group A is said to be $L_{\infty\kappa}$-free if $A \equiv_{\infty\kappa} F$ for a free group F. (It is easy to prove - using 1.3 - that if F and F' are free and $|F|, |F'| \geq \kappa$ then $F \equiv_{\infty\kappa} F'$). By the above remarks, if A is $L_{\infty\kappa}$-free, then A is κ-free. Moreover, we can prove the following.

2.1 Theorem. (Eklof [4]) A group A is $L_{\infty\kappa}$-free if and only if A is κ-free and every subset of A of cardinality $< \kappa$ is contained in a κ-pure subgroup of cardinality $< \kappa$.

Remarks on the proof. The necessity of the condition follows from the remarks above together with the observation that the property that any subset of cardinality $< \kappa$ is contained in a κ-pure subgroup of cardinality $< \kappa$ is also expressible as a sentence of $L_{\kappa\kappa}$. (Recall that B is a κ-pure subgroup of A if whenever $B \subseteq C \subseteq A$ and $|C/B| < \kappa$, then B is a direct summand of C). The sufficiency of the condition is proved using the Back and Forth Theorem (1.3). (Mekler has proved [19] that if κ is a regular limit cardinal, A is $L_{\infty\kappa}$-free if and only if A is κ-free.)

Using 2.1 one can show that there are groups of cardinality $\aleph_1$ which are $L_{\infty\omega_1}$-equivalent but not isomorphic. The significance of this to the problem of finding complete systems of invariants for certain classes of groups is discussed in [6].

2.2 Definition. Let μ be a limit ordinal. A subset C of μ is called closed and unbounded (cub) if

(1) for every $C' \subseteq C$, if $\sup C' < \mu$, then $\sup C' \in C$; and

(2) $\sup C = \mu$

If E_1 and E_2 are subsets of μ define $E_1 \sim E_2$ if and only if there is a cub C such that $E_1 \cap C = E_2 \cap C$; this defines an equivalence relation on subsets of κ. (It is transitive because the intersection of two cubs is a cub.) Let $\tilde{E}$ denote the equivalence class of E, and let $\mathscr{J}(\mu)$ denote the set of equivalence classes of subsets of μ. It is easy to see that $\subseteq$ induces a complemented, distributive, complete lattice structure on $\mathscr{J}(\mu)$ with $0 = \tilde{\emptyset}$, $1 = \tilde{\mu}$ = the equivalence class of any cub. A subset E of κ is called <u>stationary</u> (in μ) if $\tilde{E} \neq 0$. We shall use the following non-trivial facts about stationary sets:

<u>2.3 Theorem</u>. Let κ be a regular cardinal.

(1) (Solovay [22]) If E is stationary, there exist κ disjoint stationary sets E_ν, $\nu < \kappa$, such that $\bigcup_{\nu<\kappa} E_\nu = E$.

(2) (Shelah [20]) If E is stationary, there exist 2^κ different $\tilde{E}'$ with $\tilde{E}' \subseteq \tilde{E}$.

We are going to define an invariant of κ-free groups. Since by an important result of Shelah [21], κ-free implies κ^+-free when κ is singular, <u>we restrict our attention from now on to regular cardinals</u> κ.

<u>2.4 Definition</u>. Let A be a κ-free group of cardinality κ. Choose a κ-filtration $\Delta = \{A_\nu | \nu < \kappa\}$ of A i.e. $A = \bigcup_{\nu<\kappa} A_\nu$ and $\{A_\nu | \nu < \kappa\}$ is an increasing chain of subgroups of A of cardinality $< \kappa$ such that for any limit ordinal $\sigma < \kappa$, $A_\sigma = \bigcup_{\nu<\sigma} A_\nu$ (cf. [7]). Define

$$E(\Delta) = \{\nu < \kappa \mid \text{there exists } \mu > \nu \text{ such that } A_\mu / A_\nu \text{ is not free}\}$$

and let $\Gamma_\kappa(A) = \widetilde{E(\Delta)}$, the equivalence class of $E(\Delta)$.

Most of the time we will omit the subscript on Γ, since there will be no ambiguity in context.

<u>2.5 Theorem</u>. Γ is a well-defined function from the collection of κ-free groups of cardinality κ to $\mathscr{J}(\kappa)$ satisfying (for all A in the domain of Γ)

(1) if $A_1 \cong A_2$, then $\Gamma(A_1) = \Gamma(A_2)$

(2) A is free if and only if $\Gamma(A) = 0$

(3) if $\Gamma(A) \neq 1$, then A is $L_{\infty\kappa}$-free.

Proof. We must show that the definition of $\Gamma(A)$ does not depend on the choice of a κ-filtration of A. Suppose $\Delta = \{A_\nu | \nu < \kappa\}$ and $\Delta' = \{A'_\nu | \nu < \kappa\}$ are two κ-filtrations of A. Then if we let

$$C = \{\nu < \kappa | A_\nu = A'_\nu\}$$

we claim that C is a cub. Now C satisfies 2.2.1 because of the behavior of a κ-filtration at limit ordinals. As for 2.2.2, we must prove that for any $\mu < \kappa$, there exists $\nu \in C$ with $\nu \geq \mu$. By the definition of a κ-filtration and because of the regularity of κ, we can construct a chain

$$B_0 \subseteq B_1 \subseteq B_2 \subseteq \cdots$$

of subgroups of A of cardinality $< \kappa$ such that for n even (resp. odd) $B_n = A_{\nu_n}$ (resp $B_n = A'_{\nu_n}$) for some $\nu_n \geq \mu$. If we let $B = \bigcup_{n\in\omega} B_n$ and $\nu = \sup\{\nu_n | n \text{ even}\} = \sup\{\nu_n | n \text{ odd}\}$, we have that $A_\nu = B = A'_\nu$ so $\nu \in C$. Thus C is a cub.

Now we claim that $E(\Delta) \cap C = E(\Delta') \cap C$. Indeed, if $\nu \in E(\Delta) \cap C$, then by definition of $E(\Delta)$, there exists $\mu > \nu$ such that A_μ/A_ν is not free. Moreover, since C satisfies 2.2.2, and a subgroup of a free group is free, we may assume that $\mu \in C$. But then $A_\mu/A_\nu = A'_\mu/A'_\nu$ so $\nu \in E(\Delta')$. By symmetry, the claim is proved and hence Γ is well-defined. Property (1) follows immediately. As for (2), if A is free, then $\Gamma(A) = 0$ since clearly A has a κ-filtration in which each member is a direct summand of A. Conversely if $\Gamma(A) = 0$, then for any κ-filtration $\Delta = \{A_\nu | \nu < \kappa\}$ of A there is a cub C with $E(\Delta) \cap C = \emptyset$; thus $A = \bigcup \{A_\nu | \nu \in C\}$, and for any $\nu \in C$, if ν' is the next largest element of C, then $A_{\nu'}/A_\nu$ is free; therefore A is free.

Finally, to prove (3), note that if $\Gamma(A) \neq 1$ then for any κ-filtration $\Delta = \{A_\nu | \nu < \kappa\}$, $\kappa - E(\Delta)$ is cofinal in κ, so any subset of A of cardinality $< \kappa$ is contained in A_μ for some $\mu \notin E(\Delta)$; therefore by 2.1, A is $L_{\infty\kappa}$-free.

The converses of 2.5.1 and 2.5.3 are not always true. Indeed, there exist $\aleph_1$-free groups A_1 and A_2 of cardinality $\aleph_1$ such that $\Gamma(A_1) = \Gamma(A_2) = 1$ and A_1 is $L_{\infty\omega_1}$-free and A_2 is not $L_{\infty\omega_1}$-free (see [4]). We shall have more to say later about the converse of 2.5.1.

If κ is a cardinal such that κ-free does not - or may not - imply κ^+ free, it is natural to seek more information about the situation in terms of properties of the function Γ ; this we shall do, for the remainder of the paper.

First, let us note that several people (Mekler, Shelah, Gregory, Kueker) have observed that if κ is weakly compact, then κ-free implies κ^+-free. Since the proof has not yet been published elsewhere, we will give it here. (A stronger result appears in [19]).

2.6 Theorem. If κ is weakly compact (resp. strongly compact) and A is a κ-free group of cardinality κ (resp. of arbitrary cardinality), then A is free.

Proof. We strengthen our vocabulary by adding a constant symbol c_a for every element of A, and also we add a unary predicate symbol U. We consider sentences of the language $L'_{\kappa\kappa}$ defined as before except that we allow atomic formulas of the forms $\sum_{i=1}^{r} n_i\gamma_i = 0$ or $U(\gamma)$, where γ, γ_i are either variable symbols x_i or constant symbols c_{a_i}. Consider the following sentences, θ_1 and θ_2, of $L'_{\kappa\kappa}$. (Actually $\theta_1, \theta_2 \in L'_{\omega_1\omega} \subseteq L'_{\kappa\kappa}$.)

$$\theta_1 : \bigwedge_{r\in\omega} \forall\{x_1,\cdots,x_r\}[\bigwedge_{i=1}^{r} U(x_i) \rightarrow \bigwedge\{\sum_{i=1}^{r} m_i x_i \neq 0 \mid (m_1,\cdots,m_r) \neq (0,0,\cdots,0)\}]$$

$$\theta_2 : \forall y \bigvee_{r\in\omega} \exists\{x_1,\cdots,x_r\}[\bigwedge_{i=1}^{r} U(x_i) \wedge \bigvee\{y = \sum_{i=1}^{r} m_i x_i \mid (m_1,\cdots,m_r) \in Z^r\} .$$

These sentences express, respectively, the properties that (the interpretation of) U is a linearly independent set and a generating set. Let $\tilde{T}$ be the following set of sentences of $L'_{\kappa\kappa}$ (actually $\tilde{T}$ is a subset of $L'_{\omega\omega}$).

$$\tilde{T} = \{\sum_{i=1}^{k} m_i c_{a_i} = 0 \mid k \in \omega;\ a_1,\cdots,a_k \in A \text{ s.t.} \sum_{i=1}^{k} m_i a_i = 0\} .$$

Finally, let $T = \tilde{T} \cup \{\theta_1,\theta_2\}$. Note that T has cardinality $|A|$. Now if T' is a subset of T of cardinality $< \kappa$ (which we might as well assume contains

θ_1 and θ_2), then T' involves fewer than κ constant symbols c_a ; say all the constant symbols in T' correspond to elements of A' , where A' is a subgroup of A of cardinality $< \kappa$. Since A is κ-free, A' is a free group and hence we can choose a basis of A' as the interpretation of U and thus obtain a model of T' . Hence we have shown that every subset of T of cardinality $< \kappa$ has a model, and so by the definition of weakly compact or strongly compact, T has a model. In other words, there is a group B , a subgroup U^B of B and an assignment of an element b_a to each constant symbol c_a which form a model of $T = \tilde{T} \cup \{\theta_1, \theta_2\}$. Then by the definition of θ_1 and θ_2 , B is a free group (with basis U^B) and by the definition of $\tilde{T}$, the map $f : A \to B$ which takes a to b_a is an embedding. Consequently A is isomorphic to a subgroup of a free group and therefore is free.

Because of Theorem 2.6 and Shelah's result, we confine our attention now to regular uncountable cardinals which are not weakly compact.

<u>2.7 Theorem</u>. Let κ be a regular uncountable cardinal which is not weakly compact.

(1) If $\{\tilde{E}_i | i \in I\} \subseteq \mathscr{S}(\kappa)$, $|I| < \kappa$, and $\Gamma^{-1}(\tilde{E}_i) \neq \phi$ for all $i \in I$, then $\Gamma^{-1}(\bigcup_{i\in I} \tilde{E}_i) \neq \phi$.

(2) (Mekler) If $\tilde{E} \in \mathscr{S}(\kappa)$ such that $\Gamma^{-1}(\tilde{E}) \neq \phi$, then for any $\tilde{E}' \subseteq \tilde{E}$, $\Gamma^{-1}(\tilde{E}') \neq \phi$.

(3) (Eklof) If $\Gamma_\kappa^{-1}(\tilde{E}) \neq \phi$ for some $E \in \mathscr{S}(\kappa)$, then $\Gamma_{\kappa^+}^{-1}(\tilde{E}') \neq \phi$ for $E' = \{\nu < \kappa^+ | \text{ cofinality of } \nu = \kappa\} \in \mathscr{S}(\kappa^+)$.

(4) (Hill) If $\kappa = \aleph_{n+1}$ for any $n \in \omega$, then $\Gamma_\kappa^{-1}(1) \neq \phi$.

<u>Remarks on the proof</u>. Part (1) is completely elementary; in fact, if $\Gamma(A_i) = \tilde{E}_i$ then $\Gamma(\bigoplus_{i\in I} A_i) = \bigcup_{i\in I} \tilde{E}_i$. The clever proof of (2) is to be found elsewhere in this volume. The proof of (3) is to be found in [5]; the idea of the proof is that we inductively define a κ^+-filtration $\{A_\nu | \nu < \kappa^+\}$ so that A_ν will be κ-pure in the resulting group $A = \bigcup_{\nu<\kappa} A_\nu$ if and only if $\nu \notin E'$. We use the

hypothesis to define $A_{\nu+1}$ if ν belongs to E'. At limit ordinals $\mu < \kappa^+$, $A_\mu = \bigcup_{\nu<\mu} A_\nu$ is free because $E' \cap \mu$ is not stationary in μ. Notice that (3) immediately implies that if $\kappa = \aleph_{n+1}$ for $n < \omega$, then $\Gamma_\kappa^{-1}(\tilde{E}) \neq \emptyset$ for some $\tilde{E} \neq 0$, but it does not yield the additional information given by (4). On the other hand, Hill's proof (see [14]) does not seem to extend to give a result like (3). Combining (2) and (4) we obtain:

2.8 Corollary. If $\kappa = \aleph_{n+1}$ for some $n \in \omega$, then for any $\tilde{E} \in \mathscr{S}(\kappa)$, we have $\Gamma_\kappa^{-1}(\tilde{E}) \neq \emptyset$.

The following results show that the converse of 2.5.1 does not hold.

2.9 Theorem. (1) For every $\tilde{E} \in \mathscr{S}(\omega_1)$ with $\tilde{E} \neq 0$ there exist 2^{ω_1} pairwise non-isomorphic groups $\{A_i \mid i < 2^{\omega_1}\}$ with $\Gamma(A_i) = \tilde{E}$ for all i.

(2) If $\kappa = \aleph_{n+1}$ for some $n \in \omega$ then there exist 2^κ different $\tilde{E} \in \mathscr{S}(\kappa)$ such that for each such $\tilde{E}$ there exist 2^κ pairwise non-isomorphic groups $\{A_i \mid i < 2^\kappa\}$ with $\Gamma(A_i) = \tilde{E}$ for all i.

Proof. (1) Without loss of generality, we may assume E contains only limit ordinals. For any $\tilde{E} \in \mathscr{S}(\omega_1)$ there exist by 2.3.2, 2^{ω_1} different $\tilde{E}_i$, $i < 2^{\omega_1}$, such that for all i, $\tilde{E}_i \subseteq \tilde{E}$. Fix two primes, $p \neq q$; define

$$Z_{(p)} = \{ \tfrac{m}{n} \mid m \in \mathbb{Z},\ n = p^k \ \text{some}\ k \geqq 0\} \subseteq \mathbb{Q}$$

and define $Z_{(q)}$ analogously. For each $i < 2^{\omega_1}$, construct an ω_1-free group A_i with ω_1-filtration $\{A_{i,\nu} \mid \nu < \omega_1\}$ satisfying:

(i) for $\nu \notin E$, $A_{i,\nu}$ is ω_1-pure in A_i;

(ii) for $\nu \in E_i$, $A_{i,\nu+1}/A_{i,\nu} \cong Z_{(p)}$

(iii) for $\nu \in E - E_i$, $A_{i,\nu+1}/A_{i,\nu} \cong Z_{(q)}$.

Suppose that there is an isomorphism $f : A_i \to A_j$ for $i \neq j$. Then as in the proof of 2.5, there is a cub C such that for all $\nu \in C$, $f(A_{i,\nu}) = A_{j,\nu}$. Consider $\nu \in E_i \cap C$. Choose $\mu > \nu$ such that $\mu \in C$. Then f induces an isomorphism of $A_{i,\mu}/A_{i,\nu}$ with $A_{j,\mu}/A_{j,\nu}$. But for $k = i,j$,

$$A_{k,\mu}/A_{k,\nu} \cong (A_{k,\nu+1}/A_{k,\nu}) \oplus F$$

where F is a free group of countable rank ($\cong A_{k,\mu}/A_{k,\nu+1}$). Now since $\nu \in E_i$, (ii) and (iii) above imply that $\nu \in E_j$. Hence $E_i \cap C \subseteq E_j$ and, by symmetry, $E_i \cap C = E_j \cap C$. But this contradicts the definition of the $\tilde{E}_k$'s and thus proves (1).

Part (2) is proved by an induction on $n \in \omega$ using the methods of 2.7.3 and a stronger form of (1). The details will appear elsewhere.

We obtain more information if we assume the additional set theoretic hypothesis $V = L$. Results proved using the hypothesis $V = L$ are relatively consistent with ZFC, but may or may not be provable from ZFC. The universe of all sets is denoted by V. Let us call a class of sets, C, transitive if for any $a \in C$ we have $a \subseteq C$, i.e. for all $b \in V$, if $b \in a$, then $b \in C$. The constructible universe, L, is then characterized by the property that it is the smallest transitive class which contains all the ordinals of V and is a model of ZF. However, it is from the explicit construction of L as the union of a chain of classes $\{L_\alpha \mid \alpha \text{ an ordinal}\}$ that one derives its useful properties. (Briefly, $L_{\alpha+1}$ consists of all the subsets of L_α which are definable by a first-order formula - built up from atomic formulas of the form $x \in y$, $a \in y$ or $x \in b$ where $a,b \in L_\alpha$; if σ is a limit ordinal, $L_\sigma = \bigcup\{L_\alpha \mid \alpha < \sigma\}$). Gödel proved that the Axiom of Choice and the Generalized Continuum Hypothesis (i.e. $2^\kappa = \kappa^+$ for all κ) both hold in L. Moreover he proved that it is relatively consistent with ZF to assume that $V = L$, i.e. every set is constructible.

While L has exactly the same ordinals as V, an ordinal may be a cardinal in L but not in V - but every cardinal in V is a cardinal in L. For example, it may be the case that the first uncountable cardinal ω_1^L, in L is a countable ordinal from the point of view of V (that is, V contains a bijection between ω and ω_1^L which is not in L) and the first uncountable cardinal in V may be a very large cardinal from the point of view of L.

In recent years, R. Jensen ([16] and [17]) has made a detailed study of the consequences of the hypothesis $V = L$ and formulated some general principles which have proved very useful in analyzing the structure of abelian groups.

2.10 Theorem. Assume $V = L$. Let κ be a regular uncountable cardinal which is not weakly compact.

(1) (Gregory) There exists $\widetilde{E} \in \mathcal{D}(\kappa)$ such that $\widetilde{E} \neq 0,1$ and $\Gamma^{-1}(\widetilde{E}) \neq \emptyset$.

(2) (Mekler) There exist 2^{κ} pairwise non-isomorphic κ-separable groups of cardinality κ.

(3) (Eklof-Mekler) There exist 2^{κ} pairwise non-isomorphic indecomposable $L_{\infty\kappa}$ - free groups of cardinality κ.

Remarks on the proof. Results (1) - (3) make use of the following principle (sometimes called $E(\kappa)$) discovered by Jensen (see [16], Theorem 6.1 and its proof):

There is a stationary subset E_o of κ such that:

(i) every $\nu \in E_o$ has cofinality ω; and (ii) for every limit $\sigma < \kappa$ $E_o \cap \sigma$ is not stationary in σ.

Gregory's result (announced in [13]) then follows from the following lemma.

2.11 Lemma. Assume κ is regular and $E(\kappa)$ holds. Then there is a κ-free group A of cardinality κ with $\Gamma(A) = E_o$.

Proof of lemma. (See [5]). A is constructed as the union of a κ-filtration $\{A_\nu \mid \nu < \kappa\}$ of free groups of cardinality $< \kappa$ such that for $\nu \in E_o$, A_μ/A_ν is free for all $\mu > \nu$ and such that for all $\nu \in E_o$, $A_{\nu+1}/A_\nu$ is not free. Hypothesis (i) allows one to construct $A_{\nu+1}$ if $\nu \in E_o$, using the fact that ω-free does not imply ω_1-free. Hypothesis (ii) insures that at limit ordinals σ, $A_\sigma = \bigcup_{\nu<\sigma} A_\nu$ is free.

Result (2) -proved in [19]- involves a more elaborate construction to obtain a κ-free A with a κ-filtration $\{A_\nu \mid \nu < \kappa\}$ such that for $\nu \notin E_o$, A_ν is a direct summand of A. Result (3) - proved in [11]- uses not only $E(\kappa)$ as above, but also $\Diamond_\kappa$ [see Nunke's paper in this volume, or [7], Theorem 0.2, for a statement] to construct a κ-free A such that A has no direct summands.

Given κ-filtrations $\Delta = \{A_\nu | \nu < \kappa\}$ of A and $\Delta'_\nu = \{A'_\nu | \nu < \kappa\}$ of A' (where A and A' are κ-free groups of cardinality κ) let us say that Δ and Δ are quotient-equivalent if for all ν, $A_{\nu+1}/A_\nu$ is isomorphic to $A'_{\nu+1}/A'_\nu$. Theorem 2.9 produced non-isomorphic groups A and A' with $\Gamma_\kappa(A) = \Gamma_\kappa(A')$ by constructing A and A' so that they did not have quotient-equivalent filtrations. A natural question to ask is whether specifying the quotients of a filtration of A is enough to determine A up to isomorphism. Assuming $V = L$, we can give a negative answer (cf. a similar remark by Shelah regarding p-groups in the introduction to [20]).

2.12 Corollary. Assume $V = L$. If κ is a regular cardinal which is not weakly compact then there are 2^κ different $\tilde{E} \in \mathscr{D}(\kappa)$ such that there exist non-isomorphic κ-free groups A and A' of cardinality κ with: (i) $\Gamma(A) = \tilde{E} = \Gamma(A')$; and (ii) there are κ-filtrations Δ and Δ' of A and A', respectively, which are quotient-equivalent.

Proof. Let A be any indecomposable $L_{\infty\kappa}$-free group of cardinality κ (see 2.10.3), and let $A' = A \oplus \mathbb{Z}$. Then A' is not isomorphic to A since A' is not indecomposable. However if $\Delta = \{A_\nu | \nu < \kappa\}$ is any κ-filtration of A, then $\Delta' = \{A_\nu \oplus \mathbb{Z} | \nu < \kappa\}$ is a κ-filtration of A' which is quotient-equivalent to Δ (A fortiori, $\Gamma(A) = \Gamma(A')$).

2.13 Remarks. (1) Other methods yield a stronger result than 2.12 viz. assuming $V = L$, for each κ as above there are 2^κ different $\tilde{E} \in \mathscr{D}(\kappa)$ with 2^κ κ-free groups $\{A_i, i < 2^\kappa\}$ satisfying $\Gamma(A_i) = \tilde{E}$ for all i, and such that for $i \neq j$ A_i and A_j are non-isomorphic but have quotient-equivalent filtrations.

(2) Results on the splitting problem for mixed groups under the hypothesis $V = L$ are presented in [7] and [8].

We have been considering the question of when κ-free implies κ^+-free. Assuming $V = L$ we have a complete answer: κ-free implies κ^+-free if and only if κ is singular or weakly compact. Thus it is consistent with ZFC that there is no cardinal κ such that every κ-free group (of arbitrary size) is free. On the other hand, Theorem 2.6 shows that if it is consistent with ZFC that a strongly compact cardinal exists, then it is consistent with ZFC that there is a cardinal κ such that κ-free implies free. Also, Shelah (unpublished) has proved that if it is consistent with ZFC that a certain very large ("supercompact") cardinal exists, then it is consistent with ZFC that $2^{\aleph_0}$-free implies free. The following results (pointed out to us by M. Magidor) show that some assumption about the consistency of the existence of a large cardinal is necessary to prove results like those just cited. Jensen's "Covering Principle" [17] implies the following:

__2.14 Theorem__. Assume that there are no weakly compact cardinals in L. Then for every singular cardinal κ, $E(\kappa^+)$ is true (in V).

Hence by the methods of proof of 2.10 and 2.11 (and the fact that L is a model of ZFC) we obtain:

__2.15 Corollary__. Suppose it is consistent with ZFC that

(i) there is a cardinal κ such that κ-free implies free;

__or__ (ii) $\aleph_{\omega+1}$-free implies $\aleph_{\omega+2}$-free .

Then it is consistent with ZFC that there exists a weakly compact cardinal.

(The "proper" (and weaker) hypothesis - which we will not attempt to explain - for 2.14 is that "$0^{\#}$ does not exist", which yields a stronger conclusion to 2.15).

We conclude with some open questions about κ-free groups. Unless otherwise specified the question is asked in either ZFC or ZFC + V = L.

__2.16 Questions__. (1) Is it provable in ZFC that $\aleph_{\omega+1}$-free does not imply $\aleph_{\omega+2}$-free?

(2) If $\kappa > \aleph_\omega$ is such that κ-free does not imply κ^+-free, for which $\widetilde{E}$ is $\Gamma_\kappa^{-1}(\widetilde{E})$ non-empty?*

(3) If κ is not a limit cardinal and $\kappa \neq \aleph_1$, is there a κ-free group of cardinality κ which is not $L_{\infty\kappa}$-free?

*Recently Mekler has shown that if κ is weakly inaccessible, then $\Gamma_\kappa^{-1}(1) = \emptyset$.

(4) If A is a κ-free group of cardinality κ which is not free, are there 2^κ groups A' in $\Gamma_\kappa^{-1}\left(\Gamma_\kappa(A)\right)$ with κ-filtrations quotient-equivalent to a filtration of A?

(5) If $\Gamma^{-1}(\tilde{E}_i) \neq \emptyset$ for $i = 1,2$ does there exist $A \in \Gamma^{-1}(\tilde{E}_1 \cup \tilde{E}_2)$ which is not of the form $A_1 \oplus A_2$ where $\Gamma(A_i) = E_i$, $i = 1,2$? [Theorem 2.10.3 yields a partial affirmative answer assuming $V = L$].

(6) Is it provable in ZFC + $V = L$ that there is a rigid system of $L_{\infty\kappa}$-free groups for κ regular and not weakly compact?

(7) If A is an ω_1-free group of cardinality ω_1 which is not $L_{\infty\omega_1}$-free, is A a possibility I group (in the sense of [20])? [The converse is true].

REFERENCES

1. J. Barwise, Back and forth through infinitary logic, Studies in Model Theory, MAA Studies in Math. 8 (1973), 5-34.

2. J. Barwise, An introduction to first order logic, Handbook of Mathematical Logic, North Holland, Amsterdam, 1977.

3. J. Barwise and P. Eklof, Infinitary properties of abelian torsion groups, Ann. Math. Logic 2 (1970), 25-68.

4. P. Eklof, Infinitary equivalence of abelian groups, Fund. Math. 81 (1974), 305-314.

5. P. Eklof, On the existence of κ-free abelian groups, Proc. Amer. Math. Soc. 47 (1975), 65-72.

6. P. Eklof, Categories of local functions, Model Theory and Algebra, Lecture Notes in Math. 498 (1975), Springer-Verlag, Berlin, 91-116.

7. P. Eklof, Homological algebra and set theory, Trans. Amer. Math. Soc., to appear.

8. P. Eklof, Applications of logic to the problem of splitting abelian groups, Proceedings of Logic Colloquium '76, to appear.

9. P. Eklof, Ultraproducts for algebraists, Handbook of Mathematical Logic, North Holland, Amsterdam, 1977.

10. P. Eklof and E. Fisher, The elementary theory of abelian groups, Ann. Math. Logic 4 (1972), 115-171.

11. P. Eklof and A. Mekler, On constructing indecomposable groups in L, Jour. of Algebra, to appear.

12. E. Ellentuck, Categoricity regained, Jour. of Symbolic Logic 41 (1976), 639-643.

13. J. Gregory, Abelian groups infinitarily equivalent to free ones, Notices Amer. Math. Soc. 20 (1973), A-500.

14. P. Hill, New criteria for freeness in abelian groups II, Trans. Amer. Math. Soc. 196 (1974), 191-201.

15. P. Hill, A special criterion for freeness, Symposia Math. XIII (1974), 311-314.

16. R. Jensen, The fine structure of the constructible hierarchy, Ann. Math. Logic 4 (1972), 229-308.

17. R. Jensen and K. Devlin, Marginalia to a theorem of Silver, Logic Conference, Kiel 1974, Lecture Notes in Math. 499 (1975), Springer-Verlag, Berlin, 115-142.

18. I. Kaplansky, Infinite Abelian Groups, rev. ed., University of Michigan Press, Ann Arbor, 1969.

19. A. Mekler, Applications of logic to group theory, Ph.D. Dissertation, Stanford University, 1976.

20. S. Shelah, Infinite abelian groups, Whitehead problem and some constructions, Israel Jour. Math. 18 (1974), 243-256.

21. S. Shelah, A compactness theorem for singular cardinals, free algebras, Whitehead problem and transversals, Israel Jour. Math 21 (1975), 319-349.

22. R. Solovay, Real-valued measurable cardinals, Axiomatic Set Theory, A.M.S. Proc. of Symp. in Pure Math XIII (part I) (1971), Providence, 397-428.

23. W. Szmielew, Elementary properties of abelian groups, Fund. Math 41 (1955), 203-271.

ABELIAN STRUCTURES I

Edward R. Fisher

The subject matter of this paper is designed to cover the theory of modules over rings as well as sheaves of modules, modules with distinguished submodules, and "reasonable" expansions of such structures. Our treatment basically covers quasi-varieties of module diagrams with distinguished additive relations. Although the basis for our definitions lies in the model theory of modules, we feel that the stability of our notions under important algebraic and category-theoretic constructions lends credence to a sense that our subject will yield interesting applications in areas beyond our present concerns; e.g., that of "valuated groups", in which the natural language would seem to be $L_{\omega_1\omega}$ as opposed to our restriction to $L_{\omega\omega}$.

In the present Part I of this paper, we restrict ourselves to the more purely algebraic aspects of abelian structures. We begin (Section 1) with a discussion of many-sorted logic in which the conventions are chosen specifically to facilitate discussion of sheaves (or categories) of structures and to allow for natural iteration of constructions. We believe that these conventions transcend the needs of this particular work and intend to expound their general virtues in another place [5]. No results are listed as theorems, and the reader is encouraged to use this section as a reference for notation and folklore. The specific subject matter properly begins with a discussion (Section 2) of the basic features of abelian structures.

Our discussion (Section 3) of injectives is on an extremely general plane; in essence, we have included some heretofore unpublished material of a universal algebraic nature which benefited greatly from discussions with P. Bacsich at the Cambridge Summer School in Cambridge, England in 1971. As far as abelian structures are concerned, we prove the existence and uniqueness of injective and (as a special case) pure-injective hulls under quite general hypotheses. Our handling of pure-injectivity as a special case of injectivity seems to be novel--in fact our definition of "pure-essential" seems to be nontraditional, but from the present point-of-view, eminently correct; in any case, pure-injective hulls turn out to be the traditional ones (for modules, at least).

In a sequel we intend to discuss criteria for elementary equivalences and elementary embeddings, a Löwenheim-Skolem theorem for $\infty\eta$-substructures of injectives, yielding proofs of the existence of homogeneous-universal and saturated models in cardinalities η such that $\eta^{\lambda} = \eta$, $\lambda = \text{card}(L) + \aleph_0$, and hence of stability, Noetherian and coherent theories, irreducible models and Azumaya's theorem, the role of notions such as "commutative", "radical", "completely reducible", and an outline of the Mitchell Embedding Theorem from our point of view.

In the sequel, our debt to the seminal work of Eklöf and Sabbagh [2, 4, 7] will become clearer as will the relationship of the present work to that of [3]; in particular, our general machinery leads to much more direct proofs of many of the results in [3], which were there based on algebraic lemmas due to Kaplansky.

1. Many-Sorted Logic. In order to avoid what would in fact be a pointless discussion of the foundations of syntax in a system of set theory such as ZFC, we will simply assume that everything to do with "symbols" can be handled rigorously by the reader in his favorite system--a set theory with "atoms" would greatly simplify such a program. Basically, we assume that expressions are unambiguous, i.e., uniquely decipherable. Otherwise we follow rather standard set-theoretic practice in letting "A^B" denote the set of all functions from B to A and "card(A)" denote the cardinality of A, which in turn is defined to be an initial ordinal. An ordinal is identified with the set of its predecessors; it is called initial when it cannot be put in $1 - 1$ correspondence with any of its elements; it is called regular if it is not a sup of fewer earlier ordinals. If η and γ are cardinals, then $\eta^{\underline{\gamma}}$ will be defined to be $\sum_{\alpha<\gamma} \eta^{\alpha}$. We write ι_A for the identity function with domain A, leaving the codomain open, so that ι_A can be considered an inclusion map in any bigger set. Thus, if $F : B \to C$ and $A \subset B$, then $F \circ \iota_A$ is the

<u>restriction</u> of F to A, and the usual trivialities about restrictions become trivialities about compositions of maps. We follow sound category-theoretic practice only when we view sets as elements of a category. Lastly, we will write $\begin{pmatrix} a_1 & \cdots & a_n \\ b_1 & \cdots & b_n \end{pmatrix}$ to denote the $f : \{a_1, \ldots, a_n\} \to \{b_1, \ldots, b_n\}$ such that $f(a_i) = b_i$ for each i.

Given a set of symbols S called sorts, we define an <u>S-set</u> to be a pair $\langle A, s_A \rangle$, where A is a set and $s_A : A \to S$. Our intention is to be able to handle "situations" in model theory as if they were models. For example, given a homomorphism $F : G \to H$ between groups, we can think of the underlying set of this "structure" as an S-set A in which $S = \{s_1, s_2\}$, $A = \{s_1\} \times G \cup \{s_2\} \times H$, and $s_A = \pi_1$ (e.g., $s_A(\langle s_1, g \rangle) = s_1$). Intuitively, we have "sorted out" the roles of elements of G and H by disjunctification; this conforms to the healthy category-theoretic practice of ignoring the "incidental" fact that $G \cap H$ may be non-empty. Of course, we are thereby deliberately giving up some of the features of the original situation in order to develop an "abstract" discussion of it. Knowing when to be abstract (and how much to ignore) is what good mathematical practice is all about.

Fixing S, we define an <u>S-map</u> F between S-sets A and B to be a map $F : A \to B$ such that $s_A = s_B \circ F$. The category of S-sets and S-maps has most of the familiar properties of the category of sets and arbitrary maps. One notable exception is that a monomorphism, which is simply a 1 - 1 S-map, from A to B may fail to be left invertible even when A is not the initial object, which is the empty S-set. A may be half-empty, so that some elements of B (viz., of the sort for which A is empty) may have no place to go under an S-map returning to A. We will only pause briefly to show what products look like in our category, leaving other standard constructions to the diligent reader. Given a family A_i, $i \in I$, of S-sets, we define $A = \prod_{i \in I} A_i$ with projections π_i, $i \in I$, as follows:

$$A = \bigcup_{s \in S} \{s\} \times \{f \in \underset{i \in I}{\times} A_i : \forall i,\ s_{A_i}(f_i) = s\},$$

$$s_A(\langle s, f\rangle) = s,$$

$$\pi_i(\langle s, f\rangle) = f_i,$$

where $\underset{i \in I}{X} A_i$ is the ordinary cartesian product of the A_i's in which sorts are ignored. It is a trivial matter to verify that each $\pi_i : A \to A_i$ is an S-map and that we have indeed produced a product in our category. Note that when $I = \emptyset$ we get the terminal object of our category, viz., $A = \{\langle s, \emptyset\rangle : s \in S\}$, in which there is a unique element of each sort! [Of course, the empty set "$\emptyset$" in "$\langle s, \emptyset\rangle$" is the empty map $\emptyset \in \underset{i \in \emptyset}{X} A_i$.]

A vocabulary L is a collection of symbols of three kinds: sorts; relation symbols; and function symbols. We will let S be the set of sorts in L. Each relation symbol R comes equipped with a set of arguments $\alpha(R)$ and a non-empty set of admissible sort assignments $Ad_R \subset S^{\alpha(R)}$. Among the relation symbols in L we always have a distinguished truth symbol T with $\alpha(T) = \emptyset$ and, if S is non-empty, a distinguished equality symbol $=$ with two arguments ℓ and r (for left and right) for which $Ad = \left\{\begin{pmatrix} \ell & r \\ s & s \end{pmatrix} : s \in S\right\}$.

Each function symbol f comes equipped with a set of arguments $\alpha(f)$, a set of admissible sort assignments $Ad_f \subset S^{\alpha(f)}$, and a sort valuation map $Val_f : Ad_f \to S$. L will be called purely functional if and only if L has no relation symbols besides T and $=$.

The motivation for these definitions should become clear following our next definitions. If S is the set of sorts in a vocabulary L, A is an S-set, and P is a relation or function symbol of L, we define the set Ad_P^A of admissible assignments for P in A as follows:

$$Ad_P^A = \{\underset{\sim}{a} \in A^{\alpha(P)} : s_A \circ \underset{\sim}{a} \in Ad_P\}.$$

Intuitively, these are the assignments to the arguments of P at which we intend P to be "defined" in A. An L-structure $\mathcal{A}$ is a pair $\langle A, I\rangle$ in which A is an S-set and I is an interpretation in A for each of the remaining symbols in L subject to the following conditions:

(i) If R is a relation symbol in L, then I(R), which will henceforth be written R^A, is a map with domain Ad_R^A and codomain $2 = \{0, 1\}$;

(ii) if $\underset{\sim}{a} = \begin{pmatrix} \ell & r \\ a_1 & a_2 \end{pmatrix}$ is an admissible assignment for = in A (i.e., $s_A(a_1) = s_A(a_2)$), then $=^A(\underset{\sim}{a}) = 1$ if and only if $a_1 = a_2$; also, $T^A(\emptyset) = 1$;

(iii) if f is a function symbol in L, then I(f), written f^A, is a map with domain Ad_f^A and codomain A and such that $\underset{\sim}{a} \in Ad_f^A \Rightarrow s_A(f^A(\underset{\sim}{a})) = Val_f(s_A \circ \underset{\sim}{a})$.

If R is a relation symbol and $\underset{\sim}{a} \in Ad_R^A$, we say that R <u>holds at</u> $\underset{\sim}{a}$ in A and write $A \models R(\underset{\sim}{a})$ just in case $R(\underset{\sim}{a}) = 1$. Following the usual custom, if R has a single argument (i.e., is <u>unary</u>), we write $A \models R(a)$ whenever R holds at that assignment which assigns a to the argument of R; similarly, if R is <u>binary</u> and has its two arguments identified as left and right, we write $A \models a_1 R a_2$ whenever $A \models R(\underset{\sim}{a})$ for $\underset{\sim}{a} = \begin{pmatrix} \ell & r \\ a_1 & a_2 \end{pmatrix}$. For example, $A \models a = a'$ iff $a = a'$, and $A \not\models a = a'$ just in case $s_A(a) = s_A(a')$ and $a \neq a'$--we do not write expressions which do not make good sense.

If f is a function symbol and $\underset{\sim}{a} \in Ad_f^A$, then the stipulation in (iii) above says that $f^A(\underset{\sim}{a})$ should have the sort specified by Val_f. Unary and binary function symbols will be treated according to tradition, and 0-ary function symbols will be thought of as <u>individual constant symbols</u>. Incidentally, 0-ary relation symbols are thought of as <u>propositional constants</u> which are simply true or false in the structure.

L is called <u>finitary</u> if all function and relation symbols have finite sets of arguments. Finitary vocabularies are subject to most of the general theorems of model theory, and much of pure mathematics proceeds by trading in nice infinitary properties for equivalent finitary ones so that maximality principles and the like can be invoked. For example, a metrization theorem shows that, under appropriate conditions, infinitary topological structures can be viewed as two-sorted finitary structures through the introduction of a metric $d : X \times X \to R$. Here maximality principles elude us since the crucial Archimedean property of R is not subject to compactness arguments, but inductive limits are well-behaved. One studies locally compact groups by showing that there is a close relationship between the continuous Hilbert space representations of such groups and the purely finitary representations of their group algebras to which maximality principles do apply. Finally, as we will see below, finitary structures such as modules can be studied by trading in the compactness theorem to which they are subject for the existence of structures (viz., their injective hulls) with infinitary properties, which in turn have natural finitary descriptions in reasonable categories of modules; and, as a final twist, we will define a theory of abelian structures to be "Noetherian" just in case the injectives have an overtly finitary character.

A homomorphism F from one L-structure $\mathcal{A}$ to another $\mathcal{B}$ is an S-map $F : A \to B$ such that:

(i) for each relation symbol R in L and each $\underset{\sim}{a} \in Ad_R^A$,

$$\mathcal{A} \models R(\underset{\sim}{a}) \Rightarrow \mathcal{B} \models R(F \circ \underset{\sim}{a});$$

(ii) for each function symbol f in L and each $\underset{\sim}{a} \in Ad_f^A$,

$$F(f^{\mathcal{A}}(\underset{\sim}{a})) = f^{\mathcal{B}}(F \circ \underset{\sim}{a}).$$

We will write $F : \mathcal{A} \to \mathcal{B}$ or $F \in \mathcal{B}^{\mathcal{A}}$ whenever F is a homomorphism from $\mathcal{A}$ to $\mathcal{B}$. F is called an embedding, written $F : \mathcal{A} \xrightarrow{\text{emb}} \mathcal{B}$, if (i) above holds in the stronger form:

$$\mathcal{A} \models R(\underset{\sim}{a}) \Longleftrightarrow \mathcal{B} \models R(F \circ \underset{\sim}{a}).$$

Such an F is necessarily $1 - 1$, as can be seen by taking for R the equality symbol. We say that F is onto and write $F : \mathcal{A} \xrightarrow[\text{onto}]{} \mathcal{B}$ whenever F is onto as a mapping, i.e., for each $b \in B$ there exists $a \in A$ such that $F(a) = b$. F is an isomorphism if $F : \mathcal{A} \xrightarrow[\text{onto}]{\text{emb}} \mathcal{B}$, and the usual arguments show that this is the case if and only if F is an isomorphism in the category of L-structures and homomorphisms. We say that $F : A \to B$ is left-(right-)invertible if there exists $G : B \to A$ such that $G \circ F = \iota_A$ $(F \circ G = \iota_B)$, where ι_A is the isomorphism from A to A which fixes all elements.

For any category K of L-structures, if $F : A \to B$ belongs to the category, then F is called a monomorphism if and only if for all C, G, H such that $G : C \to A$ and $H : C \to A$ both belong to K, we have:

$$F \circ G = F \circ H \Rightarrow G = H,$$

i.e., F is left-cancellable. Similarly, F is called an epimorphism if it is right-cancellable.

Endomorphisms and automorphisms are defined in the usual way, and we assume the reader can check that the usual trivialities still hold. For example, the following implications hold in any category of L-structures:

(i) F is left-invertible $\Rightarrow F : A \xrightarrow{\text{emb}} B \Rightarrow F : A \xrightarrow{\text{1-1}} B \Rightarrow F$ is a monomorphism;

(ii) F is right-invertible $F : A \xrightarrow[\text{onto}]{} B \Rightarrow F$ is an epimorphism.

In both cases, the final implication is reversible in the full category of all L-structures. Also, if L is purely functional, then F 1 - 1 $\Rightarrow F$ is an embedding. Products are defined in the usual way by basing the construction on the product of the underlying S-sets.

An L-structure A is called a substructure of an L-structure B, written $A \subset B$, if $A \subset B$ and $\iota_A : A \to B$ is an embedding of A into B. It should be clear that if B is an L-structure, then a subset A of B is the underlying set of a substructure $A \subset B$ if and only if A is closed under the operations f^B of B; moreover, A is uniquely determined by A and B, viz., the sort-function, operations, and relations of A are obtained from those of B by restriction. From these observations it follows that the intersection $\bigcap_{i \in I} A_i$ of a family of substructures is again a substructure, so that the partially ordered set of all substructures of B forms a complete lattice in which the notion of infimum is simply that of intersection. If B is an L-structure and $A \subset B$, then the substructure of B generated by A is the intersection of all substructures A' of $\mathcal{B}$ such that $A \subset A'$. If $F : \mathcal{B} \to \mathcal{C}$ and $G : \mathcal{B} \to \mathcal{C}$, and $F \circ \iota_A = G \circ \iota_A$, then F and G agree on the substructure generated by A, since the <u>equalizer</u> of F and G, $\{b \in B : F(b) = G(b)\}$, is clearly a substructure containing A. $\mathcal{B}$ is called <u>η-generated</u> if and only if there exists $A \subset B$ with $\mathrm{card}(A) < \eta$ such that A generates $\mathcal{B}$; $\mathcal{B}$ is cyclic if and only if $\mathcal{B}$ is generated by a singleton. Clearly, images and inverse images of substructures under homomorphisms are substructures.

Given a class K of L-structures, we will call the members of K <u>models</u> and treat K as a category in which the morphisms are <u>all</u> the homomorphisms between models; we will generally assume that K is <u>abstract</u> (i.e., $\mathcal{A} \cong \mathcal{B} \in K \Rightarrow \mathcal{A} \in K$). K will be called <u>hereditary</u> if $\mathcal{A} \subset \mathcal{B} \in K \Rightarrow \mathcal{A} \in K$; K is <u>productive</u> if the product of any family of models is again a model. If K is hereditary then it has the <u>first isomorphism property</u> (<u>FIP</u>), viz., every $F : \mathcal{A} \to \mathcal{B}$ can be factored (uniquely up to isomorphism) as $F = G \circ H$ where H is onto and G is an embedding, since the range of F is clearly a substructure of $\mathcal{B}$. Now we note that in <u>any</u> category, if $F : \mathcal{A} \to \mathcal{B}$ has a left inverse $G : \mathcal{B} \to \mathcal{A}$, then $H = F \circ G$ is an idempotent endomorphism of $\mathcal{B}$, i.e., $H \circ H = H$. This is proved by writing: $H \circ H = (F \circ G) \circ (G \circ F) = F \circ (G \circ F) \circ G = F \circ \iota_A \circ G = F \circ G = H$. Conversely, if K has FIP and $H : \mathcal{B} \to \mathcal{B}$ is idempotent, then there exist $\mathcal{A}$, $F : \mathcal{A} \to \mathcal{B}$ and $G : \mathcal{B} \to \mathcal{A}$ such that $G \circ F = \iota_A$ and $H = F \circ G$. This is proved by factoring $H = F \circ G$ with $G : \mathcal{B} \xrightarrow[\text{onto}]{} \mathcal{A}$ and $F : \mathcal{A} \xrightarrow{\text{emb}} \mathcal{B}$, and then noting that $F \circ (G \circ F) \circ G = (F \circ G) \circ (F \circ G) = H \circ H = H = F \circ \iota_A \circ G$, so that by left- and right-cancellation, $G \circ F = \iota_A$.

An L-structure is called <u>null-generated</u> if it is generated by $\emptyset$, i.e., if it contains no proper substructures. Since L is a set, it is obvious that there are at most a set of null-generated L-structures up to isomorphism. Given a class K of L-structures, a model $\mathcal{A}$ is called <u>initial</u> if and only if for every $\mathcal{B} \in K$ there exists a unique $F : \mathcal{A} \to \mathcal{B}$. If K is hereditary and productive, there exists a (unique) initial model. This can be seen by forming the product $\prod_{i \in I} A_i$ of a representative set of null-generated models of K, and letting $A \subset \prod_{i \in I} A_i$ be null-generated. Given any B in K, there exists a (unique) null-generated substructure of B, and hence a $G : A_i \xrightarrow{\text{emb}} B$. Thus, $F = G \circ \pi_i \circ \iota_A$ is the desired homomorphism. F is unique, since A is null-generated. Finally, A is unique up to isomorphism by the usual abstract nonsense.

If $L_1 \subset L_2$ are vocabularies and A is an L_2-structure, then the <u>L_1-reduct</u> A' of A, written <u>$\text{red}_{L_1}(A)$</u>, is obtained by setting $A' = s_A^{-1}(S_1)$, where S_1 is the set of sorts of L_1, $s_{A'} = s_A \circ \iota_{A'}$, and $P^{A'}(\underset{\sim}{a}) = P^A(\underset{\sim}{a})$ for each relation or function symbol $P \in L_1$ and each $\underset{\sim}{a} \in Ad_P^{A'} = Ad_P^A \cap (A')^{\alpha(P)}$. In plain English, the reduct is obtained by simply forgetting those aspects of A which are not relevant to L_1. Note that, by demanding the elimination of all elements of A having sorts not in L_1, we are in effect subsuming the more general notion of "relativized reduct" under our one notion of reduct. In any case one should think of the L_1-reduct as that L_1-structure which best approximates the original L_2-structure. A is called an <u>L_2-expansion</u> of A'.

It is a simple matter to check that if A and B are L_2-structures, A' and B' are their L_1-reducts, and $F : A \to B$, then $F \circ \iota_{A'} : A' \to B' \cdot F \circ \iota_{A'}$ will be called the L_1-reduct of F. Further verifications then show that the act of taking reducts is functorial, i.e., compositions of homomorphisms go to compositions and identity maps go to identity maps, and that, like all decent forgetful functors, the reduct of an embedding is an embedding, the reduct of a surjective homomorphism is surjective, the reduct of a product is the product of the reducts, etc. Thus, if K_2 is a hereditary or productive class of L_2-structures, then the class $K_1 = \text{red}_{L_1}(K_2)$ of L_1-reducts of members of K_2 is also hereditary or productive, respectively. Conversely, if K_1 is a hereditary or productive class of L_1-structures, then the class K_2 of all L_2-structures A such that $\text{red}_{L_1}(A) \in K_1$ is hereditary or productive, respectively.

Before discussing "universal constructions" which are intimately related to the "coadjoints" of our reduct functors, we first turn to a quick survey of the languages which are derivable from a vocabulary.

Let S be the set of sorts of our vocabulary L. Given an S-set V of new elementary syntactic entities called variables, we will now define the set of *L-terms*, *$\mathrm{Term}_L(V)$*, and the class of *L-formulas*, *$\mathrm{Form}_L(V)$*, based on V. For each term τ, we will also define the *set of variables of* τ, *$\mathrm{Var}(\tau)$*, and the *sort of* τ, *$\mathrm{Sort}(\tau)$* (so that the terms will actually form an S-set); for each formula ϕ, we will also define the *set of free variables of* ϕ, *$\mathrm{Var}(\phi)$*, and the *set of subformulas of* ϕ, *$\mathrm{Subform}(\phi)$*. We will carry out our inductive definition by defining $\mathrm{Term}_L^{\alpha}(V)$ and $\mathrm{Form}_L^{\alpha}(V)$ for each ordinal α such that $\alpha \le \beta \Rightarrow \mathrm{Term}_L^{\alpha}(V) \subset \mathrm{Term}_L^{\beta}(V)$ and $\mathrm{Form}_L^{\alpha}(V) \subset \mathrm{Form}_L^{\beta}(V)$. It is to be understood that the auxiliary functions Sort, Subform, etc., are extended from each stage to the next (compatibility being a trivial matter), and that $\mathrm{Term}_L(V)$ and $\mathrm{Form}_L(V)$ will simply be the unions of these sets.

1. $\mathrm{Term}_L^{0}(V) = V$, $\mathrm{Var}(v) = \{v\}$, and $\mathrm{Sort}(v) = s_V(v)$.
2. $\mathrm{Term}_L^{\alpha+1}(V)$ consists of the members of $\mathrm{Term}_L^{\alpha}(V)$ together with all expressions of the form $f(\underset{\sim}{\tau})$, where $f \in L$ is a function symbol, $\underset{\sim}{\tau} : \alpha(f) \to \mathrm{Term}_L^{\alpha}(V)$, and $\mathrm{Sort} \circ \underset{\sim}{\tau} \in \mathrm{Ad}_f$; for such a new term τ, we define $\mathrm{Var}(\tau) = \bigcup_{i \in \alpha(f)} \mathrm{Var}(\underset{\sim}{\tau}_i)$ and $\mathrm{Sort}(\tau) = \mathrm{Val}_f(\mathrm{Sort} \circ \underset{\sim}{\tau})$.
3. $\mathrm{Term}_L^{\lambda}(V) = \bigcup_{\beta<\lambda} \mathrm{Term}_L^{\beta}(V)$, for λ a limit ordinal.
4. If γ is the least infinite regular cardinal $> \mathrm{card}(\alpha(f))$ for each function symbol $f \in L$, it is easy to see that $\mathrm{Term}_L^{\gamma+1}(V) = \mathrm{Term}_L^{\gamma}(V)$, and we can take $\mathrm{Term}_L(V) = \mathrm{Term}_L^{\gamma}(V)$; it also follows that $\mathrm{card}(\mathrm{Term}_L(V))$ is $\le$ the least η such that $\mathrm{card}(V) \le \eta$, $\mathrm{card}(\{f \in L : f \text{ is a function symbol}\}) \le \eta$, and $\eta^{\gamma} = \eta$.
5. $\mathrm{Form}_L^{0}(V)$ consists of those expressions of the form $R(\underset{\sim}{\tau})$, where $R \in L$ is a relation symbol, $\underset{\sim}{\tau} : \alpha(R) \to \mathrm{Term}_L(V)$ and $\mathrm{Sort} \circ \underset{\sim}{\tau} \in \mathrm{Ad}_R$; for such a formula ϕ, we define $\mathrm{Var}(\phi) = \bigcup_{i \in \alpha(R)} \mathrm{Var}(\underset{\sim}{\tau}_i)$ and $\mathrm{Subform}(\phi) = \{\phi\}$.
6. $\mathrm{Form}_L^{\alpha+1}(V)$ consists of the members of $\mathrm{Form}_L^{\alpha}(V)$ together with all

expressions of the forms:

(i) $\neg\phi$, for $\phi \in \mathrm{Form}_L^{\alpha}(V)$,

(ii) $\wedge\Phi$ or $\vee\Phi$, for $\Phi \in \mathrm{Form}_L^{\alpha}(V)$,

(iii) $\exists X\phi$ or $\forall X\phi$, for $\phi \in \mathrm{Form}_L^{\alpha}(V)$ and $X \subset V$;

we define:

(i') $\mathrm{Var}(\neg\phi) = \mathrm{Var}(\phi)$, $\mathrm{Subform}(\neg\phi) = \{\neg\phi\} \cup \mathrm{Subform}(\phi)$,

(ii') $\mathrm{Var}(\wedge\Phi) = \mathrm{Var}(\vee\Phi) = \bigcup_{\phi\in\Phi} \mathrm{Var}(\phi)$,

$\mathrm{Subform}(\wedge\Phi) = \{\wedge\Phi\} \cup \bigcup_{\phi\in\Phi} \mathrm{Subform}(\phi)$,

$\mathrm{Subform}(\vee\Phi) = \{\vee\Phi\} \cup \bigcup_{\phi\in\Phi} \mathrm{Subform}(\phi)$,

(iii') $\mathrm{Var}(\exists X\phi) = \mathrm{Var}(\exists X\phi) = \mathrm{Var}(\phi) - X$,

$\mathrm{Subform}(\exists X\phi) = \{\exists X\phi\} \cup \mathrm{Subform}(\phi)$,

$\mathrm{Subform}(\forall X\phi) = \{\forall X\phi\} \cup \mathrm{Subform}(\phi)$.

7. $\mathrm{Form}_L^{\lambda}(V) = \bigcup_{\beta<\lambda} \mathrm{Form}_L^{\beta}(V)$, for λ a limit ordinal.

8. The L-formulas do not form a set; however, if $\eta_1 \geq \eta_2$ are infinite cardinals such that η_1 is regular and $\eta_2 > \mathrm{card}(\alpha(P))$ for each function or relation symbol $P \in L$, we can define $L_{\eta_1\eta_2}(V)$ to be the <u>language</u> consisting of all members of $\mathrm{Term}_L(V)$ and all those L-formulas $\phi \in \mathrm{Form}_L(V)$ which satisfy:

(a) $\mathrm{Card}(\mathrm{Var}(\phi)) < \eta_2$,

(b) $\exists X\, \psi \in \mathrm{Subform}(\phi) \Rightarrow \mathrm{card}(X) < \eta_2$,

(c) $\mathrm{card}(\mathrm{Subform}(\phi)) < \eta_1$.

It is easy to see that (i) if ψ is a subformula of $\phi \in L_{\eta_1\eta_2}(V)$, then ψ belongs to $L_{\eta_1\eta_2}(V)$, (ii) all formulas in $L_{\eta_1\eta_2}(V)$ are contained in $\mathrm{Form}_L^{\eta_1}(V)$, and (iii) if η is the least infinte cardinal such that $\mathrm{card}(\mathrm{Form}_L^0(V)) \leq \eta$ and $\eta^{\underset{\smile}{\eta_1}} = \eta$, then $\mathrm{card}(L_{\eta_1\eta_2}(V)) \leq \eta$.

In the important case where L is finitary, we note that $\mathrm{card}(L_{\omega\omega}(V)) \leq \mathrm{card}(V) + \mathrm{card}(L) + \omega$. It is also important to note that the set of <u>atomic formulas</u>, $\mathrm{Form}_L^0(V)$, has cardinality at most the least infinite η such that $\mathrm{card}(V) \leq \eta$, $\mathrm{card}(L) \leq \eta$, and $\eta^{\underset{\smile}{\gamma}} = \eta$, where γ is the least infinite regular cardinal $> \mathrm{card}(\alpha(P))$ for each $P \in L$. Finally, it should be pointed out that with finer distinctions some improvements in our upper bounds can be given; for example, if V is finite, L is finitary and finite, and $\alpha(f) = \emptyset$ for each function symbol $f \in L$, then $\mathrm{Term}_L(V)$ and each $\mathrm{Form}_L^n(V)$, $n \in \omega$, is finite.

$L_{\infty\eta}(V)$ is the class of formulas which belong to some $L_{\eta_1\eta}(V)$. A formula σ is a sentence if and only $\mathrm{Var}(\sigma) = \emptyset$. If S is the set of sorts of L, A is an S-set, $\tau \in \mathrm{Term}_L(V)$, and $\phi \in \mathrm{Form}_L(V)$, we define Ad_τ^A to be the set of S-maps from $\mathrm{Var}(\tau)$ to A, and Ad_ϕ^A to be the set of S-maps from $\mathrm{Var}(\phi)$ to A.

Now, given an L-structure A, we define $\tau^A : \mathrm{Ad}_\tau^A \to A$ by induction on the complexity of τ as follows:

(a) if τ is $v \in V$, then $\tau^A(\underset{\sim}{a}) = \underset{\sim}{a}(v)$;

(b) if τ is $f(\underset{\sim}{\tau})$, then $\tau^A(\underset{\sim}{a}) = f^A(\underset{\sim}{a}')$,

where $\underset{\sim}{a}'_i = \underset{\sim}{\tau}_i^A(\underset{\sim}{a} \circ \iota_{\mathrm{Var}(\underset{\sim}{\tau}_i)})$ for each $i \in \alpha(f)$. Treating $2 = \{0, 1\}$ as a complete Boolean algebra in the standard way, we define $\phi^A : \mathrm{Ad}_\phi^A \to 2$ by induction on the complexity of ϕ as follows:

(i) if ϕ is $R(\underset{\sim}{\tau})$ then $\phi^A(\underset{\sim}{a}) = R(\underset{\sim}{a}')$, where $\underset{\sim}{a}'_i = \underset{\sim}{\tau}_i(\underset{\sim}{a} \circ \iota_{\mathrm{Var}(\underset{\sim}{\tau}_i)})$ for each $i \in \alpha(R)$;

(ii) $(\neg\phi)^A(\underset{\sim}{a}) = \phi^A(\underset{\sim}{a})^c$;

(iii) $(\wedge\Phi)^A(\underset{\sim}{a}) = \inf\{\phi^A(\underset{\sim}{a} \circ \iota_{\mathrm{Var}(\phi)} : \phi \in \Phi\}$;

(iv) $(\vee\Phi)^A(\underset{\sim}{a}) = \sup\{\phi^A(\underset{\sim}{a} \circ \iota_{\mathrm{Var}(\phi)} : \phi \in \Phi\}$;

(v) $(\exists X\phi)^A(\underset{\sim}{a}) = \sup\{\phi^A(\underset{\sim}{a}') : \underset{\sim}{a}' \in \mathrm{Ad}_\phi^A$ extending $\underset{\sim}{a}\}$;

(vi) $(\forall X\phi)^A(\underset{\sim}{a}) = \inf\{\phi^A(\underset{\sim}{a}') : \underset{\sim}{a}' \in \mathrm{Ad}_\phi^A$ extending $\underset{\sim}{a}\}$.

We will henceforth write $A \models \phi(\underset{\sim}{a})$ (read: <u>$\underset{\sim}{a}$ satisfies ϕ in A</u>) if and only if $\phi(\underset{\sim}{a}) = 1$. If σ is a sentence, we will write $A \models \sigma$ for $A \models \sigma(\emptyset)$, and say that <u>$\sigma$ holds in A</u>, or that <u>A is a model of σ</u>. More generally, if Σ is a set of sentences, we will say that <u>A is a model of Σ</u> if and only if $A \models \wedge\Sigma$. [Warning: Note that if $\Phi = \emptyset$, then $A \models \wedge\Phi$ and $A \not\models \vee\Phi$. Furthermore, if A is empty, then $A \models \forall x(x = x)$ (since $\inf \emptyset = 1 \in 2$), $A \not\models \exists x(x = x)$ (since $\sup \emptyset = 0 \in 2$) while $A \models \exists x\forall y(y = y)$ (since $\mathrm{Var}(\forall y(y = y)) = \emptyset = \mathrm{Var}(\exists x\forall y(y = y))$.]

It will be convenient to extend our notation slightly. Thus, if A is an L-structure, $a : X \to A$ is an S-map, $\tau \in \mathrm{Term}_L(V)$, and $\phi \in \mathrm{Form}_L(V)$ with $\mathrm{Var}(\tau) \subset X$ and $\mathrm{Var}(\phi) \subset X$, we will write $\tau^A(a)$ for $\tau^A(a \circ \iota_{\mathrm{Var}(\tau)})$ and $A \models \phi(a)$ for $A \models \phi(\underset{\sim}{a} \circ \iota_{\mathrm{Var}(\phi)})$. Finally, a trivial induction shows that if $L_1 \subset L_2$, A_2 is an L_2-structure, $A_1 = \mathrm{red}_{L_1}(A_2)$, τ is an L_1-term, and ϕ is an L_1-formula, then:

(i) $\tau^{A_1}(\underset{\sim}{a}) = \tau^{A_2}(\underset{\sim}{a})$, for $\underset{\sim}{a} \in \mathrm{Ad}_\tau^{A_1} = \mathrm{Ad}_\tau^{A_2} \cap A_1^{\mathrm{Var}(\tau)}$;

(ii) $A_1 \models \phi(\underset{\sim}{a}) \Longleftrightarrow A_2 \models \phi(\underset{\sim}{a})$ for $\underset{\sim}{a} \in \mathrm{Ad}_\phi^{A_1} = \mathrm{Ad}_\phi^{A_2} \cap A_1^{\mathrm{Var}(\phi)}$.

[This further exemplifies the idea that $\mathrm{red}_{L_1}(A_2)$ is a "best approximation" to A_2.]

If V is an S-set of variables, A is an L-structure, and $\underset{\sim}{a} : V \to A$ is an S-map, then one can check that the L-structure generated by $\mathrm{Range}(\underset{\sim}{a})$ is given by:

$$\{\tau^A(\underset{\sim}{a}) : \tau \in \mathrm{Term}_L(V)\}.$$

If this is A itself, we will say that $\underset{\sim}{a}$ <u>generates</u> A. It is also straightforward (and very useful) to check that given an S-set of variables V, L-structures A and B, and S-maps $\underset{\sim}{a} : V \to A$ and $\underset{\sim}{b} : V \to B$, if $\underset{\sim}{a}$ generates A, then there exists a unique homomorphism $H : A \to B$ such that $H \circ \underset{\sim}{a} = \underset{\sim}{b}$ if and only if for each atomic $\phi \in \mathrm{Form}_L(V)$,

$$(*) \qquad A \models \phi(\underset{\sim}{a}) \Rightarrow B \models \phi(\underset{\sim}{b}).$$

We will refer to this result as the <u>Fundamental Theorem on Defining Homomorphisms</u>. It is immediate from the proof that H is an embedding if and only if the implication in (*) holds as a bi-implication and that H is onto if and only if $\underset{\sim}{b}$ generates B. These results are normally applied under the guise of "diagram argument". For example, if A is an L_1-structure, $L_1 \subset L_2$, C is a set of individual constants such that $L_2 \cap C = \emptyset$, $\underset{\sim}{a} : C \to A$ is a generating S-map, and A' is the unique expansion of A to an $L_1 \cup C$ structure in which the new symbols are interpreted according to $\underset{\sim}{a}$, then for any $L_2 \cup C$ structure B' satisfying every atomic sentence true in A', there exists a unique homomorphism $H' : A' \to \mathrm{red}_{L_1 \cup C}(A')$. It of course follows trivially that by taking reducts, we have a homomorphism $H : A \to \mathrm{red}_{L_1}(B)$, where $B = \mathrm{red}_{L_2}(B')$; conversely, if such an H exists for an L_2-structure B, there exists a unique expansion B' to $L_2 \cup C$ such that H is the reduct of an $H' : A' \to \mathrm{red}_{L_1 \cup C}(B')$. If all the basic sentences (i.e., atomic and negated atomic sentences) are used, we get an embedding of C into B.

We now list a series of definitions according to the following scheme: we first give a name to a class of formulas, then we indicate certain simple formulas which belong to the class followed by the closure conditions satisfied by the class. [For example, our first definition says that a formula ϕ is positive-primitive if it belongs to the smallest class of formulas containing the atomic formulas and satisfying: if for all $\phi \in \Phi$, ϕ belongs to the class, then $\wedge\Phi$ belongs to the class, and if ϕ belongs to the class, $\exists X\phi$ belongs.] $\phi_1 \to \phi_2$ is defined to be $\neg(\phi_1 \wedge \neg\phi_2)$, i.e., $\neg\wedge\{\phi_1, \neg\phi_2\}$.

(i) positive-primitive includes atomic and closed under $\wedge$ and $\exists$;

(ii) positive-existential includes atomic and closed under $\wedge$, $\vee$, and $\exists$;

(iii) basic includes atomic and negated atomic;

(iv) primitive includes basic and closed under $\wedge$ and $\exists$;

(v) existential includes basic and closed under $\wedge$, $\vee$, and $\exists$.

(vi) universal includes basic and closed under $\wedge$, $\vee$, $\forall$;

(vii) positive includes atomic and closed under $\wedge$, $\vee$, $\exists$ and $\forall$;

(viii) universal-Horn includes formulas of the form $\phi_1 \to \phi_2$, where ϕ_1 is positive-existential and ϕ_2 is atomic, and closed under $\wedge$, $\forall$;

(ix) identity includes atomic and closed under $\wedge$ and $\forall$.

The importance of the above notions derives from the following easy results:

(a) If $F : A \to B$ and $A \models \phi(\underset{\sim}{a})$, then $B \models \phi(F \circ \underset{\sim}{a})$ provided ϕ is positive existential, or ϕ is existential and F is an embedding, or ϕ is positive and F is onto, or ϕ is arbitrary and F is an isomorphism.

(b) if $F : A \xrightarrow{\text{emb}} B$ and ϕ is universal, then $B \models \phi(F \circ \underset{\sim}{a}) \Rightarrow A \models \phi(\underset{\sim}{a})$;

(c) if $A = \prod_{i \in I} A_i$ and ϕ is universal-Horn, then $A \models \phi(\underset{\sim}{a})$ provided $A_i \models \phi(\pi_i \circ \underset{\sim}{a})$ for all $i \in I$.

We will outline the proof of (c). First let ϕ be $\phi_1 \to \phi_2$ with ϕ_1 positive-extential and ϕ_2 atomic. If $\underset{\sim}{a} \in Ad^A_\phi$ and $A \models \phi_1(\underset{\sim}{a})$, then (since positive-existential formulas are preserved by homomorphisms) $A_i \models \phi_1(\pi_i \circ \underset{\sim}{a})$ for each $i \in I$. But for each $i \in I$, $A_3 \models (\phi_1 \to \phi_2)(\pi_i \circ \underset{\sim}{a})$, so that for each $i \in I$, $A_i \models \phi_2(\pi_i \circ \underset{\sim}{a})$. Thus (since it is easily seen that an atomic formula holds in a product if and only it holds in each factor), $A \models \phi_2(\underset{\sim}{a})$, as desired. The general case is handled by an easy induction. Note that a universal-Horn formula is logically equivalent to a universal formula (since $\phi_1 \to \phi_2$ is logically equivalent to $\neg\phi_1 \vee \phi_2$, and $\neg\exists X\phi$ is logically equivalent to $\forall X\neg\phi$); thus, (b) above is applicable. Finally note that $\prod_{i \in \emptyset} A_i$ is the <u>terminal structure</u> in which every identity $\forall X\phi$ holds, since this sentence (which is equivalent to $\forall X(\wedge\emptyset \to \phi)$) holds in each A_i, $i \in \emptyset$.

Various converses of the above preservation results are classic theorems of model theory. For example, if A and B are L-structures such that, for every positive sentence σ, $A \models \sigma \Rightarrow B \models \sigma$, then there exists $F : A \xrightarrow[\text{onto}]{} B$. Of course more interesting results give an <u>a priori</u> bound on the "complexity" of sentences which must be checked (i.e., in which $L_{\eta_1\eta_2}$ one should look) before the conclusion follows.

We will examine briefly the situation regarding universal-Horn sentences. If K is a class of L-structures and σ is an L-sentence, we say that σ <u>is valid in K</u>, or <u>is K-valid</u>, if and only if $A \in K \Rightarrow A \models \sigma$. We will now show that if K is hereditary and productive, and A is an L-structure satisfying every universal-Horn sentence valid in K, then $A \in K$. Given such an A, take an S-set of variables X with a generating assignment $\underset{\sim}{a} : X \to A$. Let Φ_1 be the set of atomic $\phi \in Form_L(X)$ such that $A \models \phi(\underset{\sim}{a})$, and let Φ_2 be the set of atomic $\phi \in Form_L(X)$ such that $A \models \neg\phi(\underset{\sim}{a})$. For each $\phi \in \Phi_2$, there exists $B_\phi \in K$ and an S-map $\underset{\sim}{b} : X \to B_\phi$ such that: $B_\phi \models \neg\phi(\underset{\sim}{b})$ and $B_\phi \models \wedge\Phi_1(\underset{\sim}{b})$; for otherwise, every $B \in K$ would satisfy $\forall X(\wedge\Phi_1 \to \phi)$, while A does not. By the Fundamental Theorem on Defining Homomorphisms, there exists $F_{\phi} : A \to B_{\phi}$ such that $F_{\phi} \circ \underset{\sim}{a} = \underset{\sim}{b}$, and thus in particular, $B_{\phi} \models \neg\phi(F_{\phi} \circ \underset{\sim}{a})$. Let $B = \prod_{\phi\in\Phi_2} B_{\phi} \in K$. Then the induced $F : A \to B$ is an embedding since:

$$A \models \neg\phi(\underset{\sim}{a}) \Rightarrow B_{\phi} \models \neg\phi(F_{\phi} \circ \underset{\sim}{a}) \Rightarrow B \models \neg\phi(\pi_{\phi} \circ F \circ \underset{\sim}{a}) \Rightarrow$$
$$B \models \neg\phi(F \circ \underset{\sim}{a}).$$

Thus A is embeddable in a model, and hence a model, since K is hereditary.

The careful reader will have noted that the universal-Horn sentences arising in the above proof are of a special form, viz., the hypotheses are simply conjections of atomic formulas. This is not surprising in $L_{\omega\omega}$ where every universal-Horn formula is equivalent to a conjection of such simpler formulas. But if K is the class of models of a sentence $\sigma \in L_{\omega_1\omega}$, then to show (using the Model Existence Theorem) that σ is equivalent to a universal-Horn sentence <u>in</u> $L_{\omega_1\omega}$ <u>again</u>, we must allow the more general definition.

Given a finitary vocabulary L, a class K of L-structures is called <u>compact</u> if and only if for every finitary $L' \supset L$ and every set Σ of sentences in $L'_{\omega\omega}$, there exists an L'-structure $A' \models \wedge\Sigma$ with $\mathrm{red}_L(A') \in K$, provided that for each $L^0 \subset L'$ and $\Sigma^0 \subset L^0_{\omega\omega} \cap \Sigma$ with $L \subset L^0$, $L^0 - L$ finite, and Σ^0 finite, there exists an L^0-structure $A^0 \models \wedge\Sigma^0$ with $\mathrm{red}_L(A^0) \in K$. Through a use of new individual constants one can readily demonstrate the following (useful) strengthening: if K is a compact class of L-structures, $L' \supset L$ is finitary, and $\Phi \subset L'_{\omega\omega}$ is a set of formulas with $\mathrm{Var}(\wedge\Phi) \subset X$, then there exists an L'-structure $A' \models \exists X \wedge\Phi$ with $\mathrm{red}_L(A') \in K$, provided that for each $L^0 \subset L'$ and $\Phi^0 \subset L^0_{\omega\omega} \cap \Phi$ with $L \subset L^0$, $L^0 - L$ finite, and Φ^0 finite, there exists an L^0-structure $A^0 \models \exists X^0 \wedge\Phi^0$ with $\mathrm{red}_L(A^0) \in K$, where $X^0 = \mathrm{var}(\wedge\Phi^0)$. [Note that $\exists X^0 \wedge\Phi^0 \in L_{\omega\omega}$.] Our major use for this version will be in cases where $L' = L$. The <u>compactness theorem</u> (for which a proof using "ultraproducts" still works in our general framework) states that if $L^* \supset L$ is finitary, $\Sigma \subset L^*_{\omega\omega}$, and K is the class of L-reducts of models of Σ, then

K is compact. [Note that much of the fuss in the above wording is due to our desire to tolerate "emptiness" in structures. Thus, we must define ultraproducts, not as homomorphic images of products, but rather in such a way that, if $\exists x(x = x)$ is true in "almost all factors", then it is true in the ultraproduct--hence, an ultraproduct is defined to consist of equivalence classes of partial functions defined in "almost all factors". This makes it a colimit of a directed family of products. Much of this unusual delicacy will generally be unnecessary in our later developments due to the fact that our vocabularies will normally have individual constants of each sort.]

Now if K is hereditary, productive, and compact, then every A which satisfies $\Sigma = \{\sigma \in L_{\omega\omega} : \sigma$ is a universal-Horn sentence valid in $K\}$ belongs to K, so that K is simply the class of models of Σ. This is obtained by noting that in the stage of the above argument where we needed a $B \in K$, if none existed then no member of K would be a model of $\exists X(\wedge\Phi_1 \wedge\neg\Phi)$ so that by compactness there would be a finite $\Phi_0 \subset \Phi$, with $X_0 = \mathrm{Var}(\wedge\Phi_0)$ such that no member of K is model of the sentence $\exists X_0(\wedge\Phi_0 \wedge\neg\Phi)$; thus every member of K would satisfy $\forall X_0(\wedge\Phi_0 \to \phi) \in L_{\omega\omega}$, while A does not. [There is a subtlety here, which we intend to leave to the reader in further "diagram arguments". Namely, in appealing to compactness, we have implicitly assumed that $\wedge\Phi_1 \wedge\neg\phi$ (i.e., $\wedge\{\wedge\Phi_1, \neg\phi\}$) is logically equivalent to $\wedge\{\phi_1 \wedge\neg\phi : \phi_1 \in \Phi\}$. This is justifiable, since Φ_1 is non-empty (in particular, $T \in \Phi_1$).] An even better-known argument shows that the class of substructures of members of a compact class K of L-structures is axiomatized by $\{\sigma \in L_{\omega\omega} : \sigma$ is universal and K-valid$\}$.

We will finish our discussion of many sorted logic by indicating how universal constructions fit into our framework. In most cases the following procedure suffices. We are given $L_1 \subset L_2$, an L_1-structure A_1, and a hereditary and productive class K of L_2-structures. We want to show the existence of a pair $\langle F, B\rangle$ with $B \in K$ and $F : A \to \mathrm{red}_{L_1}(B)$ such that for every other such pair $\langle F', B'\rangle$ there exists a unique $H : B \to B'$ with $F' = H_1 \circ F$, where H_1 is the L_1-reduct of H. The essential uniqueness of such a "universal" pair is clear by category theory. For existence, we let C be a set of individual constants with $C \cap L_2 = \emptyset$ and $\underset{\sim}{a} : C \to A$ a generating S-map. Let A^* be the unique expansion of A to an $L_1 \cup C$ structure with interpretation according to $\underset{\sim}{a}$. Finally, let K^* be the class of $L_2 \cup C$ structures which satisfy $\{\sigma \in \mathrm{Form}_{L \cup C}(\emptyset) : \sigma$ is atomic and $A^* \models \sigma\}$ and whose reducts belong to K. Trivially, by our previous discussion concerning reducts together with the observation that atomic sentences are universal-Horn, K^* is the intersection of two hereditary and productive classes and hence is itself such a class. An initial object $B^* \in K^*$ immediately provides us with a homomorphism $F^* : A^* \to \mathrm{red}_{L_1 \cup C}(B^*)$, and taking $B \in K$ to be $\mathrm{red}_{L_2}(B^*)$ and F to be the L_1-reduct of F^*, we get $F : A \to \mathrm{red}_{L_1}(B)$. Universality is now immediate.

As a special case of the above, we have an existence and uniqueness result for free models in hereditary and productive classes, where the free generators form an S-set (S-sets being L-structures in which only "=" occurs in L). Notice that, in general, we have many free cyclic models (freely generated by a singleton), viz., one for each sort in S.

Most other universal constructions fit into obvious variations on the above theme. For example, given two appropriate modules A and B, their tensor product can be obtained by first introducing constants $a \otimes b$ for each pair $\langle a, b\rangle \in A \times B$ together with the usual bilinearity axioms, which are clearly universal-Horn. Alternatively one introduces the class of many-sorted structures with three sorts $\{s_1, s_2, s_3\}$, the module vocabulary in each sort, and a new binary function symbol taking admissible arguments $\begin{pmatrix} l & r \\ a_1 & a_2 \end{pmatrix}$ with sorts $\begin{pmatrix} l & r \\ s_1 & s_2 \end{pmatrix}$ and giving outputs of sort s_3. The axioms on f, for bilinearity, are then universal-Horn, and we take K to be the class of $L \cup C_1 \cup C_2$-structures in which the reducts to each sort are modules, and which satisfy the atomic sentences true of A and B (through some $\underset{\sim}{a} : C_1 \to A$ and $\underset{\sim}{b} : C_2 \to B$).

We intend to provide a more precise notational framework for handling such many-sorted expansions in another place [5]. We only remark that our conventions are expressly designed to handle iterations of concepts. For example, a homomorphism between "situations" $F_1 : A_1 \to B_1$ and $F_2 : A_2 \to B_2$ makes sense (as above), conforms to the usual notion, and gives rise to a new structure in which the set of sorts is four times what it is in A_1 (and is organized into a natural hierarchy via "super-sorts").

In this discussion (and some others below), we assume that the reader can check the eminently trivial facts concerning "similar" vocabularies. Two vocabularies L_1 and L_2 are called <u>similar</u> if there exists an isomorphism $I : L_1 \to L_2$ respecting their basic syntactic structures. Given an S_1-set of variables X, an S_2-set of variables Y, and a map $\chi : X \xrightarrow[\text{onto}]{1-1} Y$ such that $s_Y \circ \chi = I \circ s_X$, we are led to a natural induced isomorphism from each $L_{\eta\eta}(X)$ to $L_{\eta\eta}(Y)$ (viz., via substitution). I also induces a natural "large" isomorphism from the full category of L_2-structures to that of L_1-structures "co-respecting" all semantic notions. Thus, in the second approach to tensor products, the three "modules" which are reducts of the whole situation actually have three different vocabularies <u>similar</u> to that for modules. In [5], we combine similarity and reduction into one notion called <u>interpretation</u>, which amounts to an embedding of one vocabulary in another.

Finally, we note that the set of variables V used in forming $\mathrm{Form}_L(V)$ is frequently larger than necessary in the following sense. If L, for example, is finitary, then every $\phi \in L_{\omega\omega}(V)$ actually belongs to some $L_{\omega\omega}(V')$, where $V' \subset V$ is finite. Thus, by employing a part of the above discussion concerning similarity (viz., that which is concerned with "change of variables"), it is easy to see that the "expressive power" of $L_{\omega\omega}(V)$ is essentially no greater than that of $L_{\omega\omega}(X)$, where X is countably infinite. When we desire economy of expression, we will write $L_{\omega\omega}$, implicitly assuming that an <u>economical</u> choice of variables is made.

2. <u>Basic Features of Abelian Structures</u>.

Throughout the remainder of this paper, we will normally assume that L is a fixed finitary vocabulary containing among other symbols a distinguished binary function symbol $+$, a distinguished unary function symbol $-$, and distinguished individual constants

0_s (one for each sort $s \in S$, the (necessarily non-empty) set of sorts of L). We assume furthermore that $Ad_+ = \left\{\begin{pmatrix}1 & r\\ s & s\end{pmatrix} : s \in S\right\}$, $Val_+\left(\begin{pmatrix}1 & r\\ s & s\end{pmatrix}\right) = s$, $Ad_- = \{s : s \in S\}$, $Val_-(s) = s$, $Val_{0_s} = s$. In other words, + and - work independently in each sort, and each 0_s has the obvious sort. We will call such a vocabulary abelian.

If L is abelian, then an L-structure will be called abelian if and only if it satisfies all sentences in $L_{\omega\omega}$ of the following forms:

(i) $\forall x[x + 0_s = x \wedge 0_s + x = x]$;

(ii) $\forall x[-x + x = 0_s]$;

(iii) $\forall X, Y(f(\underset{\sim}{\tau})) = f(\underset{\sim}{x}) + f(\underset{\sim}{y})$, where $f \in L$ is a function symbol, $\underset{\sim}{x} \in X^{\alpha(f)}$, $\underset{\sim}{y} \in Y^{\alpha(f)}$, and for each $i \in \alpha(f)$, $\underset{\sim}{\tau}_i$ is $\underset{\sim}{x}_i + \underset{\sim}{y}_i$;

(iv) $\forall X, Y[R(\underset{\sim}{x}) \wedge R(\underset{\sim}{y}) \rightarrow R(\underset{\sim}{\tau})] \wedge \forall X[R(\underset{\sim}{x}) \rightarrow R(\underset{\sim}{\tau}')] \wedge R(\underset{\sim}{0})$, where $R \in L$ is a relation symbol, $\underset{\sim}{x} \in X^{\alpha(R)}$, $\underset{\sim}{y} \in Y^{\alpha(R)}$, and for each $i \in \alpha(R)$, $\underset{\sim}{\tau}_i$ is $\underset{\sim}{x}_i + \underset{\sim}{y}_i$, $\underset{\sim}{\tau}'$ is $-\underset{\sim}{x}_i$, and $\underset{\sim}{0}_i \in \{0_s : s \in S\}$.

Note that we do not need to mention any side conditions, since we are only taking those expressions of the above forms which are sentences (so that sorts and numbers of arguments are automatically correct). Also note that the instances of (iii) where f is + or - imply and are implied by the associative and commutative laws for + (using (i) and (ii); also, the instances at (iv) involving the symbol T or the equality symbol are simply logically valid. If f is a function symbol with no arguments, then (iii) implies $f + f = f$, so that by (i), $f + f = f + 0_s$ (where $s = Val_f$), and by the usual arguments $f = 0_s$. Thus, if there are individual constants besides the 0's, they are redundant. Similarly, any propositional constant must always be true. [We will thus ignore any propositional constants in abelian structures.]

We will find it convenient to adopt some conventions concerning the substitution of terms into other terms and atomic formulas according to an assignment. To this end, let V be an S-set of variables, $\tau \in Term_L(V)$, $\phi \in Form^0_L(V)$. If X is another S-set of variables and $\underset{\sim}{\tau} : V \rightarrow Term_L(X)$ is an S-map (i.e., a map such that $Sort(\underset{\sim}{\tau}_v) = s_V(v)$), then the results of substituting into τ and ϕ according to $\underset{\sim}{\tau}$ (defined in the usual way, by induction) will be written $\tau(\underset{\sim}{\tau})$ and $\phi(\underset{\sim}{\tau})$, respectively. If $\underset{\sim}{\tau}' : V \to \mathrm{Term}_L(X)$ is another S-map, then $\underset{\sim}{\tau} + \underset{\sim}{\tau}'$ and $-\underset{\sim}{\tau}$ will be defined by:

$$(\underset{\sim}{\tau} + \underset{\sim}{\tau})_v = \underset{\sim}{\tau}_v + \underset{\sim}{\tau}'_v;$$

$$(-\underset{\sim}{\tau})_v = -(\underset{\sim}{\tau}_v).$$

We will use this notation most commonly in situations where $\underset{\sim}{\tau}$ assigns variables to variables, i.e., $\underset{\sim}{\tau} = \underset{\sim}{x} : V \to X$. In such contexts we will let V be fixed and chosen to include the variables occuring in the terms and formulas on which substitution is being performed; X, Y, etc., will be Range($\underset{\sim}{x}$), Range($\underset{\sim}{y}$), etc.; and $\underset{\sim}{0}$ will be the unique S-map from V to $\{0_s : s \in S\}$. We will occasionally use a similar "vector" notation for function or relation symbols P, where V should be replaced by $\alpha(P)$ in the above discussion.

As a useful illustration, it is now a simple matter to check that all sentences of the following forms are consequences of our axioms:

(*) $\forall X, Y[\tau(\underset{\sim}{x} + \underset{\sim}{y}) = \tau(\underset{\sim}{x}) + \tau(\underset{\sim}{y})]$, τ an L-term;

(**) $\forall X, Y[\phi(\underset{\sim}{x}) \wedge \phi(\underset{\sim}{y}) \to \phi(\underset{\sim}{x} + \underset{\sim}{y})] \wedge \forall X[\phi(\underset{\sim}{x}) \to \phi(-\underset{\sim}{x})] \wedge \phi(\underset{\sim}{0})$,

ϕ an atomic L-formula. Also, all instances of (iii) and (iv) follow from (i), (ii), and (**). This is obvious for (iv), while (iii) follows from:

$$\forall X, Y, u, v \quad (f(\underset{\sim}{x}) = u \wedge f(\underset{\sim}{y}) = v \to f(\underset{\sim}{x} + \underset{\sim}{y}) = u + v).$$

We will refer to (v) and (vi) or (*) and (**) as the <u>additivity assumptions</u>. [Note that we are <u>not</u> treating the case of bilinear maps, and thus tensor products do not fall under our treatment.] The class of all abelian L-structures is clearly compact, hereditary, and productive (since it is axiomatized using universal-Horn sentences in $L_{\omega\omega}$). It is closed under homomorphic images if L is purely functional (since the axioms (i)-(iii) are identities).

A class K will be said to have the <u>Homomorphism Extension Property</u> (<u>HEP</u>), if and only if, for any three models A, B, $C \in K$ and $F : A \xrightarrow{\text{emb}} B$, $G : A \to C$, there exists a $D \in K$ and $F' : C \xrightarrow{\text{emb}} D$, $G' : B \to D$ such that

$F' \circ G = G' \circ F$. In other words, every diagram:

$$\begin{array}{c} B \\ \text{emb} \uparrow \\ A \longrightarrow C \end{array}$$

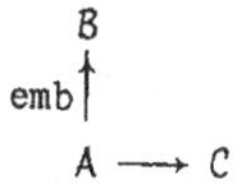

in K can be completed to a commutative diagram:

$$\begin{array}{ccc} B & \longrightarrow & D \\ \text{emb} \uparrow & & \uparrow \text{emb} \\ A & \longrightarrow & C \end{array}$$

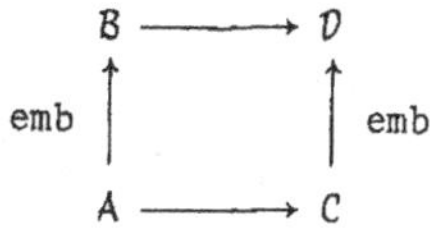

in K. If K is hereditary and productive, and hence closed under limits, then it <u>has</u> arbitrary colimits, and the pushout will always work for D. This is seen as follows; if P is the pushout, and D is as above, then the universal properties of B yield the commutative diagram:

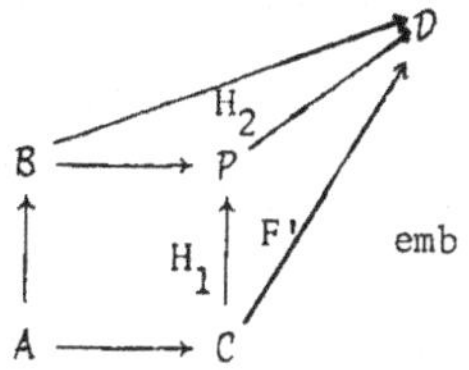

But quite generally, if $H_2 \circ H_1$ is an embedding of L-structures, then H_1 is, since:

$$B \models \phi(H_1 \circ \underline{c}) \Rightarrow D \models \phi(H_2 \circ H_1 \circ \underline{c}) \Rightarrow C \models \phi(\underline{c}).$$

A class K of abelian structures will be called <u>abelian</u> if and only if K is compact, hereditary, and productive and has HEP. With a slight abuse of language, we will say that a class K of abelian structures is an <u>additive</u> variety if and only if K is axiomatized by the additivity assumptions plus a set of identities.

For an example of a compact, hereditary, and productive class of abelian structures not satisfying HEP, we may take the class K of abelian groups which are models of $\forall x(4 \cdot x = 0 \to 2 \cdot x = 0)$. The following diagram in K cannot be completed:

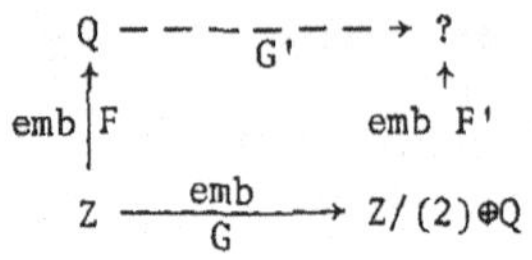

where Z is the free cyclic group, F is the natural embedding into the rationals Q, and G(1) = <1, 1>. [In ? we would have: $2 \cdot (G'(1/2) - F'(\langle 0, 1/2\rangle)) = G' \cdot F(1) - F'(\langle 0, 1\rangle) = F'(G(1) - \langle 0, 1\rangle) = F'(\langle 1, 0\rangle) \neq 0$, and $2 \cdot F'(1, 0) = 0$.] Another example is provided by the class of structures of the form:

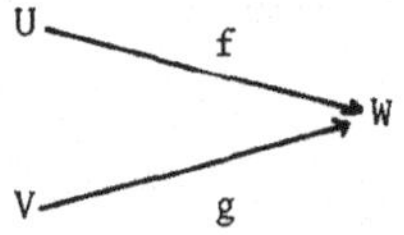

where U, V, and W are vector spaces over Z/(2), f and g are linear transformations, and the following axiom is satisfied:

$$\forall x, y[f(x) = g(y) \rightarrow x = 0].$$

[Indeed, a non-zero member of W may be a f-image in one extension and (incompatibly) a g-image in another.]

Theorem 2.1. If K is an additive variety, then K is abelian. In particular, the class of all abelian L-structures is abelian.

Proof. If K is an additive variety, then K is clearly hereditary and productive. Since L has an individual constant of each sort, every identity is logically equivalent to a conjunction of identities in $L_{\omega\omega}$. [$\forall X \wedge \Phi \rightarrow \wedge\{\forall X \phi : \phi \in \Phi\}$ is not otherwise logically valid by our semantic conventions.] Thus, K is compact. In verifying HEP, we will show that the pushout works.

First note that, under the assumption, every substructure of a model is the kernel of a homomorphism. Namely, if $A \subset B \in K$, then we define an equivalence relation $\sim$ by: $b_1 \sim b_2$ if and only if $b_2 - b_1 \in A$. This is a congruence, since $\underset{\sim}{b}_1 \sim \underset{\sim}{b}_2 \Rightarrow \exists \underset{\sim}{a} \ni \underset{\sim}{b}_2 = \underset{\sim}{b}_1 + \underset{\sim}{a} \Rightarrow f(\underset{\sim}{b}_2) = f(\underset{\sim}{b}_1 + \underset{\sim}{a}) = f(\underset{\sim}{b}) + f(\underset{\sim}{a})$, and since $f(\underset{\sim}{a}) \in A$, $f(\underset{\sim}{b}_1) \sim f(\underset{\sim}{b}_2)$. If q is the quotient S-map onto the S-set C of equivalence classes, we can define a structure C on C according to: $f^{\mathcal{C}}(q \circ b) = q(f^{\mathcal{B}}(b))$, by the congruence property, and $\mathcal{C} \models R(q \circ b)$ if and only if $\exists \underset{\sim}{a} \in Ad_R^A \ni \mathcal{B} \models R(b + a)$. Since $\underset{\sim}{0} \in Ad_R^A$ q is clearly a homomorphism, and the additivity assumptions (besides identities, which are obvious) follow from:

$$\mathcal{C} \models R(q \circ \underset{\sim}{b}_1) \wedge R(q \circ \underset{\sim}{b}_2)$$

$$\Rightarrow \exists \underset{\sim}{a}_1, \underset{\sim}{a}_2 \in Ad_R^A \ni \mathcal{B} \models R(\underset{\sim}{b}_1 + \underset{\sim}{a}_1) \wedge R(\underset{\sim}{b}_2 + \underset{\sim}{a}_2)$$

$$\Rightarrow \exists \underset{\sim}{a}_1, \underset{\sim}{a}_2 \in Ad_R^A \ni \mathcal{B} \models R(\underset{\sim}{b}_1 + \underset{\sim}{b}_2 + \underset{\sim}{a}_1 + \underset{\sim}{a}_2) \wedge R(-\underset{\sim}{b}_1 + (-\underset{\sim}{a}_1))$$

$$\Rightarrow \mathcal{C} \models R(q \circ (\underset{\sim}{b}_1 + \underset{\sim}{b}_2)) \wedge R(q \circ (-\underset{\sim}{b}_1))$$

$$\Leftrightarrow \mathcal{C} \models R(q \circ \underset{\sim}{b}_1 + q \circ \underset{\sim}{b}_2) \wedge R(-q \circ \underset{\sim}{b}_1).$$

Now given $F : A \xrightarrow{emb} \mathcal{B}$ and $G : A \to \mathcal{C}$, we can construct the pushout by forming the quotient of $\mathcal{B} \oplus \mathcal{C} = \mathcal{B} \times \mathcal{C}$ by the substructure $\{\langle F(a), -G(a)\rangle : a \in A\}$. [That this is a substructure is immediate; we will see below that $\mathcal{B} \times \mathcal{C}$ happens to be a coproduct.] If $\iota_1 : \mathcal{B} \to \mathcal{B} \oplus \mathcal{C}$ and $\iota_2 : \mathcal{C} \to \mathcal{B} \oplus \mathcal{C}$ are the natural inclusion homomorphisms given by $\iota_1(b) = \langle b, 0\rangle$ and $\iota_2(c) = \langle 0, c\rangle$ and q is the quotient map onto the quotient $\mathcal{D}$, then

$$\mathcal{D} \models R(q \circ \iota_2 \circ \underset{\sim}{c}) \Rightarrow \exists \underset{\sim}{a} \in Ad_R^A \ni \mathcal{B} \oplus \mathcal{C} \models R(\iota_2 \circ \underset{\sim}{c} + \iota_1 \circ F \circ \underset{\sim}{a} - \iota_2 \circ G \circ \underset{\sim}{a})$$

$$\Rightarrow \exists \underset{\sim}{a} \in Ad_R^A \ni \mathcal{B} \models R(\underset{\sim}{a}) \text{ and } \mathcal{C} \models R(\underset{\sim}{c} - G \circ \underset{\sim}{a}), \text{ by projection,}$$

$$\Rightarrow \exists \underset{\sim}{a} \in Ad_R^A \ni A \models R(\underset{\sim}{a}) \wedge \mathcal{C} \models R(\underset{\sim}{c} - G \circ \underset{\sim}{a}), \text{ since } F \text{ is an embedding}$$

$$\Rightarrow \exists \underset{\sim}{a} \in A \ni \mathcal{C} \models R(G \circ \underset{\sim}{a}) \wedge R(\underset{\sim}{c} - G \circ \underset{\sim}{a})$$

$$\Rightarrow \mathcal{C} \models R(\underset{\sim}{c}), \text{ by additivity.}$$

Thus, $F' = q \circ \iota_2$ is an embedding, and obviously $F' \circ G = G' \circ F$.

We will now give examples of classes fulfilling the above conditions. Since the class of modules over a fixed ring is an obvious well-known example we will describe a slightly more sophisticated case: the class of chain complexes of modules over a fixed ring. L has a sort s_n for each integer, the function symbols $+$, $-$, 0_{s_n} as above, a unary function symbol f_r for each element r of the ring R, and a unary function symbol ∂ for the boundary map. $Ad_{f_r} = S$,

$Val_{f_r}(s_n) = s_n$, $Ad_\partial = S$, $Val_\partial(s_{n+1}) = s_n$. K is then axiomatized by all sentences of the forms:

(i)-(iii) above (i.e., the additivity assumptions);

(iv) $\forall x[f_1(x) = x]$;

(v) $\forall x[f_{r_1+r_2}(x) = f_{r_1}(x) + f_{r_2}(x)]$;

(vi) $\forall x[f_{r_1 \cdot r_2}(x) = f_{r_1}(f_{r_2}(x))]$;

(vii) $\forall x[\partial(f(x)) = f(\partial(x))]$;

(viii) $\forall x[\partial(\partial x)) = 0]$.

K is thus clearly an additive variety, and the homomorphisms between models are exactly the "chain transformations" which are normally of primary interest.

We assume that the reader knows how to axiomatize the class of (small) categories using two sorts: one for objects and the other for morphisms. A nicer axiomatization results if we choose to fix the set of objects $\mathcal{O}$ and vary only the morphisms; of course, this radically changes the class of models and the notion of homomorphism: instead of arbitrary functors, we take only those which map the objects according to the prescription of sorts. In this approach the sorts s correspond to pairs of objects $<o_1, o_2>$, we have an individual constant e_o for each $o \in \mathcal{O}$ with $Val_{e_o} = <o, o>$, and we have a binary function symbol $\circ$ with $Ad_\circ = \left\{ \left(\overset{\ell}{<o_2, o_3>} \quad \overset{r}{<o_1, o_2>} \right) : o_1, o_2, o_3 \in \mathcal{O} \right\}$ and $Val_\circ \left(\left(\overset{\ell}{<o_2, o_3>} \quad \overset{r}{<o_1, o_2>} \right) \right) = <o_1, o_3>$. The axioms then read like those for a monoid:

$$\forall x, y, z[x \circ (y \circ z) = (x \circ y) \circ z];$$

$$\forall x[x \circ e_o = x \wedge e_o \circ x = x].$$

There is no need to mention "when defined", since the sorts are such that L-terms correspond exactly to "defined" expressions.

However the subject is axiomatized, we assume that the reader knows what a small abelian category C is [6], and we show that the class G^C of additive group-valued functors is a K satisfying our hypotheses. We now take for the sorts of L the objects of C, we take $+$, $-$, 0_s as above, and a unary function symbol f for each morphism $f \in C$ with $Ad_f = \{\text{domain of } f\}$ and $Val_f(\text{dom}(f)) = \{\text{codomain of } f\}$. The axioms are all sentences of the forms:

(i)-(iii), the additivity assumptions;

(iv) $\forall x[f(g(x)) = f \circ g(x)]$;

(v) $\forall x[f(x) = x]$, f an identity morphism $\in C$;

(vi) $\forall x[(f + g)(x) = f(x) + g(x)]$;

(vii) $\forall x[f(x) = 0_s]$, f a zero morphism $\in C$.

It is now obvious that G^C is an additive variety.

Of course, there is no reason why we could not equally well take any of our abelian K's instead of the category of abelian groups, and consider the class K^C of all additive K-valued functors; the axioms besides those of K in each "supersort", corresponding to an object in K, are all universal-Horn; for example, "$\forall X(R(\underset{\sim}{x}) \Rightarrow R(f \circ \underset{\sim}{x}))$" is a typical part of the assertion that a morphism f corresponds to a K-homomorphism. HEP is seen by forming the pushout in each "supersort" (according to K) and taking the natural induced homomorphisms for the interpretations of the f's $\in C$. [If this sounds a bit vague, it is because we have no need in this paper for such generality, but only mention the idea as evidence for the stability of our concepts under natural constructions and iterations. G^C plays a special role in our discussion of the Mitchell Embedding Theorem in the sequel.]

Theorem 2.2. Let K be an arbitrary compact class of L-structures. If for every positive primitive formula $\phi_1 \in L_{\omega\omega}$, there exists an atomic formula ϕ_2 such that for every $A \in K$ there exists an $A' \in K$ such that $A \subset A' \models X(\phi_1 \leftrightarrow \phi_2)$, then K has HEP.

Proof. If A, B, C, F, and G are given as above, we take sets of variables X, Y, Z such that $Y \cap Z = X$ with generating S-maps $\underset{\sim}{a} : X \to A$, $\underset{\sim}{b} : Y \to B$, $c : Z \to C$, such that $F \circ \underset{\sim}{a} = \underset{\sim}{b} \circ \iota_X$ and $G \circ \underset{\sim}{a} = \underset{\sim}{c} \circ \iota_X$. Now if $\Phi_1 = \{\phi : \phi$ is atomic and $B \models \phi(\underset{\sim}{b})\}$, and $\Phi_2 = \{\phi : \phi$ is basic and $C \models \phi(\underset{\sim}{c})\}$, then any member D of K with $d : Y \cup Z \to D$ satisfying

$\mathcal{D} \models (\wedge\Phi_1 \wedge \wedge\Phi_2)(\underset{\sim}{d})$ will be as desired, by the Fundamental Theorem on Defining Homomorphisms. Since Φ_1 and Φ_2 are non-empty, it suffices (by compactness) to consider finite subsets Φ_1^* and Φ_2^* of Φ_1 and Φ_2. Now $\exists Y, Z(\wedge\Phi_1^* \wedge \wedge\Phi_2^*)$ is logically equivalent to $\exists X(\exists Y' \wedge \Phi_1^* \wedge \exists Z' \wedge\Phi_2^*)$, where $Y' = Y - X$ and $Z' = Z - X$. Clearly, since $\Phi_1^* \subset \Phi_1$, $\mathcal{B} \models \wedge\Phi_1^*(\underset{\sim}{b})$, and since $F \circ \underset{\sim}{a} = \underset{\sim}{b} \circ \iota_X$, $\mathcal{B} \models \exists Y' \wedge \Phi_1^*(F \circ \underset{\sim}{a})$. But $\phi_1 = \exists Y' \wedge \Phi_1^*$ is logically equivalent to a positive-primitive formula in $L_{\omega\omega}$, and is thus equivalent to an atomic formula ϕ_2 in extensions $\mathcal{B}'$ of $\mathcal{B}$ and $\mathcal{C}'$ of $\mathcal{C}$. Now we see that:

$$\mathcal{B} \models \phi_1(F \circ \underset{\sim}{a}) \Rightarrow \mathcal{B}' \models \phi_1(F \circ \underset{\sim}{a}) \Rightarrow \mathcal{B}' \models \phi_2(F \circ \underset{\sim}{a}) \Rightarrow \mathcal{A} \models \phi_2(\underset{\sim}{a}),$$

since F is an embedding and $\mathcal{B}'$ is an extension of $\mathcal{B}$. Finally, $\mathcal{C} \models (\phi_2 \wedge \exists Z' \wedge\Phi_2^*)(G \circ \underset{\sim}{a})$, since G is a homomorphism and $G \circ \underset{\sim}{a} = \underset{\sim}{c} \circ \iota_X$, so that $\mathcal{C}' \models (\exists Y' \wedge\Phi_1^* \wedge \exists Z' \wedge\Phi_2^*)(G \circ \underset{\sim}{a})$ and $\mathcal{C}' \models \exists X(\exists Y' \wedge\Phi_1^* \wedge \exists Z' \wedge\Phi_2^*)$, as desired.

We will now apply Theorem 2.2 to get a wealth of classes satisfying HEP. Let K be an arbitrary compact class of L-structures not necessarily satisfying HEP. Then if we are willing to restrict our notion of embedding, we can always obtain it cheaply. We call an embedding $F : \mathcal{A} \to \mathcal{B}$ <u>pure</u> if for every positive primitive $\phi \in L_{\omega\omega}$,

$$\mathcal{A} \models \phi(\underset{\sim}{a}) \Leftrightarrow \mathcal{B} \models \phi(F \circ \underset{\sim}{a}).$$

We can discuss the category of models in K with arbitrary homomorphisms and <u>pure embeddings</u> by taking $L^* \supset L$ to have the same sorts as L and a new relation symbol R_ϕ for each positive primitive $\phi \in L_{\omega\omega}$ with $\alpha(R_\phi) = \mathrm{Var}(\phi)$ and $\mathrm{Ad}_{R_\phi} = \{\underset{\sim}{s}\}$, where $\underset{\sim}{s}(x) = \mathrm{Sort}(x)$. We let K^* be the class of L^*-structures whose L-reducts belong to K (i.e., satisfying the axioms for K), and which satisfy:

$$(*) \qquad \forall X(R_\phi(\underset{\sim}{x}) \leftrightarrow \phi),$$

where $\phi \in L_{\omega\omega}$ is positive-primitive and $\underset{\sim}{x} = \iota_{\mathrm{Var}(\phi)}$. Every member of K has a unique expansion to a member of K^*, every homomorphism between members of

K is the L-reduct of a homorphism between their expansions in K^* (since homomorphisms preserve positive primitive formulas), while a homomorphism $F : A^* \to B^*$ is an embedding if and only if its reduct is a pure embedding if and only if F itself is a pure embedding. K^* is a compact class, since it is axiomatizable relative to K. Moreover, if K is axiomatizable in $L_{\omega\omega}$, then K^* is axiomatizable in $L^*_{\omega\omega}$.

If K is productive, then in checking the axioms (*) for $A^* = \prod_{i\in I} A^*_i$, $A^*_i \in K^*$, we see:

$$\begin{aligned} & A^* \models R_\phi(\underset{\sim}{a}) \\ \iff & A^* \models R_\phi(\pi_i \circ \underset{\sim}{a}), \text{ for all } i \in I, \\ \iff & A^*_i \models \phi(\pi_i \circ \underset{\sim}{a}), \text{ for all } i \in I \text{ (by (*))}, \\ \iff & A^* \models \phi(\underset{\sim}{a}). \end{aligned}$$

The last equivalence holds from right to left since the homomorphisms π_i preserve positive primitive formulas. It holds from left to right, since if $\phi = Y(\wedge\Phi)$, $Var(\phi) = X$, $\underset{\sim}{a} : X \to A$, and for each $i \in I$, $A^*_i \models \phi(\pi_i \circ \underset{\sim}{a})$, then for each $i \in I$ there exists $\underset{\sim}{a}^i : X \cup Y \to A_i$ with $\underset{\sim}{a}^i \circ \iota_X = \pi_i \circ \underset{\sim}{a}$ and $A^*_i \models \wedge\Phi(\underset{\sim}{a}^i)$, and letting $\underset{\sim}{a}' : X \cup Y \to A$ be the induced map with $\pi_i \circ \underset{\sim}{a}' = \underset{\sim}{a}^i$, we have $A^*_i \models \wedge\Phi(\pi_i \circ \underset{\sim}{a}')$, hence $A^* \models \wedge\Phi(\underset{\sim}{a}')$; finally, $A^* \models \exists Y(\wedge\Phi)(\underset{\sim}{a}' \circ \iota_X)$ and thus $A^* \models \phi(\underset{\sim}{a})$, since $\pi_i \circ \underset{\sim}{a}' \circ \iota_X = \underset{\sim}{a}^i \circ \iota_X = \pi_i \circ \underset{\sim}{a}$ and (by the uniqueness of induced maps into products) $\underset{\sim}{a}' \circ \iota_X = \underset{\sim}{a}$. Thus, if K is productive, so is K^*.

K^* is not hereditary, but if we let K^P be the class of substructures of members of K^*, then K^P is still compact--in fact, axiomatized by the set of universal $\sigma \in L^*_{\omega\omega}$ which are valid in K^*. If K (and hence K^*) is productive, then K^P is also productive (since, as is readily checked, a homomorphism into a product induced by embeddings is an embedding). K^P is also clearly hereditary. K^P has HEP by Theorem 2.2, since every member of K^P has an extension in K^*, where, by the axioms (*), every positive primitive $\phi_1 \in L^*_{\omega\omega}$ is equivalent to one in $L_{\omega\omega}$ and thus to an atomic formula $\phi_2 \in L^*_{\omega\omega}$.

The price we have paid in moving to K^* is that of restricting our notion of embedding. The further price paid for the hereditary nature of K^P is that of having to tolerate "dishonest" models (in which an R_ϕ may not imply ϕ). However, every dishonest model is embeddable in an honest one (in K^*). Let us define a member A of a class K to be <u>absolutely pure</u> if for all $F : A \xrightarrow{\text{emb}} B \in K$, F is a pure embedding. It is easy to check that, if K is hereditary, then the absolutely pure members of K^P are exactly the members of K^*, which is an axiomatizable class.

Every claim so far (back to Theorem 2.2) is generally-valid model-theoretic lore. Now suppose that K consists of abelian structures. Note that the set of formulas $\phi \in L_{\omega\omega}$ satisfying the additivity assumptions includes (by assumption) the atomic formulas and is closed under $\wedge$ and $\exists$, so that by (*) the additivity assumptions hold for all symbols in K^* and thus in K^P, since these assumptions are universal in nature. Thus K^P is also a class of abelian structures. Finally, our remarks so far are sufficient to demonstrate the following.

<u>Theorem 2.3</u>. If K is a compact, hereditary, and productive class of abelian structures, then K^P is abelian (even if K does not have HEP).

We note every abelian K is closed under limits, has free models and colimits, and that the initial object is the terminal object, viz., the model of all sentences of the form $\forall x(x = 0_s)$, called the <u>zero-model</u> (or trivial model). Given A, $B \subset C \in K$, the smallest submodel $\mathcal{D}$ of C containing A and B is given by:

$$D = A + B = \{a + b : a \in A,\ b \in B\}.$$

The closure properties should be clear by additivity. We will write $\mathcal{D} = A + B$. Note that even when $A \cap B$ is trivial, the canonical maps given by: $F(a + b) = a$ and $G(a + b) = b$ need not be homomorphisms (unless L is purely functional). If one is, say F, then so is the other: $\mathcal{D} \models \phi(\underset{\sim}{a} + \underset{\sim}{b}) \Rightarrow \mathcal{D} \models \phi(\underset{\sim}{a})$ (since F is a homomorphism and $A \subset \mathcal{D}) \Rightarrow \mathcal{D} \models \phi(\underset{\sim}{b})$ (by additivity--actually "subtractivity") $\Rightarrow B \models \phi(\underset{\sim}{b})$ (since $B \subset \mathcal{D}$). In fact we have just deduced a criterion for F (and G) to be a homomorphism, viz.,

$$\mathcal{D} \models \phi(\underset{\sim}{a} + \underset{\sim}{b}) \Rightarrow \mathcal{D} \models \phi(\underset{\sim}{a}), \quad \phi \text{ atomic}.$$

[This criterion even includes the well-definedness of F (i.e., the triviality of $A \cap B$) by the Theorem on Defining Homomorphisms, or more simply, by noting the special case:

$$\mathcal{D} \models a + 0 = 0 + b \Rightarrow \mathcal{D} \models a = 0.]$$

When this is the case, we write $\mathcal{D} = A \oplus B$ and call F and G the <u>natural homomorphisms</u> $\pi_1 : \mathcal{D} \to A$ and $\pi_2 : \mathcal{D} \to B$. π_1 and π_2 are clearly left-inverses for the respective inclusion maps.

If F and G are homomorphisms from $A \in K$ to $B \in K$, then the sum $F + G : A \to B$ given by $(F + G)(a) = F(a) + G(a)$ is again clearly a homomorphism, and the usual properties pertain (e.g., $\mathrm{End}(A)$ is a ring). Also, if $F : A \to B$, then $\mathrm{Ker}(F) = \bigcup_{s \in S} \{a \in A : F(a) = 0_s\}$ is easily seen to be a substructure of A (viz., the inverse image of the trivial submodel of B). We can now reconstrue the above discussion of $\oplus$, by noting that $\mathrm{Ker}(\pi_1) = B$ and $\mathrm{Ker}(\pi_2) = A$. Also, if π_1 and π_2 are considered as idempotent endomorphisms of $\mathcal{D}$, then $\pi_1 + \pi_2 = \iota_A$. Conversely, if $F \in \mathrm{End}(A)$ is idempotent, then so is $\iota_A - F$; also, $\mathrm{Ker}(F) = \mathrm{Range}(\iota_A - F)$, $\mathrm{Ker}(\iota_A - F) = \mathrm{Range}(F)$, and $A = \mathrm{Range}(F) \oplus \mathrm{Ker}(F)$.

Given A, $B \in K$, we have natural embeddings $\iota_1 : A \to A \times B$ and $\iota_2 : B \to A \times B$ given by: $\iota_1(a) = \langle a, 0_{s_A(a)} \rangle$ and $\iota_2(b) = \langle 0_{s_B(b)}, b \rangle$. Making the usual identifications, $A \times B$ becomes $A \oplus B$, and the fact that $A \oplus B$ is a coproduct is trivial. More generally, a coproduct of A_i, $i \in I$, is given by $A = \bigoplus_{i \in I} A_i \subset \prod_{i \in I} A_i$ with:

$$A = \{a \in \prod_{i \in I} A_i : \{i \in I : \pi_i(a) \text{ is non-zero}\} \text{ is finite}\}.$$

In particular, every abelian class is closed under the formation of coproducts in the class of all abelian structures.

Although every homomorphism has a kernel, a substructure is not generally the kernel of a homomorphism. Also, with relations present, a homomorphism onto is not generally determined by its kernel (e.g., F is $1 - 1$ if and only if

Ker(F) is trivial, but F may be 1 - 1 and onto without being an isomorphism). The role of ideals in our theory will be played by presentations. A presentation is a pair $\langle X, \Phi\rangle$, where X is a set of variables and Φ is a set of atomic formulas based on X. $\langle X, \Phi\rangle$ is called a complete presentation if there exists $A \in K$ and a generating S-map $\underset{\sim}{a} : X \to A$ such that $\Phi = \{\phi \in \mathrm{Form}_L^0(X) : A \models \phi(\underset{\sim}{a})\}$. $\langle X, \Phi, \underset{\sim}{a}\rangle$ is said to be a presentation of A. The complete presentations based on a fixed set X are closed under intersections, since given A_i, $i \in I$, and $\underset{\sim}{a}^i : X \to A_i$, if $A = \prod_{i \in I} A_i$ and $\underset{\sim}{a} : X \to A$ is the induced map, the presentation of $A' \subset A$ given by $\underset{\sim}{a}$ yields the desired intersection. Since the complete presentations form a complete lattice under intersections, it makes sense to speak of the complete presentation generated by an arbitrary presentation. If γ is a cardinal, $A \in K$ is said to be γ-presentable if there exists a presentation $\langle X, \Phi, \underset{\sim}{a}\rangle$ of A with $\mathrm{card}(X) < \gamma$ and Φ generated by a subset Φ' with $\mathrm{card}(\Phi') < \gamma$. A is called finitely presentable if it is ω-presentable.

Note that a complete presentation presents a unique model by the Fundamental Theorem on Defining Homomorphisms. Also, by the compactness of K, if $\langle X, \Phi\rangle$ is a presentation, then the complete presentation generated by Φ is given by:

$$\{\phi \in \mathrm{Form}_L^0(X) : \text{ for some finite } \Phi' \in \Phi,\ \forall X(\wedge\Phi' \to \phi) \text{ is valid in } K\}.$$

[In other words, we have an algebraic closure operator on $\mathrm{Form}_L^0(X)$.] It follows from this, that if γ is infinite, A is γ-presentable, $\mathrm{card}(X) < \gamma$, and $\underset{\sim}{a} : X \to A$ generates A, then $\Phi = \{\phi \in \mathrm{Form}_L^0(X) : A \models \phi(\underset{\sim}{a})\}$ is generated by a subset of cardinality $< \gamma$. To see this, take any $\underset{\sim}{a}' : X' \to A$ in which the presentation of A is generated by Φ' of cardinality $< \gamma$. Each $\underset{\sim}{a}'(x') \in A$ can be written in the form $\tau^A(\underset{\sim}{a})$; we choose such a term $\tau_{x'}$ for each $x' \in X'$, let T be the set (of cardinality $< \gamma$) of terms so obtained, and let $X_0 = \bigcup_{\tau \in T} \mathrm{Var}(\tau)$. Also, each $\underset{\sim}{a}(x) \in A$ can be written in the form $\tau^A(\underset{\sim}{a}')$; we choose such a term τ_x for each $x \in X_0$. By our choices, for each $x \in X_0$ there exists a formula $\phi_x \in \Phi$ of the form $x = \tau_x(\tau_1(\underset{\sim}{x}), \ldots, \tau_n(\underset{\sim}{x}))$ with $\tau_1, \ldots, \tau_n$ T. Finally, let $\Phi_0 \subset \Phi$ be

$\{\phi_x : x \in X_0\} \cup \{\phi'(\tau_{x'_1}(\underset{\sim}{x}), \ldots, \tau_{x'_n}(\underset{\sim}{x})) : \phi'(x'_1, \ldots, x'_n) \in \Phi'\}$. Clearly, $\text{card}(\Phi_0) < \gamma$ and by the Theorem on Defining Homomorphisms Φ_0 generates Φ, as desired.

Similarly, if γ is infinite, $\langle X, \Phi\rangle$ is a presentation generating a complete presentation $\langle X, \Phi'\rangle$, and Φ' is generated by a subset $\Phi'' \subset \Phi'$ with $\text{card}(\Phi'') < \gamma$, then there exists a subset Φ_0 of Φ with $\text{card}(\Phi_0) < \gamma$ such that Φ_0 also generates Φ'. The proof is similar to the one just given, except easier; namely, we need only choose finitely many members of Φ for each member of Φ'' (i.e., enough to imply it).

A <u>cyclic</u> presentation is a presentation $\langle X, \Phi\rangle$ in which $\text{card}(X) = 1$. Obviously, each cyclic presentation generates a presentation of a cyclic model, and conversely every cyclic model has a cyclic presentation.

Given a presentation $\langle X, \Phi\rangle$, we will frequently speak of the model presented by Φ; by this we will mean the essentially unique model having a complete presentation generated by $\langle X, \Phi\rangle$.

Note that (by the proof of Theorem 2.1), if K is an additive variety, then every submodel of a model is a kernel. More trivially, if L is purely functional and K is an abelian class of L-structures, then the presentations in K correspond to those submodels of free models which are kernels; in this case, a presentation is γ-generated if its corresponding kernel is γ-generated (as a substructure). Finally, if L is purely functional and K is an additive variety of L-structures, then a presentation is γ-generated if and only if its corresponding kernel is γ-generated.

3. <u>Injectives and Pure-Injectives</u>. Given a class K of L-structures, a model A is called <u>injective</u> if and only if for all $B, C \in K$ with $F : B \to A$ and $G : B \xrightarrow{\text{emb}} C$, there exists $H : C \to A$ such that $H \circ G = F$, i.e., every diagram:

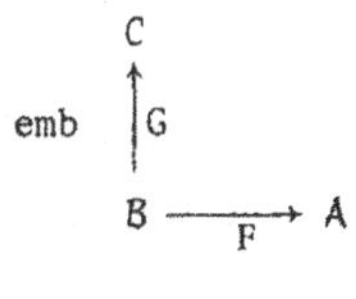

can be completed to a commutative diagram:

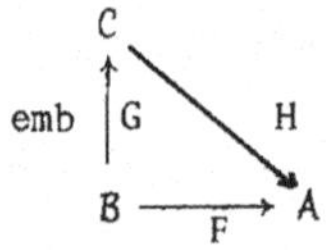

$A \in K$ is called an absolute retract if and only if A is a retract of every extension in K; i.e., if and only if every $F : A \xrightarrow{\text{emb}} B \in K$ is left-invertible. Every absolute retract is obviously absolutely pure.

The following propositions are well-known and trivial.

Proposition 3.1. Every injective is an absolute retract. If $A, B \in K$, B is injective, and $F : A \xrightarrow{\text{emb}} B$, then A is injective if and only if F is left-invertible.

Proposition 3.2. If each A_i, $i \in I$, is injective and $A = \prod_{i \in I} A_i \in K$, then A is injective.

Proposition 3.3. If K has HEP, then $A \in K$ is injective if and only if A is an absolute retract.

We will say that K has enough injectives (or, enough absolute retracts) if every model can be embedded in an injective (respectively, absolute retract).

Proposition 3.4. K has enough injectives if and only if K has enough absolute retracts and HEP.

We say that $A \in K$ is pure-injective if and only if for all $B, C \in K$ with $F : B \to A$ and $G : B \xrightarrow{\text{pure emb}} C$, there exists $H : C \to A$ such that $H \circ G = F$. Obviously, every injective model is pure-injective, but pure-injectivity can even be treated as a special case of injectivity by virtue of the next two propositions (in which K^* is the class of positive-primitive expansions of members of K and K^P is the class of substructures of members of K^*).

Proposition 3.5. $A \in K$ is pure-injective if and only if $A^* \in K^*$ is injective.

Proof. Trivial.

Proposition 3.6. If K is hereditary and compact, then an L^*-structure is an injective member of K^* if and only if it is an injective member of K^P.

Proof. Since K^* and K^P have HEP, their injective models are just those which are absolute retracts. Now if $A \in K^*$ is an absolute retract and $F : A \xrightarrow{\text{emb}} B \in K^P$, then there exists $G : B \xrightarrow{\text{emb}} C \in K^*$, so that there exists a left-inverse H for $G \circ F$, and consequently $H \circ G$ is a left-inverse for F.

On the other hand, if $A \in K^P$ is an absolute retract, then A is absolutely pure, and hence a member of K^*. It is then obvious that A is an absolute retract there, and the proof is complete.

An entirely straightforward check shows that a model is injective if and only if it is pure-injective and absolute pure.

$A \in K$ is called algebraically compact if and only if for all S-sets of variables X, Y, all $\Phi \subset \text{Form}_L^0(X \cup Y)$ and S-maps $\underset{\sim}{a} : X \to A$,

$$A \models \exists Y \wedge \Phi(\underset{\sim}{a}),$$

provided that for each finite $\Phi^0 \subset \Phi$,

$$A \models \exists Y \wedge \Phi^0(\underset{\sim}{a}).$$

Proposition 3.7. If K is compact, then $A \in K$ is pure-injective if and only if it is algebraically compact.

Proof. Suppose that $A \in K$ is pure-injective, $\Phi \subset \text{Form}_L^0(X \cup Y)$, $\underset{\sim}{a} : X \to A$, and that for every finite $\Phi^0 \subset \Phi$, $A \models \exists Y \wedge \Phi^0(\underset{\sim}{a})$. Then an easy diagram argument (using compactness) shows that there exists a pure-embedding (indeed, an elementary embedding) $F : A \to B \in K$ with $B \models \exists Y \wedge \Phi(F \circ \underset{\sim}{a})$. Since A is pure-injective, there exists $G : B \to A$ with $G \circ F = \iota_A$. Thus $A \models \exists Y \wedge \Phi(G \circ F \circ \underset{\sim}{a})$, and so $A \models \exists Y \wedge \Phi(\underset{\sim}{a})$.

On the other hand, if A is algebraically compact and $F : A \to B$ is a pure-embedding, then we can choose X, Y and generating S-maps $\underset{\sim}{a} : X \xrightarrow[\text{onto}]{} A$, $\underset{\sim}{b} : X \cup Y \to B$ such that $F \circ \underset{\sim}{a} = \underset{\sim}{b} \circ \iota_X$. Let $\Phi = \{\phi \in \text{Form}_L^0(X \cup Y) : B \models \phi(\underset{\sim}{b})\}$. If $\Phi^0 \subset \Phi$ is finite, then

$$\mathcal{B} \models \wedge\Phi^0(\underset{\sim}{b}) \Rightarrow \mathcal{B} \models \exists Y \wedge \Phi^0(\underset{\sim}{b} \circ \iota_X) \Rightarrow \mathcal{B} \models \exists Y \wedge \Phi^0(f \circ \underset{\sim}{a})$$

$$(\text{since } F \text{ is a pure-embedding}) \quad A \models \exists Y \wedge \Phi^0(\underset{\sim}{a}).$$

Now since A is algebraically compact, $A \models \exists Y \wedge \Phi(\underset{\sim}{a})$, so that there exists $\underset{\sim}{a}' : X \cup Y \to A$ with $\underset{\sim}{a}' \circ \iota_X = \underset{\sim}{a}$ and $A \models \wedge\Phi(\underset{\sim}{a}')$. Now by the Fundamental Theorem on Defining Homomorphisms, there exists (a unique) $H : \mathcal{B} \to A$ such that $H \circ \underset{\sim}{b} = \underset{\sim}{a}'$. But then $H \circ F \circ \underset{\sim}{a} = H \circ \underset{\sim}{b} \circ \iota_X = \underset{\sim}{a}' \circ \iota_X = \underset{\sim}{a} = \iota_A \circ \underset{\sim}{a}$, so that by the uniqueness half of the Fundamental Theorem again, $H \circ F = \iota_A$, as desired.

The next proposition, though trivial, is fundamental to our development and relies heavily on the finitarity of L (which has played no role so far, except indirectly where compactness was assumed). $\mathcal{B} \supset A$ is called a <u>cyclic extension of</u> A if and only if there exists $b \in \mathcal{B}$ such that $A \cup \{b\}$ generates $\mathcal{B}$. Thus $\mathcal{B}$ is cyclic (i.e., generated by some $b \in \mathcal{B}$) if and only if $\mathcal{B}$ is a cyclic extension of its null-generated substructure. $F : A \xrightarrow{\text{emb}} \mathcal{B}$ is called cyclic if $\mathcal{B}$ is a cyclic extension of Range(F).

<u>Proposition 3.8</u>. If L is finitary and K is hereditary, then $A \in K$ is injective if and only if for all $\mathcal{B}, C \in K$ with $F : \mathcal{B} \to A$ and a cyclic $G : \mathcal{B} \xrightarrow{\text{emb}} C$, there exists $H : C \to A$ with $H \circ G = F$.

<u>Proof</u>. Every extension of a model can be viewed as a union (direct limit) of a well-ordered sequence of cyclic extensions. The homomorphism at the bottom can be extended successively by assumption, since a union of homomorphisms is again one (by finitarity).

<u>Corollary 3.9</u>. If L is finitary and K is hereditary with HEP, then $A \in K$ is injective if and only if for all S-maps $\underset{\sim}{a} : X \to A$, variables z, and $\Phi \subset \mathrm{Form}_L^0(X \cup \{z\})$,

$$A \models \exists z \wedge \Phi(\underset{\sim}{a})$$

provided that there exists $F : A \xrightarrow{\text{emb}} \mathcal{B} \in K$ such that

$$\mathcal{B} \models \exists z \wedge \Phi(F \circ \underset{\sim}{a}).$$

<u>Proof:</u> The syntactic condition is clearly equivalent (by the Fundamental Theorem

on Defining Homomorphisms) to the existence of a left-inverse for every cyclic embedding of A into a member of K. But HEP guarantees that given $B, C \in K$, $F : B \to A$, and cyclic $G : B \xrightarrow{\text{emb}} C$, there exists $F' : C \to D$ and $G' : A \xrightarrow{\text{emb}} D$ such that $F' \circ G = G' \circ F$. We may assume (by replacing D by the Range(F')), that G' is cyclic, and then the result follows from the theorem.

We now return to our usual situation in which L is finitary and K is abelian. Let $\lambda = \text{card}(L) + \omega$, and let γ = the least <u>infinite</u> <u>regular</u> cardinal such that every cyclic model is γ-presentable. Clearly, $\gamma \le \lambda^+$.

If η is an infinite cardinal, we will say that $A \in K$ is <u>η-injective</u> if and only if for all S-maps $\underset{\sim}{a} : X \to A$, variables z, and $\phi \subset \text{Form}^0_L(X \cup \{z\})$ <u>with $\text{card}(\Phi) < \eta$</u>.

$$A \models \exists z \wedge \Phi(\underset{\sim}{a})$$

provided that there exists $F : A \xrightarrow{\text{emb}} B \in K$ such that

$$B \models \exists z \wedge \Phi(F \circ \underset{\sim}{a}).$$

The following theorem (for the case of modules) is due to Eklof and Sabbagh [4]. We feel that our proof amounts to a serious modification of theirs; indeed, even Bass's characterization of Noetherian--that a direct limit of injectives by injective--is given a new twist.

<u>Theorem 3.10</u>. Let K be an abelian class, and let η be an infinite regular cardinal. Then the following are equivalent:

(i) $\eta \ge \gamma$;

(ii) every η-injective model is injective;

(iii) a η-directed direct limit of injective models is injective;

(iv) the limit of an increasing η-sequence of injective models is injective.

<u>Proof</u>. (i) $\Rightarrow$ (ii): Suppose that A is η-injective, and hence γ-injective. Let $\underset{\sim}{a} : X \to A$ be an S-map, z a variable, and $\Phi \subset \text{Form}^0_L(X \cup \{z\})$. Let $F : A \xrightarrow{\text{emb}} B \in K$ be such that $B \models \exists z \wedge \Phi(F \circ \underset{\sim}{a})$. Now

$\Phi^* = \{\phi(\underset{\sim}{0}, z) : \phi(\underset{\sim}{x}, z) \in \Phi\}$, treated as a cyclic presentation, has a subset of cardinality $< \gamma$ which generates the same complete presentation (since γ is infinite). That is, there exists $\Phi_0 \subset \Phi$ with $\mathrm{card}(\Phi_0) < \gamma$ such that if $\Phi_0^* = \{\phi(\underset{\sim}{0}, z) : \phi(\underset{\sim}{x}, z) \in \Phi_0\}$, then:

(*) $$\forall z(\wedge\Phi_0^* \to \wedge\Phi^*)$$

is valid in K.

Since $\Phi_0 \subset \Phi$, $B \models \exists z \wedge \Phi_0(F \circ \underset{\sim}{a})$, so that by γ-injectivity, $A \models \exists z \wedge \Phi_0(\underset{\sim}{a})$. Let $a' \in A$ be a witness to this existential assertion, i.e., $A \models \wedge\Phi_0(\underset{\sim}{a}, a')$. Then

(**) $$B \models \wedge\Phi_0(F \circ \underset{\sim}{a}, F(a')).$$

Now if $b \in B$ is such that:

(***) $$B \models \wedge\Phi(F \circ \underset{\sim}{a}, b),$$

then by invoking additivity in relation to (**) and (***), we see that: $B \models \wedge\Phi_0(\underset{\sim}{0}, F(a') - b)$. But then by (*), $B \models \wedge\Phi(\underset{\sim}{0}, F(a') - b)$, and by addition to (***), $B \models \wedge\Phi(F \circ \underset{\sim}{a}, F(a'))$. Finally, since F is an embedding, $A \models \wedge\Phi(\underset{\sim}{a}, a')$.

An application of Corollary 3.9 shows that A is injective as desired.

Since γ-injectives can be constructed by a chain argument (as for γ-saturated models), we see already that K has enough injectives. [This assertion is stated explicitly as Corollary 3.11 below.] (ii) $\Rightarrow$ (iii): A η-directed direct limit is a direct limit over a partially ordered set P having the property that every subset of P with cardinality $< \eta$ has an upper bound. More simply, once the direct limit is in hand, it becomes a union of isomorphic copies of the original injective models of which it was a limit. By hypothesis, we need only check that it is η-injective. But given $\underset{\sim}{a} : X \to A$ and $\Phi \subset \mathrm{Form}_L^0(X \cup \{z\})$ with $\mathrm{card}(\Phi) < \eta$, $\mathrm{card}(\bigcup_{\phi\in\Phi} \mathrm{Var}(\phi)) < \eta$ (since each set $\mathrm{Var}(\phi)$ is finite). Thus, it suffices to suppose that $\mathrm{card}(X) < \eta$. But then $\mathrm{Range}(a)$ is contained in one of the injective models (by η-directedness), and the desired witness can be found there, so certainly in the union. [Note that we have actually used less than η-directedness: it suffices that every subset of the union having cardinality $< \eta$ be contained in one of the terms of union.]

(iii) $\Rightarrow$ (iv): Trivial, since a η-sequence is η-directed. (iv) $\Rightarrow$ (i): We will first give the proof (of the contrapositive) in the case of $\eta = \omega$, where things are simpler. If η is $< \gamma$, since η is regular and infinite, there exists a complete presentation $<z, \Phi>$ of a cyclic model which is not generated by any finite subset, and we can choose $\phi_0, \phi_1, \ldots, \phi_n, \ldots$ such that each $\phi_n \in \Phi$ and $\forall z(\bigwedge_{i<n} \phi_i \to \phi_n)$ is not valid in K. Let A_n, $n \in \omega$, be a sequence of cyclic models with elements $a_n \in A_n$ such that $A_n = (\bigwedge_{i<n} \phi_i \wedge \neg\phi_n)(a_n)$. Since (by (i) $\Rightarrow$ (ii)) K has enough injectives, each A_n may be supposed to be injective, provided we give up the claim of cyclicity.

Now the chain $A_0 \to A_0 \oplus A_1 \to A_0 \oplus A_1 \oplus A_2 \to \ldots$, with the natural inclusions, is a chain of injectives, since finite direct sums are direct products. Every finite subset of $\Phi = \{\phi_n(x - (a_0 + a_1 + \ldots + a_n)) : n \in \omega\}$ is satisfiable in the union $A = \bigoplus_{n\in\omega} A_n$, since if N = the maximum n such that ϕ_n belongs to the finite subset, $a_0 + a_1 + \ldots + a_N$ is a witness by additivity. By compactness, Φ is satisfiable in an extension of A belonging to K. But Φ is not satisfiable in A itself, since if a satisfies Φ, $a \in \bigoplus_{n<N} A_n$, and F is the natural projection onto A_N,

$$\begin{aligned} & A \models \phi_N(a - (a_0 + a_1 + \ldots + a_N)) \\ \Rightarrow\ & A_N \models \phi_N(F(a) - (F(a_0) + \ldots + F(a_N)) \\ \Rightarrow\ & A_N \models \phi_N(-a_N) \\ \Rightarrow\ & A_N \models \phi_N(a_N), \text{ a contradiction.} \end{aligned}$$

Thus A is not injective, as desired.

In the general case, if $\eta < \gamma$, there exists a presentation $<z, \Phi>$ of a cyclic which is not generated by any subset of cardinality $< \eta$, and we can choose ϕ_ξ, $\xi < \eta$, such that each $\phi_\xi \in \Phi$ and $\forall z(\wedge\{\phi_\eta : \eta < \xi\} \to \phi_\xi)$ is not valid in K. Let A_ξ, $\xi < \eta$, be a sequence of injective models with elements $a_\xi \in A_\xi$ such that $A_\xi \models (\wedge\{\phi_\eta : \eta < \xi\} \wedge \phi_\xi)(a_\xi)$. We now construct an increasing sequence of injectives, B_ξ, $\xi < \eta$, as follows. Let $B_0 = A_0$, $B_{\xi+1} = B_\xi \oplus A_\xi$, and let B_λ, λ a limit ordinal, be obtained by taking any injective containing $\operatorname{Lim}_{\xi<\lambda} B_\xi$ and forming a direct sum with A_λ.

Again, the chain $B_0 \to B_1 \to \ldots$ is a chain of injectives. We can now show that there exists a sequence of elements $b_\xi \in B_\xi$ such that:

(i) $b_0 = 0$,

(ii) $b_{\xi+1} = b_\xi + a_\xi$,

(iii) $\xi_0 < \xi \Rightarrow B \models \wedge\{\phi_\eta : \eta < \xi_0\}(b_\xi - b_{\xi_0})$,

where $B = \operatorname{Lim}_{\xi<\eta} B_\xi$. [Intuitively, $b_\xi = \Sigma a_\eta$.] The proof is by transfinite induction. $b_0 = 0$ is clearly all right. If b_ξ satisfies (iii), then for $\xi_0 \leq \xi$ (the case where $\xi = \xi$ being trivial), we have:

$$B \models \wedge\{\phi_\eta : \eta < \xi_0\}(b_\xi - b_{\xi_0}).$$

But $B \models \wedge\{\phi_\eta : \eta < \xi_0\}(a_\xi)$, so that (using (ii) with additivity):

$$B \models \wedge\{\phi_\eta : \eta < \xi_0\}(b_{\xi+1} - b_{\xi_0})$$

i.e., $b_{\xi+1}$ satisfies (iii). If λ is a limit ordinal, we note that $\{\wedge\{\phi_\eta : \eta < \xi_0\}(x - b_{\xi_0}) : \xi_0 < \lambda\}$ is finitely satisfiable (in fact, by any b_{ξ^*} with $\xi^* < \lambda$ beyond the last member of the finite subset of λ). Thus, by compactness, we have satisfiability in an extension of B_λ, and hence by injectivity, in B_λ itself.

Now $\Phi = \{\wedge\{\phi_\eta : \eta < \xi\}(x - b_\xi),\ \xi < \eta\}$ is again finitely satisfiable and, thus, satisfiable in an extension of $B = \operatorname{Lim}_{\xi<\eta} B_\xi$. But Φ is not satisfiable in B itself, since if b satisfies Φ, $b \in B_\xi$, and F is a retraction onto $B_{\xi+1} = B_\xi \oplus A_\xi$,

$$\begin{aligned} &B \models \wedge\{\phi_\eta : \eta < \xi + 1\}(b - b_{\xi+1}) \\ \Rightarrow\ &B \models \phi_\xi(b - b_{\xi+1}) \\ \Rightarrow\ &B_{\xi+1} \models \phi_\xi(F(b) - F(b_{\xi+1})) \\ \Rightarrow\ &B_{\xi+1} \models \phi_\xi(b - b_{\xi+1}), \quad \text{since } b,\ b_{\xi+1} \in B_{\xi+1} \end{aligned}$$

Finally, if $\pi_2 : B_{\xi+1} \in A_\xi$ is the natural projection (annihilating B_ξ),

$$B_{\xi+1} \models \phi\,(b - b_{\xi+1})$$

$$\Rightarrow A \models \phi(-\pi_2(b_{\xi+1})), \quad \text{since} \quad b \in B_\xi$$

$$\Rightarrow A_\xi \models \phi_\xi(-a_\xi), \quad \text{since} \quad b_{\xi+1} = b_\xi + a_\xi$$

$$\Rightarrow A_\xi \models \phi_\xi(a_\xi), \quad \text{a contradiction.}$$

Thus, $\mathcal{B}$ is not injective, as desired.

Corollary 3.11. Every abelian class K has enough injectives.

Corollary 3.12. The following are equivalent for any abelian K:

(i) $\gamma = \omega$;

(ii) arbitrary direct sums of injective models are injective;

(iii) direct sums of countable sequences of injective models are injective.

Proof. (i) $\Rightarrow$ (ii): If $\gamma = \omega$, then every (ω-) directed direct limit of injectives is injective. But $\bigoplus_{i\in I} A_i$ is the direct limit of $\{\bigoplus_{i\in F} A_i : F \subset I,$ F finite$\}$, where the finite subsets of I are directed by inclusion.

(ii) $\Rightarrow$ (iii): Trivial.

(iii) $\Rightarrow$ (i): By the theorem, we need only show that the direct limit of an increasing sequence of injectives is injective. But if $A_0 \subset A_1 \subset \ldots \subset A_n \subset \ldots$ is such a sequence, we can inductively define injectives: $\mathcal{B}_0, \mathcal{B}_1, \ldots$ such that $A_n \cong \bigoplus_{i\le n} \mathcal{B}_i$. But then $\operatorname{Lim}_{n<\omega} A_n = \bigoplus_{n<\omega} \mathcal{B}_n$, which is injective by assumption.

One of the major tools in our study of abelian structures will be the injective hull. We will discuss some of its more important features in a general setting. To this end let K be a class of L-structures.
$F : A \xrightarrow{\text{emb}} \mathcal{B} \in K$ is called essential if and only if for all $G : \mathcal{B} \to \mathcal{C} \in K$, if $G \circ F$ is an embedding, then G is an embedding. [Intuitively, $\mathcal{B}$ sits "tightly" over A.] An embedding which is not onto (i.e., not an isomorphism) will be called proper.

Proposition 3.13. Given $F : A \xrightarrow{\text{emb}} \mathcal{B}$ and $G : \mathcal{B} \xrightarrow{\text{emb}} \mathcal{C} \in K$, if F and G are essential, then $G \circ F$ is essential; conversely, if $G \circ F$ is essential, then G is essential. Finally, if K has HEP and $G \circ F$ is essential, then

F is essential.

Proof. First suppose that F and G are essential. If $H : \mathcal{C} \to \mathcal{D} \in K$ is such that $H \circ G \circ F$ is an embedding, then by the essentiality of F, $H \circ G$ is an embedding, and by the essentiality of G, H is an embedding. Thus, $G \circ F$ is essential.

Secondly, let $G \circ F$ be essential. If $H : \mathcal{C} \to \mathcal{D} \in K$ is such that $H \circ G$ is an embedding, then, since a composition of embeddings is again one, $H \circ G \circ F$ is an embedding. Thus, by the essentiality of $G \circ F$, H is an embedding. This shows that G is essential.

Finally, assume that K has HEP, $G \circ F$ is essential, and $H : \mathcal{B} \to \mathcal{D} \in K$ is such that $H \circ F$ is an embedding. By HEP the ingredients for the following commutative diagram exist in K:

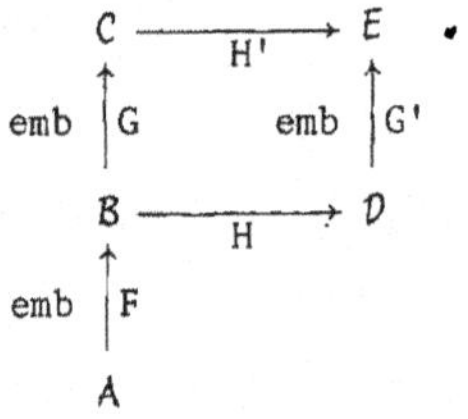

Since $H \circ F$ and G' are embeddings, $H' \circ G \circ F = G' \circ H \circ F$ is an embedding. By the essentiality of $G \circ F$, H' is an embedding. Thus, $G' \circ H = H' \circ G$ is an embedding. But if a composition is an embedding, the first map must be, so that H is an embedding as desired.

The following theorems (3.14 and 3.18) are due to Taylor [8] in the case of varieties of algebras and subsequently to P. Bacsich and the author in essentially the generality indicated.

Theorem 3.14. If $A \in K$ is an absolute retract, then A has no proper essential extensions in K. Conversely, for compact K with FIP, if $A \in K$ has no proper essential extensions in K, then A is an absolute retract.

Proof. Let $A \in K$ be an absolute retract, and let $F : A \xrightarrow{\text{emb}} \mathcal{B} \in K$ with G its left-inverse. $G \circ F = \iota_A$ is clearly an embedding. If F is essential, then G is an embedding, and thus a monomorphism. Since $G \circ (F \circ G) = \iota_A \circ G = G \circ \iota_B$, left-cancellation yields: $F \circ G = \iota_B$, i.e., F is also right-invertible, and hence an isomorphism (which is improper).

For the converse, let K be compact and $A \in K$ have no proper essential extensions. Given $F : A \xrightarrow{\text{emb}} B \in K$, we will show that F is left-invertible. to this end, choose S-sets $X \subset Y$ of variables with generating S-maps $\underset{\sim}{a} : X \to A$ and $\underset{\sim}{b} : Y \to B$ such that $F \circ \underset{\sim}{a} = \underset{\sim}{b} \circ \iota_X$. Let $\Phi_1 = \{\phi \in L_{\omega\omega}(X) : \phi$ is basic and $A \models \phi(\underset{\sim}{a})\}$ and $\Phi_2 = \{\phi \in \mathrm{Form}^0_L(Y) : B \models \phi(\underset{\sim}{b})\}$. Now $\exists Y \wedge(\Phi_1 \cup \Phi_2)$ is consistent with K, since $A \models \wedge\Phi_1(\underset{\sim}{a})$, so that $B \models \wedge\Phi_1(\underset{\sim}{b})$. By compactness, there exists a maximal $\Phi_3 \subset \mathrm{Form}^0_L(Y)$ containing Φ_2 such that $\exists Y \wedge(\Phi_1 \cup \Phi_3)$ is consistent with K.

Now choosing $C \in K$ with $\underset{\sim}{c} : Y \to C$ such that $C = \wedge(\Phi_1 \cup \Phi_3)(\underset{\sim}{c})$, the Fundamental Theorem on Defining Homomorphisms yields $G : B \to C$ with $G \circ \underset{\sim}{b} = \underset{\sim}{c}$. Since K has FIP, we may assume that G is onto. We claim that $G \circ F$ is an essential embedding. It is an embedding, since for basic ϕ, $A \models \phi(\underset{\sim}{a}) \Rightarrow \phi \in \Phi_1 \Rightarrow C \models \phi(\underset{\sim}{c}) \Rightarrow C \models \phi(G \circ \underset{\sim}{b}) \Rightarrow C \models \phi(G \circ F \circ \underset{\sim}{a})$.

It is essential, since given $H : C \to D \in K$ with $H \circ G \circ F$ an embedding, if $\Phi = \{\phi \in \mathrm{Form}^0_L(Y) : D \models \phi(H \circ \underset{\sim}{c})\}$, then

(i) $\Phi_3 \subset \Phi$, since $\phi \in \Phi_3 \Rightarrow C \models \phi(\underset{\sim}{c}) \Rightarrow D \models \phi(H \circ \underset{\sim}{c}) \Rightarrow \phi \in \Phi$, and

(ii) $D \models \wedge\Phi_1(H \circ \underset{\sim}{c})$, since $\phi \in \Phi_1 \Rightarrow D \models \phi(\underset{\sim}{a}) \Rightarrow$ (since $H \circ G \circ F$ is an embedding) $D \models \phi(H \circ G \circ F \circ \underset{\sim}{a}) \Rightarrow D \models \phi(H \circ \underset{\sim}{c})$.

But (ii) implies that $D \models \exists Y \wedge(\Phi_1 \cup \Phi)$, so that $\exists Y \wedge(\Phi_1 \cup \Phi)$ is consistent with K. Thus, by (i) and maximality, $\Phi = \Phi_3$. Hence, $D \models \phi(H \circ \underset{\sim}{c}) \Rightarrow \phi \in \Phi = \Phi_3 \Rightarrow C \models \phi(\underset{\sim}{c})$, so that H is an embedding. This demonstrates the essentiality of $G \circ F : A \to C \in K$.

But A has no proper essential extensions in K, so that $G \circ F$ is an isomorphism. If H is a (left-) inverse for $G \circ F$, then $H \circ G$ is the desired left-inverse for F.

If $F : A \xrightarrow{\text{emb}} B \in K$ is essential and $C \in K$ is injective with $G : A \xrightarrow{\text{emb}} C$, then there exists an $H : B \to C$ such that $H \circ F = G$ (by injectivity), and H is an embedding (since $H \circ F = G$ is an embedding, and F

is essential). Thus, every essential extension of A can be made to "sit in" every injective extension. Now if $F : A \xrightarrow{\text{emb}} B \in K$ is essential with B injective, then F (with its codomain B) is called an <u>injective hull</u> of A. If F is an inclusion map, we will frequently denote B by $\bar{A}$. Although, as opposed to "algebraic closures" of fields, A may have several distinct injective hulls sitting in a common extension, nevertheless, injective hulls are unique up to isomorphism. The following proposition shows this and a little more.

<u>Proposition 3.15</u>. If $F : A \xrightarrow{\text{emb}} B \in K$ and $G : A \xrightarrow{\text{emb}} C \in K$ are essential with B an absolute retract and C injective, then there exists an isomorphism $H : B \to C$ such that $H \circ F = G$.

<u>Proof</u>. By the injectivity of C, there exists $H : B \to C$ such that $H \circ F = G$. Since F is essential, H is an embedding. Since B is an absolute retract, H is actually left-invertible; let H' be such a left-inverse. We will show that it is also a right-inverse.

Since $H' \circ G = H' \circ H \circ F = \iota_B \circ F = F$ is an embedding and G is essential, H' is an embedding, hence a monomorphism. But $H' \circ (H \circ H') = (H' \circ H) \circ H' = \iota_B \circ H' = H' \circ \iota_C$, and by left-cancellation $H \circ H' = \iota_C$, as desired.

<u>Theorem 3.16</u>. Given a finitary L and a compact, hereditary class K of L-structures, if $A \in K$ has an injective extension in K, then A has an injective hull in K.

<u>Proof</u>. By our earlier remarks, every essential extension of A can be made to "sit in" any fixed injective extension of A. Thus, there is a bound on the size of essential extensions. It is easy to see (by the finitarity of L) that the direct limit of a chain of essential extensions is essential. Thus, by transfinite induction, we can construct a well-ordered increasing sequence of essential extensions, which must then terminate. If $F : A \xrightarrow{\text{emb}} B \in K$ is such a maximal essential extension, then Proposition 3.13 guarantees that B has proper essential extensions, and Theorem 3.14 guarantees that B is an absolute retract. But B is embeddable in any injective extension C of A; and, being a retract of C, it is injective. Thus $F : A \to B$ is an injective hull of A.

Corollary 3.17. If L is finitary and $\mathcal{K}$ is compact and hereditary with enough injectives, then every member of $\mathcal{K}$ has an injective hull.

The following curious theorem will enable us to extend the range of applicability of Corollary 3.11. It is not essential to later developments.

Theorem 3.18. If L is finitary and $\mathcal{K}$ is compact and hereditary with enough pure-injectives (i.e., such that every model is embeddable in a pure-injective model), then for every member $\mathcal{A}$ of $\mathcal{K}$ there exists a cardinal $\eta_{\mathcal{A}}$ such that every essential extension of $\mathcal{A}$ (in $\mathcal{K}$) has cardinality $\le \eta_{\mathcal{A}}$. In particular, every model has a maximal essential extension, and every such extension is an absolute retract. Thus, if $\mathcal{K}$ has HEP, then every member of $\mathcal{K}$ has an injective hull.

Proof. If $\mathcal{K}$ satisfies the hypotheses, then (by the Löwenheim-Skolem Theorem) for every $\mathcal{A} \in \mathcal{K}$ there exists a (small) family $F_i : \mathcal{A} \to \mathcal{B}_i \in \mathcal{K}$, $i \in I$, of embeddings such that given $G : \mathcal{A} \xrightarrow{\text{emb}} \mathcal{C} \in \mathcal{K}$ there exists an elementary embedding $H_i : \mathcal{B}_i \to \mathcal{C}$ with $G = H_i \circ F_i$. For each $i \in I$, choose a $K_i : \mathcal{B}_i \xrightarrow{\text{emb}} \overline{\mathcal{B}}_i \in \mathcal{K}$ with $\overline{\mathcal{B}}_i$ pure-injective. Note that K_i is not in general a pure-embedding!

Now if $G : \mathcal{A} \to \mathcal{C} \in \mathcal{K}$ is an essential embedding, then there exists $i \in I$ for which the following diagram is commutative:

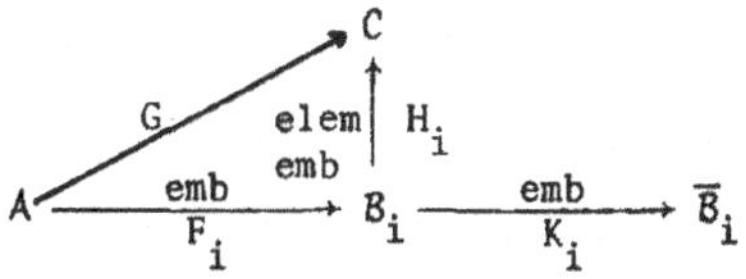

Since H_i is, in particular, a pure-embedding, and $\overline{\mathcal{B}}_i$ is pure-injective, there exists $L_i : \mathcal{C} \to \overline{\mathcal{B}}_i$ such that $L_i \circ H_i = K_i$. But then $L_i \circ G = L_i \circ H_i \circ F_i = K_i \circ F_i$ is an embedding, and since G is essential, L_i is an embedding. Thus $\operatorname{card}(C) \le \operatorname{card}(\overline{B}_i)$.

We have thus shown that $\sup_{i \in I} \operatorname{card}(\overline{B}_i)$ is an upper bound on the sizes of essential extensions of $\mathcal{A}$. Thus, as in the proof of Theorem 3.16, there exists a maximal essential extension of $\mathcal{A}$, which (by Proposition 3.13) has no proper essential extensions, and thus (by Theorem 3.14) is an absolute retract.

Returning to our concern with abelian structures, we have the following extension of Corollary 3.11.

Corollary 3.19. If K is a compact class of abelian structures which is closed under retracts, then K has enough absolute retracts. Thus, if K has HEP, then it has enough injectives.

Proof. The class K_0 of substructures of members of K is compact, hereditary, and also satisfies the additivity assumptions. [Incidentally, it is productive if K is.] The class K_0^p of substructures of positive primitive expansions of members of K_0 clearly has the same properties plus HEP.

Now note that the proof of (i) $\Rightarrow$ (ii) in Theorem 3.10, and hence the proof of Corollary 3.11 did not make use of the productiveness of K. [Indeed it did not make use of compactness for that metter; but HEP is even more elusive in non-compact classes!] Thus, K_0^p has enough injectives; i.e., K_0 has enough pure-injectives. By Theorem 3.18, then, K_0 has enough absolute retracts.

But every absolute retract in K_0 is a retract of any extension in K, and hence belongs to K (since K is closed under retracts). Thus, K itself has enough absolute retracts, and if K has HEP [while K_0 may not!], then K has enough injectives.

We tend to view this Corollary as merely an indication of techniques for extending some of our results to broader classes than we normally work in. For example, it is clear that if K_0 above has HEP, then it is the class to work in (simply noting that all injectives fall in the original class K). But in the interesting case, where K has HEP and K_0 does not, work should probably be done in K_0 (to handle chain constructions conveniently, etc.), noting carefully when models actually belong to K (where the powerful HEP applies). In such tricky situations, the technical proposition below (3.20) may be useful.

A class K is said to have the Amalgamation Property (AP) if every diagram in K of the form:

$$\begin{array}{lcl} B & & \\ {\scriptstyle \mathrm{emb}}\uparrow & & \\ A & \xrightarrow{\mathrm{emb}} & C, \end{array}$$

can be completed in K to a commutative diagram of the form:

$$\begin{array}{ccc} B & \xrightarrow{\mathrm{emb}} & D \\ {\scriptstyle \mathrm{emb}}\uparrow & & \uparrow{\scriptstyle \mathrm{emb}} \\ A & \xrightarrow{\mathrm{emb}} & C \end{array}$$

Note that if K has HEP and pushouts (as we are generally assuming), then K has AP, since D may be taken to be the pushout of the original diagram and HEP then guarantees that each new side of the square is an embedding. Note also that if K has AP, FIP, and a weak version of HEP stating that homomorphisms <u>onto</u> can be extended, then it has HEP in the usual sense, since given:

$$\begin{array}{lcl} B & & \\ {\scriptstyle \mathrm{emb}}\uparrow & & \\ A & \longrightarrow & C, \end{array}$$

we can factor (by FIP):

$$\begin{array}{lcccc} B & & & & \\ {\scriptstyle \mathrm{emb}}\uparrow & & & & \\ A & \xrightarrow[\mathrm{onto}]{} & C' & \xrightarrow{\mathrm{emb}} & C, \end{array}$$

apply "weak" HEP:

$$\begin{array}{ccccc} B & \longrightarrow & D' & & \\ {\scriptstyle \mathrm{emb}}\uparrow & & \uparrow{\scriptstyle \mathrm{emb}} & & \\ A & \xrightarrow[\mathrm{onto}]{} & C' & \xrightarrow{\mathrm{emb}} & C, \end{array}$$

and then AP:

$$\begin{array}{ccccc} B & \longrightarrow & D' & \longrightarrow & D \\ {\scriptstyle \mathrm{emb}}\uparrow & & \uparrow{\scriptstyle \mathrm{emb}} & & \uparrow{\scriptstyle \mathrm{emb}} \\ A & \xrightarrow[\mathrm{onto}]{} & C' & \xrightarrow{\mathrm{emb}} & C. \end{array}$$

The following proposition is useful, relative to the discussion following Corollary 3.19 above, in the case where K_0 does not itself have HEP, but does have AP.

Proposition 3.20. Let K have AP. If $F : A \xrightarrow{\text{emb}} B \in K$ and $G : A \xrightarrow{\text{emb}} C \in K$ are essential with B and C absolute retracts, then there exists an isomorphism $H : B \to C$ such that $H \circ F = G$.

Proof. By AP there exist $D \in K$ and $F' : C \xrightarrow{\text{emb}} D$ and $G' : B \xrightarrow{\text{emb}} D$ such that $F' \circ G = G' \circ F$. Now, since C is an absolute retract, F' has a left-inverse $L : D \to C$. If $H = L \circ G'$, then $H \circ F = L \circ G' \circ F = L \circ F' \circ G = 1_C \circ G = G$. The rest of the proof is word-for-word that of Proposition 3.15 above (starting with the second sentence).

Along these lines we cannot resist quoting the following gem due to Pierce.

Proposition 3.21. If K is closed under (finite, non-trivial) products and has enough injectives, then K has AP.

Proof. Given $F : A \xrightarrow{\text{emb}} B \in K$ and $G : A \xrightarrow{\text{emb}} C \in K$, we may assume (by taking extensions, if necessary) that B and C are injective. Thus, we can find $H_1 : B \to C$ and $H_2 : C \to B$ with $H_1 \circ F = G$ and $H_2 \circ G = F$, and then form the diagram:

$$
\begin{array}{ccc}
B & \xrightarrow{\quad G' \quad} & B \times C \\
{\scriptstyle F}\big\uparrow & & \big\uparrow{\scriptstyle F'} \\
A & \xrightarrow{\quad G \quad} & C,
\end{array}
$$

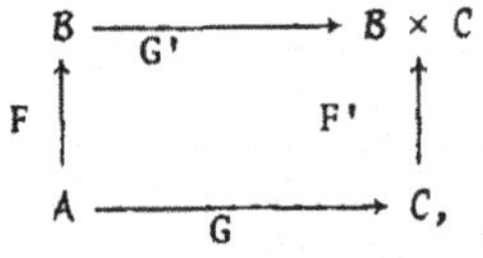

in which F' and G' are such that $\pi_1 \circ F' = H_2$, $\pi_2 \circ F' = 1_C$, $\pi_1 \circ G' = 1_B$, and $\pi_2 \circ G' = H_1$. The diagram commutes, since $\pi_1 \circ (F' \circ G) = H_2 \circ G = F = 1_B \circ F = \pi_1 \circ (G' \circ F)$ and $\pi_2 \circ (F' \circ G) = 1_C \circ G = G = H_1 \circ F = \pi_2 \circ (G' \circ F)$. F' is an embedding, since $\pi_2 \circ F' = 1_C$ is; and G' is an embedding, since $\pi_1 \circ G' = 1_B$ is.

We leave to the reader the variations required for a proof of the following.

Proposition 3.22. If K is productive, has HEP, and is inductive (i.e., closed

under direct limits), then $\mathcal{K}$ has AP.

[Hint: To get H_1 and H_2, just iterate HEP back-and-forth ω-times until they exist.]

Theorem 3.23. If $\mathcal{K}$ is a compact and productive class of abelian structures which is closed under retracts, then $\mathcal{K}$ has enough injectives if and only if $\mathcal{K}$ has AP.

Proof. By Proposition 3.21, if $\mathcal{K}$ has enough injectives, then $\mathcal{K}$ has AP. In the other direction, we know (by Corollary 3.19) that $\mathcal{K}$ has enough absolute retracts. We will show that (finite) direct products of absolute retracts are absolute retracts--an assertion which not true in general (even in varieties of algebras).

Thus, suppose that A, $\mathcal{B} \in \mathcal{K}$ are absolute retracts, and choose any absolute retract $\mathcal{D} \in \mathcal{K}$ with $A \oplus \mathcal{B} \subset \mathcal{D}$. Since A is an absolute retract, there exists $C \subset \mathcal{D}$ such that $\mathcal{D} = A \oplus C$; C is a model, since C is a retract of $\mathcal{D} \in \mathcal{K}$. Let $\pi : \mathcal{D} \to C$ be the natural projection. Then:

$$\begin{aligned} & C \models \phi(\pi \circ \underset{\sim}{b}) \\ \Rightarrow\ & \exists \underset{\sim}{a} \in \mathcal{D} \models \phi(\underset{\sim}{a} + \underset{\sim}{b}) \\ \Rightarrow\ & \mathcal{B} \models \phi(\underset{\sim}{b}), \end{aligned}$$

so that π embeds $\mathcal{B}$ into C. Thus, since $\mathcal{B}$ is an absolute retract, $\mathcal{D} = A \oplus \pi(\mathcal{B}) \oplus C_0$ (for some $C_0 \subset C$), and finally $A \oplus \mathcal{B} \cong A \oplus \pi(\mathcal{B})$ is a retract of $\mathcal{D}$. But (by AP) a retract of an absolute retract is an absolute retract, so that $A \oplus \mathcal{B}$ is an absolute retract, as claimed.

Now given a diagram in $\mathcal{K}$:

$$\begin{array}{ccc} \mathcal{B} & & \\ \text{emb} \uparrow F & & \\ A & \xrightarrow[G]{} & C, \end{array}$$

with C an absolute retract, we may suppose (by taking a further extension) that $\mathcal{B}$ is also an absolute retract. If $H : A \to \mathcal{B} \times C$ is induced by F and G, then H is an embedding, since $\pi_1 \circ H$ is. By AP, there exists a commutative diagram in K of the form:

$$
\begin{array}{ccc}
B & \xrightarrow[H']{\text{emb}} & D \\
\text{emb} \uparrow \; F & & \text{emb} \uparrow F' \\
A & \xrightarrow[H]{\text{emb}} & B \oplus C \; .
\end{array}
$$

Since $B \oplus C$ (as shown above) is an absolute retract, there exists a left-inverse K for F'. Now $\pi_2 \circ K \circ H' \circ F = \pi_2 \circ K \circ F' \circ H = \pi_2 \circ H = G$, i.e., the following diagram commutes:

$$
\begin{array}{ccc}
B & & \\
F \uparrow & \searrow \;\; \pi_2 \circ K \circ H' & \\
A & \xrightarrow[G]{} & C \; ,
\end{array}
$$

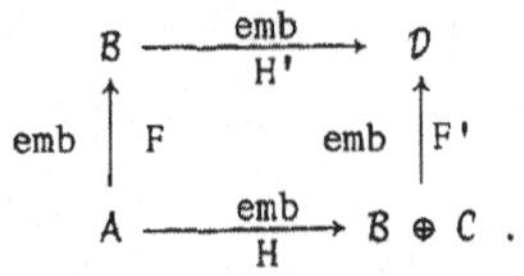

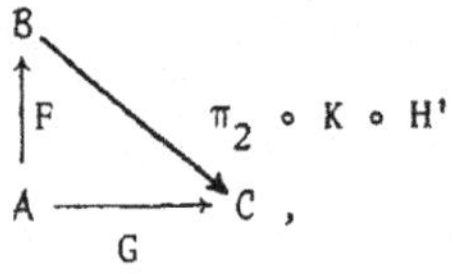

establishing the injectivity of C, as desired.

<u>Corollary 3.24.</u> If K is a compact, hereditary, and productive class of abelian structures, then K is abelian if and only K has AP.

<u>Proof.</u> Under the given assumptions, K is abelian if and only K has enough injectives if and only if (by the Theorem) K has AP.

The rest of this section will be concerned with our usual classes: L is finitary and K is an abelian class of L-structures.

<u>Theorem 3.25.</u> If $\overline{A} \in K$ is an essential extension of A and $B \in K$ is arbitrary, then $\overline{A} \oplus B$ is an essential extension of $A \oplus B$.

<u>Proof.</u> Let $F : A \oplus B \xrightarrow{\text{emb}} \overline{A} \oplus B$ be the inclusion map, and let $H : \overline{A} \oplus B \to C \in K$ be such that $H \circ F$ is an embedding. We must show that H is an embedding.

Now if $\pi_1 : A \oplus B \to A$ is the natural projection, then by HEP, we can find the ingredients for the following commutative diagram in K:

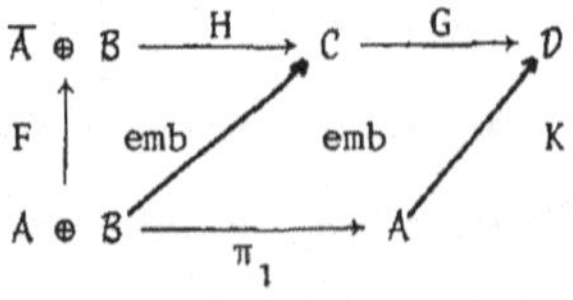

Let $\iota_1 : A \to A \oplus B$ be the inclusion map; let $\overline{\iota}_1 : \overline{A} \to \overline{A} \oplus B$ be the inclusion map; and let $\overline{\pi}_1 : \overline{A} \oplus B \to \overline{A}$ be the natural projection. Clearly, the inclusion map of A into $\overline{A}$ is given by: $\overline{\pi}_1 \circ F \circ \iota_1$; also $\gamma_1 \circ \iota_1 = \iota_A$ and $\overline{\pi}_1 \circ \overline{\iota}_1 = \iota_{\overline{A}}$. $G \circ H \circ \overline{\iota}_1 \circ (\overline{\pi}_1 \circ F \circ \iota_1) = G \circ H \circ F \circ \iota_1 = K \circ \pi_1 \circ \iota_1 = K \circ \iota_A$ is an embedding, and the inclusion map $\overline{\pi}_1 \circ F \circ \iota_1$ of A into $\overline{A}$ is essential, so that $G \circ H \circ \overline{\iota}_1$ is an embedding. We are now ready to show that H itself is an embedding, thereby establishing the essentiality of F.

If X is an S-set of variables, $\phi \in \mathrm{Form}_L^0(X)$, and $\underset{\sim}{a} : X \to \overline{A}$, $\underset{\sim}{b} : X \to B$ are S-maps, then

$$
(*) \qquad
\begin{aligned}
& \mathcal{C} \models \phi(H \circ (\underset{\sim}{a} + \underset{\sim}{b})) \\
\Rightarrow\ & \mathcal{D} \models \phi(G \circ H \circ (\underset{\sim}{a} + \underset{\sim}{b})) \\
\Rightarrow\ & \mathcal{D} \models (G \circ H \circ \underset{\sim}{a} + G \circ H \circ \underset{\sim}{b}) \\
\Rightarrow\ & \mathcal{D} \models \phi(G \circ H \circ \underset{\sim}{a}),
\end{aligned}
$$

since if $\iota_2 : B \to A \oplus B$ is the inclusion map, $G \circ H \circ \underset{\sim}{b} = G \circ H \circ F \circ \iota_2 \circ \underset{\sim}{b} = K \circ \pi_1 \circ \iota_2 \circ \underset{\sim}{b} = K \circ 0 \circ b = 0$. But we have seen that $G \circ H \circ \overline{\iota}_1$ is an embedding, and thus

$$
(**) \qquad
\begin{aligned}
& \mathcal{D} \models \phi(G \circ H \circ \underset{\sim}{a}) \\
\Rightarrow\ & \mathcal{D} \models \phi(G \circ H \circ \overline{\iota}_1 \circ \underset{\sim}{a}) \\
\Rightarrow\ & \overline{A} \models \phi(\underset{\sim}{a}) \\
\Rightarrow\ & \mathcal{C} \models \phi(H \circ \underset{\sim}{a}).
\end{aligned}
$$

Subtracting the last line from (*), we get:

$$
(***) \qquad
\begin{aligned}
& \mathcal{C} \models \phi(H \circ \underset{\sim}{b}) \\
\Rightarrow\ & \mathcal{C} \models \phi(H \circ F \circ \iota_2 \circ \underset{\sim}{b}) \\
\Rightarrow\ & B \models \phi(\underset{\sim}{b}),
\end{aligned}
$$

since $H \circ F$ is an embedding. But adding (**) and (***) yields:

$$\overline{A} \oplus B \models \phi(\underset{\sim}{a} + \underset{\sim}{b}),$$

showing that H is an embedding, as desired.

<u>Corollary 3.26</u>. If $F : A \xrightarrow{\text{emb}} \overline{A} \in K$ and $G : B \xrightarrow{\text{emb}} \overline{B} \in K$ are injective

hulls, then the induced map of $A \oplus B$ into $\overline{A} \oplus \overline{B}$ is an injective hull:

Proof. $\overline{A} \oplus \overline{B} = \overline{A} \times \overline{B}$ is clearly injective, while two successive applications of the Theorem yield the essentiality of $A \oplus B \to \overline{A} \oplus B \to \overline{A} \oplus \overline{B}$.

Corollary 3.27. Given $A \oplus B \subset C \in K$, if $A' \subset C$ and $B' \subset C$ are essential extensions of A and B (via natural inclusion maps), then the sum $A' + B' \subset C$ is direct.

Proof. Let $A' \oplus B'$ be an external direct sum, $H : A' \oplus B' \to A + B \subset C$ be the (natural) induced map. Since $A' \oplus B'$ is an essential extension of $A \oplus B$ and H restricted to $A \oplus B$ is an embedding, $H : A' \oplus B' \xrightarrow[\text{onto}]{\text{emb}} A' + B'$, as desired.

We will refer to our next theorem as the Schroeder-Bernstein Theorem for Injectives; it is due to Bumby [1] in the case of modules. We adopt the notation $A^{(\eta)}$ for the coproduct of η copies of A.

Theorem 3.28. Given an abelian class K, if A, $B \in K$ are injective and each is embeddable in the other, then A and B are isomorphic.

Proof. By the nature of left-invertible embeddings in K, there exist (injective) C, $D \in K$ such that $A \cong C \oplus B$ and $B \cong D \oplus A$. By iteration, we see that $A \cong C \oplus D \oplus A$, and hence, using induction, we can define $F_n : (C \oplus D)^{(n)} \xrightarrow{\text{emb}} A$, $n < \omega$, such that $F_n = F_{n+1} \circ \iota_n$, where ι_n denotes the natural initial inclusion map of $(C \oplus D)^{(n)}$ into $(C \oplus D)^{(n+1)}$. Thus, there exists an embedding of the limit $(C \oplus D)^{(\omega)} \cong C^{(\omega)} \oplus D^{(\omega)}$ into A.

Since A is injective, the injective hull $\overline{C^{(\omega)} \oplus D^{(\omega)}}$ is also embeddable in A, giving $A \cong \overline{C^{(\omega)} \oplus D^{(\omega)}} \oplus R$, for $R \in K$. But $C^{(\omega)} \oplus D^{(\omega)} \cong D \oplus C^{(\omega)} \oplus D^{(\omega)}$ (by absorption of another copy of D into the ω-sum), so that:

$$\begin{aligned} A &\cong \overline{C^{(\omega)} \oplus D^{(\omega)}} \oplus R \\ &\cong \overline{D \oplus C^{(\omega)} \oplus D^{(\omega)}} \oplus R \\ &\cong \overline{D} \oplus \overline{C^{(\omega)} \oplus D^{(\omega)}} \oplus R \quad \text{(by Corollary 3.26)} \\ &\cong D \oplus A \quad (\text{since } \overline{D} = D) \\ &\cong B, \quad \text{as desired.} \end{aligned}$$

Finally, we note that all our work is immediately applicable to K^p, yielding results about pure-injectives, pure-essential pure-embeddings, and pure-injective hulls. We will now simply state the translations back to K of our theorems applied to K^p.

Theorem 3.29. If K is abelian, then every member of K is purely-embeddable in a pure-injective member of K.

This is Corollary 3.11 applied to K^p. Note that the γ of Theorem 3.10 applied to K^p is $\leq \lambda^+$ where $\lambda = \text{card}(L^*) + \omega = \text{card}(\{\phi \in L_{\omega\omega} : \phi$ is positive-primitive$\}) = \text{card}(L) + \omega$. Theorem 3.10 thus yields the following.

Theorem 3.30. If $\lambda = \text{card}(L) + \omega$ and K is an abelian class of L-structures, then every λ^+-pure-injective member of K is pure-injective.

This estimate is best possible, in general, since in the case of abelian groups, for example, $\lambda = \omega$, $\gamma = \omega$ (since Z is Noetherian), while $\gamma^p = \omega_1 = \lambda^+$ (since the pure-injective hull of Z contains a copy of the p-adic integers for each prime p--see [3] for details). Of course on the other hand, for vector spaces over a field F, $\lambda^p = \lambda = \text{card}(F) + \omega$, while $\gamma = \omega$ and $\gamma^p = \gamma$ (since all models are injective and hence pure-injective).

A glance at the proof of Corollary 3.19 shows that we actually also established the following "technical" result.

Theorem 3.31. If K is a compact class of abelian structures which is closed under retracts, then every member of K is purely embeddable in a pure-injective member of K.

Given A, $B \in K$ with $F : A \to B$, F is called a pure-essential embedding if and only if F is a pure-embedding and for all $G : B \to C \in K$, if $G \circ F$ is a pure-embedding, then G is a pure-embedding. Our usual techniques readily show the following.

Theorem 3.32. If A, $B \in K$ and A^* and B^* are their positive primitive expansions in K^p, then $F : A \to B$ is a pure-essential embedding if and only if $F : A^* \to B^*$ is an essential embedding in K^p.

Given $A \in K$, $F : A \to B \in K$ is called a pure-injective hull of A if and only if B is pure-injective and F is a pure-essential embedding. Theorem 3.15 and Corollary 3.17 applied to K^p translate as follows.

Theorem 3.33. If K is abelian, then every member A of K has a pure-injective hull which is unique up to isomorphism over A.

Theorem 3.34. If K is abelian and A, $B \in K$ are pure-injectives such that each is pure-embeddable in the other, then $A \cong B$.

We will mention one further result which essentially mixes notions between K and K^p.

Proposition 3.35. If K is abelian with $A \in K$, then a pure-injective hull of A is injective if and only if A is absolutely pure.

Proof. If a pure-injective hull of A is injective, then A is absolutely pure, since a pure-substructure of an absolutely pure model is clearly absolutely pure (by compactness or AP). If A is absolutely pure, then one readily sees that an injective hull satisfies the requirements for being a pure-injective hull (since the containment is automatically pure).

REFERENCES

1. R. T. Bumby, "Modules which are isomorphic to submodules of each other", Arch. Math., 16(1965), 184-185.

2. P. C. Eklof, "Homogeneous universal modules", Math. Scand. 29(1971), 187-196.

3. P. C. Eklof and E. R. Fisher, "The elementary theory of abelian groups," Annals of Math. Logic 4(1972), 115-171.

4. P. C. Eklof and G. Sabbagh, "Model-completions and modules", Annals of Math. Logic 2(1970), 251-295.

5. E. R. Fisher, Categories of Structures (in preparation).

6. P. Freyd, Abelian Categories: An Introduction to the Theory of Functors (Harper and Row, New York, 1964).

7. G. Sabbagh, "Aspects logiques de la pureté dans les modules", C. R. Acad. Sc. Paris 271(1970), 909-912.

8. W. Taylor, "Residually small varieties", Alg. Univ. 2(1972), 33-53.

THE NUMBER OF κ - FREE ABELIAN GROUPS AND THE SIZE OF EXT

Alan Mekler

Introduction. In this paper "group" shall mean "abelian group". For uncountable cardinals κ, a group is κ - free if all subgroups of cardinality less than κ are free. A group is strongly κ - free if every subset of cardinality less than κ is contained in a κ - pure free subgroup of cardinality $< \kappa$.

In Section 1, it is shown how, given a κ - free non - free group of cardinality κ, to construct non - free strongly κ - free groups of cardinality κ. From this follows:

Theorem 1.2: If there exists a κ - free non - free group of cardinality κ, then there exists 2^{κ} strongly κ - free groups of cardinality κ.

Recall that a group, A, is a Whitehead group (W group) if $\mathrm{Ext}(A, \mathbb{Z}) = 0$. The construction can be used to show:

Theorem 1.3: If all W - groups of cardinality κ are free and A is a κ - free non - free group of cardinality κ, then $|\mathrm{Ext}(A, \mathbb{Z})| = 2^{\kappa}$.

This strengthens a result due independently to Eklof and Mekler [5] and Shelah.

In Section 2, these results are generalized to projective models.

Section 0 - Set Theoretic Preliminaries

Certain set theoretic concepts are needed. An ordinal ν is identified with the set of ordinals less than $\nu : \nu = \{\mu | \mu < \nu\}$. A cardinal is identi-

This research is funded, in part, by a grant from the National Research Council of Canada.

fied with an initial ordinal. If A is a set, $|A|$ denotes the cardinality of A.

If ν is a limit ordinal, the <u>cofinality</u> of ν ($cf(\nu)$) is the smallest ordinal σ such that there is a strictly increasing function $f: \sigma \to \nu$ such that $\sup \{f(\mu):\mu < \sigma\} = \nu$. A cardinal κ is regular if $cf(\kappa) = \kappa$.

Let ν and σ be limit ordinals. A function $f:\sigma \to \nu$ is called <u>normal</u> if f is continuous (i.e., $f(\lambda) = \sup \{f(\mu)|\mu < \lambda\}$ for every limit ordinal $\lambda < \sigma$) and strictly increasing. A subset S of σ is called <u>stationary</u> if S meets the range of every normal function $f:cf(\sigma) \to \sigma$. Equivalently, S is stationary in σ if and only if S meets every closed cofinal subset of σ. For example, $S = \{\nu|cf(\nu) = \omega, \nu < \sigma\}$ is stationary in σ if $cf(\sigma) > \omega$.

An important property of stationary sets, which will be used repeatedly in this paper, is the following theorem of Solovay's [10].

<u>Theorem:</u> If κ is a regular cardinal and S a stationary subset, then S can be partioned into κ disjoint stationary subsets.

A recent paper by Baumgartner and Prikry [1] is recommended to the reader who wishes to learn more about this important concept.

A structural notion is now defined.

<u>Definition:</u> Assume $|X| = \kappa$. A sequence of subsets $\langle X_\nu|\nu < \kappa\rangle$ is a <u>κ - filtration</u> of X if the following conditions are true:

(i) $X = \bigcup_{\nu<\kappa} X_\nu$

(ii) $|X_\nu| < \kappa$

(iii) $X_\lambda = \bigcup_{\nu<\lambda} X_\nu$ if λ is a limit ordinal.

<u>Section 1</u>

In the following κ is assumed to be a regular uncountable cardinal and A a κ - free group of cardinality κ. To investigate the structure of A

one more concept is needed.

<u>Definition</u>: A subgroup B of a group C is <u>κ - pure</u> if whenever $D \supseteq B$ and $|D/B| < \kappa$, B is a direct summand of D.

<u>Lemma 1.1</u>: There is a κ - filtration $\langle A_\nu | \nu < \kappa \rangle$ of A by free subgroups and a subset E of κ such that $A_{\nu+1}/A_\nu$ is not free if $\nu \in E$ and A_σ/A_ν is free for $\sigma > \nu$ if $\nu \notin E$. A is not free if and only if E is stationary.

<u>Proof</u>: Choose $\langle B_\nu | \nu < \kappa \rangle$ a κ - filtration by free subgroups with $B_0 = \langle 0 \rangle$. Such groups exist since every subgroup of a free group is free. Let $S = \{\nu | B_\nu$ is not κ - pure $\}$.

Define a normal function $f: \kappa \to \kappa$ by $f(0) = 0$; if $f(\nu) \in S$ choose τ such that $A_{f(\nu)}$ is not a direct summand of A_τ and set $f(\nu + 1) = \tau$; if $f(\nu) \notin S$, set $f(\nu + 1) = f(\nu) + 1$; if ν is a limit ordinal, set $f(\nu) = \cup_{\tau<\nu} f(\tau)$. Define $A_\nu = B_{f(\nu)}$ and $E = \{\nu | A_\nu$ is not κ - pure$\}$. The A_ν and E satisfy the requirements of the lemma.

If E is not stationary, a κ - filtration by κ - pure free subgroups could be found and so A would be free. Suppose E is stationary and A is free. Choose a κ - filtration of A by direct summands. By a standard argument (cf. Eklof [2]), for some $\nu \in E$, A_ν is in this κ - filtration. Hence, A_ν must be κ - pure, a contradiction.

Take the A_ν and E to be fixed. The above lemma can be extended to obtain a more complete structural lemma.

<u>Lemma 1.2</u>: There exist free groups F_ν, K_ν, for all $\nu < \kappa$ such that

(0) if $F = \Sigma_{\nu<\kappa} F_\nu$ $K = \Sigma_{\nu<\kappa} K_\nu$ then F/F_ν, K/K_ν are free for all ν and $F_\nu \cap F_\tau = K_\nu \cap K_\tau = \langle 0 \rangle$ for $\nu \neq \tau$;

(1) $A = F/K$;

(2) $A_\beta = \Sigma_{\nu<\beta} F_\nu / \Sigma_{\nu<\beta} K_\nu$, $\beta \geq 1$;

(3) $K_\nu = \langle 0 \rangle$ for $\nu \notin E$.

In the above equalities certain identifications are assumed.

Proof: Define F_ν, K_ν by induction.

Case 0 $\nu = 0$. Let F_0 be a free group isomorphic to A_1 and $K_0 = \langle 0 \rangle$.

Case 1 $\nu \notin E$. Choose F_ν isomorphic to $A_{\nu+1}/A_\nu$, and $K_\nu = \langle 0 \rangle$.

Case 2 $\nu \in E$. Choose F'_ν a complimentary summand of $\Sigma_{\tau<\nu} K_\tau$ in $\Sigma_{\tau<\nu} F_\tau$. (Such a group exists since $\Sigma_{\tau<\nu} F_\tau / \Sigma_{\tau<\nu} K_\tau = A_\nu$ is free.)

Now choose F_ν isomorphic to $A_{\nu+1}$ and let K_ν be the kernel obtained by embedding F'_ν into F_ν as A_ν is embedded into $A_{\nu+1}$.

In the following, the groups F_ν, K_ν are assumed to be fixed groups which were described in the previous lemma. If $E' \subset E$ define K(E') to be $\Sigma_{\nu\in E'} K_\nu$ and A(E') to be F/K(E'). Groups constructed this way have the following properties.

Lemma 1.3: If E' is a stationary subset of E, then A(E') is a κ - free non - free group of cardinality κ. If $(\kappa - E')$ is cofinal, A(E') is strongly κ - free. If E', E" $\subset$ E and (E' - E") $\cup$ (E" - E') is stationary, then A(E') is not isomorphic to A(E").

Proof: Define $A(E')_\nu = \Sigma_{\tau<\nu} F_\tau / \Sigma_{\tau\in E'\cap\nu} K_\tau$. Identifying the natural embeddings with inclusion maps gives $A(E') = \bigcup_{\nu<\kappa} A(E')_\nu$. By the construction it is clear that $A(E')_\lambda = \bigcup_{\nu<\lambda} A(E')_\nu$, and $A(E')_\nu / A(E')_\tau$ is free for $\nu > \tau$ if $\tau \notin E'$. If $\nu \in E'$, $A(E')_{\nu+1}/A(E')_\nu$ is not free since this group has a direct summand isomorphic to $A_{\nu+1}/A_\nu$. Hence if E' is stationary, A(E') is a κ - free non - free group of cardinality κ.

Suppose B is a subgroup of A(E') and $|B| < \kappa$. If $(\kappa - E')$ is cofinal there is a $\nu \in E$ such that $A(E')_\nu \supset B$. By the construction, $A(E')_\nu$ is κ - pure. Hence if $(\kappa - E')$ is cofinal, A(E') is strongly κ - free.

The final part of this lemma follows from an argument of Shelah's [8]. A version of this argument will be used in Theorem 2.1.

The third part of the lemma is implied by the general fact that the set E is an isomorphism invariant, where sets whose symmetric difference is non - stationary are equivalent.

Theorem 1.1: If there exists a κ - free non - free group of cardinality κ, then there exist 2^κ strongly κ - free groups of cardinality κ.

Proof: Choose $\mathcal{E} = \{E_i \mid i \in I\}$ a family of 2^κ stationary subsets of E such that $(E_i - E_j) \cup (E_j - E_i)$ is stationary if $i \neq j$. This can be done since E can be partioned into κ disjoint stationary subsets [10]. By Lemma 1.3, $\{A(E_i) \mid E_i \in \mathcal{E}\}$ are the required groups.

The relationship between the group A and the groups constructed is exploited to prove the following theorem.

Theorem 1.2: If all W - groups of cardinality κ are free and A is a κ - free non - free group of cardinality κ, then $|\mathrm{Ext}(A, \mathbb{Z})| = 2^\kappa$.

Proof: Partition E into κ disjoint stationary subsets E_ν, $\nu < \kappa$. Note that $K = \Sigma_{\nu<\kappa} K(E_\nu)$. For each $\nu < \kappa$, there exists a non - trivial extension G_ν of $\mathbb{Z}$ by $A(E_\nu)$. The following sequence is exact.

(1) $\mathrm{Hom}(F, \mathbb{Z}) \to \mathrm{Hom}(K(E_\nu), \mathbb{Z}) \xrightarrow{\delta_\nu} \mathrm{Ext}(A(E_\nu), \mathbb{Z}) \to \mathrm{Ext}(F, \mathbb{Z}) = 0$.

So there exists $f_\nu \in \mathrm{Hom}(K(E_\nu), \mathbb{Z})$ such that $\delta_\nu(f_\nu) = G_\nu$.

For each $\nu < \kappa$, choose such an f_ν. Now for each $X \subseteq \kappa$, define f_X to be the homomorphism defined by $F_X \restriction K(E_\nu) = f_\nu$, if $\nu \in X$, and $f_X \restriction K(E_\nu) = 0$ if $\nu \notin X$. By the exact sequence

$\mathrm{Hom}(F, \mathbb{Z}) \to \mathrm{Hom}(K, \mathbb{Z}) \to \mathrm{Ext}(A, \mathbb{Z}) \to \mathrm{Ext}(F, \mathbb{Z}) = 0$,

these give rise to 2^κ extensions. It now must be shown they are all different.

Assume not. Then there are X, X' and g, such that $X \neq X'$, $g \in \mathrm{Hom}(F, \mathbb{Z})$, and $f_X = f_{X'} + \bar{g}$, where $\bar{g} = g \restriction K$. Without loss of generality assume $\nu \in X$, and $\nu \notin X'$. Then $f_\nu = f_X \restriction K_\nu = f_{X'} \restriction K_\nu + g \restriction K_\nu = g \restriction K_\nu$. Since (1) is exact, this implies G_ν is trivial. This is the desired contradiction.

Shelah [8], [9] has shown if "V = L" then all W - groups are free. Eklof and Mekler [5], and Shelah have shown Theorem 1.2 follows from "V = L". In [7], Hiller has shown if $|\mathrm{Ext}(A, \mathbb{Z})| \geq \omega_1$ for all κ - free nonfree A of cardinality κ, then for any $n \geq 2$ there is no topological space X such that $H^n(X, \mathbb{Z}) = \mathbb{Q}$, the groups of rationals. This provides a partial answer to a conjecture of Kan and Whitehead [6]. It is only a partial answer since Shelah [8] has shown that it is consistent that there are W - groups which are not free.

Section 2

The results of Section 1 can be generalized to projective modules over countable Dedekind domains by substituting "projective" for "free" (cf.[4]). For modules over an arbitrary ring, a slightly different approach is needed. First κ - projective must be defined.

Definition: Assume Λ is a commutative ring and κ a regular cardinal such that $\kappa > |\Lambda| + \omega$. A Λ - module A of cardinality κ is κ - projective if there is a κ - filtration of A by projective modules.

This notion of κ - projective can be made more general by considering a more general notion of κ - filtration. Shelah [8] has given a general definition of "κ - free", where "free" is a property which satisfies certain requirements. All modules of size κ which are κ - projective in Shelah's sense are κ - projective according to the above definition.

In the next theorem, modules are constructed which are $L_{\infty\kappa}$ - equivalent to projective modules. This asserts that each such module satisfies the same sentences in the infinitary language $L_{\infty\kappa}$ as a projective module. In order to understand the proof it is enough to know that $L_{\infty\kappa}$ - equivalence is characterized by the existence of a set of κ - extendable partial isomorphisms. That is, two structures A and B are $L_{\infty\kappa}$ - equivalent if there is a non-empty set F of isomorphisms from substructures of A to substructures of B such that for all $f \in \mathcal{F}$, $X \subseteq A$ $(Y \subseteq B)$ with $|X| < \kappa$ $(|Y| < \kappa)$, there exists $g \in \mathcal{F}$, $g \supseteq f$ and $X \subseteq$ domain (g) $(Y \subseteq$ range $(g))$. Note that a group is $L_{\infty\kappa}$ - equivalent to a free group if and only if it is strongly κ - free [2].

Theorem 2.1 Assume Λ is a commutative ring, κ a regular cardinal such that $\kappa > |\Lambda| + \omega$, and A is a κ - projective module of cardinality κ which is not projective. There exist 2^κ κ - projective non - projective modules of cardinality κ which are $L_{\infty\kappa}$ - equivalent to a projective module.

Proof Mimic Lemma 1.1 to define a κ - filtration $\langle A_0 | \nu < \kappa \rangle$ of A by projective submodules and a stationary subset E such that A_ν is κ - pure if $\nu \notin E$ and $A_{\nu+1}/A_\nu$ is not projective if $\nu \in E$. As in Section 1, for each $E' \subseteq E$, a module $A(E')$ can be defined. If E' is stationary $A(E')$ is a κ - projective module which is not projective. It is not clear whether or not $A(E')$ is $L_{\infty\kappa}$ - equivalent to a projective module even if $(\kappa - E')$ is cofinal.

It will be shown in Lemma 2.1 that if $(\kappa - E')$ is cofinal, then the direct sum of κ copies of $A(E')$ is $L_{\infty\kappa}$ - equivalent to a projective module.

Assuming Lemma 2.1, in order to complete the proof it must be shown that there are 2^κ such groups. Suppose E_1, $E_2 \subseteq E$ and $(E_1 - E_2) \cup (E_2 - E_1)$ is stationary. For $i = 1, 2$ Define $A^i = \Sigma_{\nu<\kappa} A(E_i)^\nu$ where $A(E_i)^\nu \simeq A(E_i)$. By the construction we have κ - filtrations $\langle A(E_i)_\nu | \nu < \kappa \rangle$ of the $A(E_i)$. Define $A_0^i = \langle 0 \rangle$; and for $\nu > 0$, define $A_\nu^i = \Sigma_{\tau<\nu} A(E_i)_\nu^\tau$. These modules give κ - filtrations of the A^i by projective modules. Further, A_ν^i is κ - pure iff $\nu \notin E_i$.

Suppose $f: A^1 \to A^2$ is an isomorphism. If $C = \{\nu \mid f \restriction A^1_\nu$ is an isomorphism onto $A^2_\nu\}$, then C is a closed cofinal set. Choose $\nu \in C \cap ((E_1 - E_2) \cup (E_2 - E_1))$. Assume without loss of generlity that $\nu \in E_1 - E_2$ and choose $\tau > \nu$ such that $\tau \in C$ and $A^1{}_\tau/A^1{}_\nu$ is not projective. Since $\nu \notin E_2$, $A^2_\tau/A^2{}_\nu$ is projective. But, $A^2_\tau/A^2{}_\nu \simeq A^1{}_\tau/A^1{}_\nu$, a contradiction. The above argument is the one referred to in Lemma 1.3.

The proof is then completed (assuming Lemma 2.1) by choosing $\mathcal{E} = \{E_i \mid i \in I\}$ a family of 2^κ stationary subsets of E, such that $(E_i - E_j) \cup (E_j - E_i)$ is stationary if $i \neq j$.

Lemma 2.1 Assume Λ is a commutative ring, κ a regular cardinal, $\kappa > |\Lambda| + \omega$, $\{A_\nu \mid \nu < \kappa\}$ a κ - filtration of A by projective modules, and S a cofinal set such that A_ν is κ - pure if $\nu \in S$. Let $A^\nu \cong A$ and $A^\nu_\tau \simeq A_\tau$. Then $\Sigma_{\nu<\kappa} A^\nu$ is $L_{\infty\kappa}$ - equivalent to $\Sigma_{\tau\in S} (\Sigma_{\nu<\kappa} A^\nu_\tau)$.

Proof: A set of κ - extendable partial isomorphism, roughly corresponding to coordinate maps will be defined. Suppose $X \subset \kappa$, $|X| < \kappa$, $f: X \to \kappa$, $g: X \to S$, are such that for all ν, τ X $(f(\nu), g(\nu)) \neq (f(\tau), g(\tau))$ if $\nu \neq \tau$. Define $p(X,f,g)$ to be the isomorphism induced by the component isomorphisms from $\Sigma_{\nu \in X} A^\nu_{g(\nu)}$ to $\Sigma_{\nu \in X} A^{f(\nu)}_{g(\nu)}$. Define I_0 to be the set of all such $p(X,f,g)$. Suppose now that X, Y, f, g are such that $X \cap Y = 0$; $|X \cup Y| < \kappa$; $f \restriction X: X \to \kappa$; $f \restriction Y: Y \to \kappa \times \kappa$; $f(\nu)_0 \neq f(\nu)_1$; $g \restriction X: X \to S$; $g \restriction Y: Y \to S \times S$; for all $\nu \in Y$, $g(\nu)_0 < g(\nu)_1$; and for all ν, τ and $i, j = 0, 1$, $(f(\nu)_i, g(\nu)_i) \neq (f(\tau)_j, g(\tau)_j)$ if $\nu \neq \tau$. For each $\nu, \tau \in S$, $\nu < \tau$, choose $A_{\nu\tau}$ to be such that $A_\tau = A_\nu \oplus A_{\nu\tau}$. Define $p(X, Y, f, g)$ to be the map induced by the component isomorphisms from

$$\Sigma_{\nu\in X} A^\nu_{g(\nu)} \oplus \Sigma_{\nu\in Y} A^\nu_{g(\nu)_1} \quad \text{to} \quad \Sigma_{\nu\in X} A^{f(\nu)}_{g(\nu)} \oplus \Sigma_{\nu\in Y} A^{f(\nu)_0}_{g(\nu)_0} \oplus \Sigma_{\nu\in Y} A^{f(\nu)_1}_{g(\nu)_0 g(\nu)_1}$$

Define I to be the set of all such $p(X, Y, f, g)$. Similarly, I_n can be defined. Let $I = \bigcup_{n<\omega} I_n$. The back and forth criterion for $L_{\infty\kappa}$ - equivalence is satisfied by I .

Lemma 2.1 can be generalized. For example it applies to direct sums of countable groups.

REFERENCES

1 . J.E. Baumgartner and K. Prikry, Singular cardinals and the generalized continuum hypothesis, Amer. Math. Monthly 84(1977) 108-113.

2 . P.C. Eklof, Infinitary equivalence of abelian groups, Fund. Math. LXXXI (1974) 305-314.

3 . _________, On the existence of κ - free abelian groups, Proc. Amer. Math Soc. 47 (1975) 65-72.

4 . _________, Independence results in Algebra (Mimeo).

5 . _________, and Alan Mekler, Some Independence results in group theory. Notices Amer. Math. Soc. 23 (1976) A-273 Abstract #76T-A66.

6 . D. Kan and G. Whitehead, On the realizability of singular cohomology groups, Proc. Amer. Math. Soc., V.12 (1961) 24-25.

7 . Howard L. Hiller, A conjecture on cohomology in L, (preprint).

8 . S. Shelah, Infinite abelian groups, Whitehead problem and some constructions, Israel J. Math. 18 (1974), 243-256.

9 . _________, A compactness theorem for singular cardinals, free algebras, Whitehead problem and transversals, Israel J. Math 21 (1975) 319-349.

10 . R.M. Solovay, Real-valued measurable cardinals, Proceedings of Symposia in Pure Mathematics, XIII, Part I, A.M.S. Providence, R.I., 1971.

THE JACOBSON RADICAL OF SOME ENDOMORPHISM RINGS

Jutta Hausen[1]

Throughout the following, G will denote a [reduced] totally projective abelian p-group, where p is some fixed prime. Mappings will be written on the right.

The Jacobson radical, $J(\text{End } G)$, of the endomorphism ring, $\text{End } G$, of G is the quasi-regular ideal of $\text{End } G$ which contains all quasi-regular ideals of $\text{End } G$; the ideal I of $\text{End } G$ is quasi-regular if and only if the coset $1_G + I$ of the additive subgroup I of $\text{End } G$ is a subset of the automorphism group $\text{Aut } G$ of G.

It is the purpose of the present discussion to describe $J(\text{End } G)$ in terms of its action on G. It will turn out that actually $J(\text{End } G)$ can be described in terms of its action on the socle $G[p]$ of G.

In [6; p. 279], R. S. Pierce has shown that an endomorphism γ of G is one-to-one and onto if, for each non-negative integer n, the restriction of γ to $p^nG[p]$ is an automorphism. Consequently, if $\varepsilon \in \text{End } G$ has the property that its restriction to the socle is nilpotent then, for each $n < \omega$, $1_G + \varepsilon$ induces an automorphism in $p^nG[p]$ and $1_G + \varepsilon \in \text{Aut } G$. This proves

<u>Lemma 1</u>. If I is an ideal of $\text{End } G$ which induces in $G[p]$ a nil ring of endomorphisms then $I \subseteq J(\text{End } G)$.

An easy way to obtain two-sided ideals of $\text{End } G$ whose restriction to $G[p]$ is nil is the following. Let $\lambda > 0$ be an ordinal such that $p^\lambda G = 0$. Consider any finite sequence of ordinals $\sigma_0, \sigma_1, \ldots, \sigma_n$ such that

$$0 = \sigma_0 < \sigma_1 < \ldots < \sigma_{n-1} < \sigma_n = \lambda ,$$

[1]This research was partially supported by a University of Houston Faculty Development Leave Grant.

and let

$$J = \{\varepsilon \in \text{End } G : p^{\sigma_i}G[p]\varepsilon \subseteq p^{\sigma_{i+1}}G \text{ for } i = 0, 1, \dots, n-1\}.$$

Clearly, J is an ideal of End G and its restriction to G[p] is nil. Thus,

$$J \subseteq J(\text{End } G) ,$$

establishing

Lemma 2. Let I_G denote the set of all $\varepsilon \in \text{End } G$ for which there exists a finite sequence of ordinals

$$0 = \sigma_0 < \sigma_1 < \dots < \sigma_{m-1} < \sigma_m = \lambda$$

such that, for $i = 0, 1, \dots, m-1$, $p^{\sigma_i}G[p]\varepsilon \subseteq p^{\sigma_{i+1}}G$. Then $I_G \subseteq J(\text{End } G)$.

If G is a direct sum of cyclic groups then $I_G = J(\text{End } G)$ as was shown by W. Liebert in [5; p. 170]. In general, the inclusion in Lemma 2 is a proper one [6; p. 287]. However, if G is totally projective then $I_G = J(\text{End } G)$. This is an easy consequence of the following theorem and the fact that every properly decreasing sequence of ordinals terminates after finitely many steps [8; p. 270].

Theorem 3. Let G be totally projective and let $\lambda > 0$ be an ordinal such that $p^{\lambda}G = 0$. Let $\varepsilon \in J(\text{End } G)$ and let $0 < \tau \leq \lambda$ be any ordinal. Then there exists an ordinal $\sigma < \tau$ such that $p^{\sigma}G[p]\varepsilon \subseteq p^{\tau}G$.

The proof of this result is based upon two facts, the proof of which can be found in [3]. Throughout the following, λ is an ordinal such that $p^{\lambda}G = 0$ and $\lambda > 0$. If L is a reduced p-group, $\ell(L)$ denotes its length. The height of $x \in G$ is denoted by $h(x)$.

Lemma 4. If G is totally projective and $\varepsilon \in J(\text{End } G)$ then, for each ordinal $\sigma < \lambda$, $p^{\sigma}G[p]\varepsilon \subseteq p^{\sigma+1}G$.

Theorem 5 [3]. Let G be totally projective and let τ be a limit ordinal such that $0 < \tau \leq \lambda$. Let $\{x_\sigma\}_{\sigma<\lambda}$ and $\{x'_\sigma\}_{\sigma<\lambda}$ be subsets of G[p] such that, for all $\sigma < \lambda$,

$$\sigma \leq h(x_\sigma) < h(x'_\sigma) < \tau .$$

Then there exist a countable subset $\{x_{\sigma_n}\}_{n<\omega}$ of $\{x_\sigma\}_{\sigma<\tau}$ and a decomposition

$$G = \bigoplus_{n<\omega} L_n \oplus C$$

of G satisfying the following. For all $n < \omega$,

$$x'_{\sigma_n} \pi \in L_n$$

and

$$h(x'_{\sigma_n} \pi) < \ell(L_n) \leq h(x_{\sigma_{n+1}}) ,$$

where $\pi : G \twoheadrightarrow \bigoplus_{n<\omega} L_n$ denotes the canonical projection.

Proof of Theorem 3. Assume, by way of contradiction, that, for all $\sigma < \tau$, $p^\sigma G[p]\varepsilon \nsubseteq p^\tau G$. It follows from Lemma 4 that τ is a limit ordinal and that, for each $\sigma < \tau$, there exists $x_\sigma \in p^\sigma G[p]$ such that

$$\sigma \leq h(x_\sigma) < h(x_\sigma \varepsilon) < \tau .$$

Thus, Theorem 5 is applicable with $x'_\sigma = x_\sigma \varepsilon$, and there exist x_{σ_n} , $n = 0, 1, \ldots$, and a decomposition

(6) $$G = \bigoplus_{n<\omega} L_n \oplus C$$

such that, for $n = 0, 1, \ldots$,

$$x_{\sigma_n} \varepsilon\pi \in L_n$$

and

(7) $$h(x_{\sigma_n} \varepsilon\pi) < \ell(L_n) \leq h(x_{\sigma_{n+1}}) .$$

Since totally projective p-groups are detachable [7; p. 71], for each $n < \omega$, L_n has a decomposition

(8) $$L_n = A_n \oplus B_n$$

such that

$$(9) \qquad p^{\mu_n} A_n = \langle x_{\sigma_n} \varepsilon\pi \rangle \quad , \quad \mu_n = h(x_{\sigma_n} \varepsilon\pi) \; .$$

Clearly, the function mapping $x_{\sigma_n} \varepsilon\pi$ to $x_{\sigma_{n+1}}$ extends to a homomorphism φ'_n from $p^{\mu_n} A_n$ to $\langle x_{\sigma_{n+1}} \rangle$. Because of (6) and (8), A_n is totally projective [2; p. 89], and using (7) and Corollary 3.9 of [9; p. 252], φ'_n extends to a homomorphism φ''_n from A_n to G. It follows from (6) and (8) that there exists $\varphi \in \text{End } G$ such that, for each $n < \omega$, the restriction of φ to A_n equals φ''_n [1; p. 40]. Hence, for all $n < \omega$,

$$x_{\sigma_n} \varepsilon\pi\varphi = x_{\sigma_{n+1}} \quad ,$$

and

$$x_{\sigma_n} \varepsilon\pi\varphi\varepsilon\pi = x_{\sigma_{n+1}} \varepsilon\pi \; .$$

Put $w_n = x_{\sigma_n} \varepsilon\pi$ and let $\varepsilon' = \varphi\varepsilon\pi$. Since $\varepsilon \in J(\text{End } G)$ which is a two-sided ideal, $\varepsilon' \in J(\text{End } G)$ and, for each $n < \omega$,

$$w_n \varepsilon' = w_{n+1} \; .$$

Using (6), (7), and (9), one easily verifies that $1_G + \varepsilon'$ does not map G <u>onto</u> G, contradicting the fact that $\varepsilon' \in J(\text{End } G)$. This completes the proof.

Recalling the definition of I_G given in Lemma 2 we can state

<u>Corollary</u>. If G is totally projective then $J(\text{End } G) = I_G$.

REFERENCES

1. L. Fuchs, <u>Infinite Abelian Groups</u>, Vol. I., New York, 1970.

2. ________, <u>Infinite Abelian Groups</u>, Vol. II., New York, 1973.

3. J. Hausen, Quasi-regular ideals of some endomorphism rings, <u>Illinois J. Math</u>. (to appear).

4. N. Jacobson, <u>Structure of Rings</u>, Amer. Math. Soc. Colloq. Publ., Vol. 37, Revised Edition, Providence, 1968.

5. W. Liebert, The Jacobson radical of some endomorphism rings, J. Reine Angew. Math. 262/263 (1973), 166-170.

6. R. S. Pierce, Homomorphisms of primary abelian groups, Topics in Abelian Groups, Chicago, 1963, pp. 215-310.

7. F. Richman, Detachable p-groups and quasi-injectivity, Acta Math. Sci. Hungar. 27 (1976), 71-73.

8. W. Sierpinski, Cardinal and Ordinal Numbers, Warszawa, 1958.

9. E. A. Walker, The groups P_β , Symposia Math., Vol. 13, New York, 1974, pp. 245-255.

ULM VALUATIONS AND CO-VALUATIONS ON TORSION-COMPLETE P-GROUPS

Wolfgang Liebert

1. Introduction. A systematic investigation of valuated abelian groups was recently initiated in [11]. In the present paper we study the relationship between valuations and co-valuations on a torsion-complete abelian p-group and the one-sided ideals of its endomorphism ring. This continues our work in [6].

Let us fix a prime p. Throughout this paper G will denote a torsion-complete abelian p-group and $E(G)$ its endomorphism ring. For the most part our terminology and notation is taken from [4] and [5]. The most notable exception is that we write homomorphisms on the right. I denotes the non-negative integers and N the natural numbers. An element x in G has height $hx = n$ [exponent $ex = n$] if x is divisible [annihilated] by p^n but not by p^{n+1} [p^{n-1}]. Set $h0 = \infty$ and $e0 = 0$. The symbol ∞ satisfies $\infty < \infty$ and $i < \infty$ for all $i \in I$. A valuation [co-valuation] on G is a function $v : G \to I \cup \{\infty\}$ such that

(1) $vpx > vx$ [$< vx$ if $x \neq 0$ and $v0 = 0$]

(2) $vnx = vx$ if $(p,n) = 1$

(3) $v(x+y) \geq \min\{vx,vy\}$ [$\leq \max\{vx,vy\}$].

The valuations [co-valuations] on G form a partially ordered set under the natural pointwise ordering: $v \leq w$ if $vx \leq wx$ for each $x \in G$. In this partially ordered set arbitrary infs exist and are taken pointwise [not necessarily pointwise]. A model for such valuations [co-valuations] is the height [exponent] function $h[e]$. Notice that $v \geq h$ [$\geq e$] for any valuation [co-valuation] v on G. If H is a p-group then $r_p(H)$ will be the dimension of the vector space H/pH over the field of p elements. For each $i \in I$ we set

$G_v(i) = \{x \in G : vx \geq i\}$ [$= \{x \in G : vx \leq i\}$] and

$G_v(i)^* = G_v(i) \cap p^{-1}G_v(i+2)$ $[= G_v(i) + pG_v(i+2)]$. We will write $G(i)$ for $G_v(i)$ if the context makes clear which valuation [co-valuation] is meant. The i-th Ulm invariant of the valuated [co-valuated] group G is defined to be $f_v(i) = r_p(G_v(i)^*/G_v(i+1))$ $[= r_p(G_v(i+1)/G_v(i)^*]$. As usual we write $G_h(i) = p^iG$ $[G_e(i) = G[p^i]]$ and $f_h(i) = f(i)$ $[f_e(i) = f(i)]$. Note that $f(i)$ is the classical i-th Ulm invariant of the group G . For torsion-complete p-groups the classical Ulm invariants form a complete set of invariants.

By an Ulm valuation [co-valuation] on G we mean a valuation [co-valuation] v such that $f(i) \neq 0$ whenever $f_v(i) \neq 0$. The inf and sup of any number of Ulm valuations [co-valuations] on G is again an Ulm valuation [co-valuation], but one must be aware that sups [infs] are not necessarily taken pointwise. So the Ulm valuations [co-valuations] form a complete lattice. Each endomorphism α of G determines an Ulm valuation [co-valuation] v_α defined by $v_\alpha x = hx\alpha$ $[= \inf\{i \in I : x \in G[p^i]\alpha\}$ and $= \infty$ if $x \notin G\alpha]$. Notice that $v_1 = h$ [=e]. If M is a right [left] ideal of $E(G)$, we set $M^\perp = \inf\{v_\alpha : \alpha \in M\}$. And if v is any valuation [co-valuation] on G , then we let $v^\perp = \{\alpha \in E(G) : v_\alpha \geq v\}$. The correspondences $v \to v^\perp$ and $M \to M^\perp$ are inverse lattice anti-isomorphisms between Ulm valuations [co-valuations] on G and right [left] ideals of E(G) which are closed in the finite topology of E(G). It is our main purpose to characterize those Ulm valuations [co-valuations] on G which under this Galois correspondence are paired with the principal and the finitely generated right [left] ideals of E(G). This will also lead to a characterization of all one-sided ideals of $E(G)$ in terms of valuations and co-valuations on G . A remarkable duality between Ulm valuations and Ulm co-valuations on G (right and left ideals of $E(G)$) is all-round, but we can explain this phenomenon only if the classical Ulm invariants of G are finite.

Let B be a basic subgroup of G . Then $G = \bigoplus_{n \in \mathbb{N}} B_n$. where each B_n is a direct sum of cyclic groups of order p^n . The number of cyclic summands of B_n is precisely $f(n-1)$ which in turn equals $r_p(B_n)$. Such a decomposition of B is not canonical, and in general G has many basic subgroups. But for our purposes it will be very convenient to select and fix one basic subgroup of G with such a decomposition once and for all. By means of the p-adic topology of G we can assign a meaning to certain infinite sums in G. If $\{x_n\}$ is a bounded sequence of elements in G such that $hx_n \to \infty$, then the partial sums $\sum_{n<m} x_n$ converge to an element of G in the p-adic topology of G. It is convenient to denote this element by $\sum_{n\in N} x_n$. Let $B_n^* = B_{n+1} \oplus B_{n+2} \oplus \cdots$. Then we have $G = B_1 \oplus \cdots \oplus B_n \oplus (B_n^* + p^nG)$ for all $n \in N$ [4; Theorem 32.4]. Let π_n denote the projection of G onto B_n corresponding to this direct decomposition of G. Then the identity $x = \sum_{n\in N} x\pi_n$ holds for all $x \in G$ [10; p.127].

We restrict ourselves to torsion-complete p-groups because these groups possess plenty of endomorphisms. Let there be given homomorphisms $\varphi_n : G \to B_n$ for all $n \in N$. Then they define an endomorphism φ of G by $x\varphi = \sum_{n\in N} x\varphi_n$ for all $x \in G$. Dually, if homomorphisms $\varphi_n : B_n \to G$ are given, then they define an endomorphism φ of G by $x\varphi = \sum_{n\in N} x\pi_n\varphi_n$ for all $x \in G$. As far as the possibility of extending homomorphisms between subgroups of G to endomorphisms of G is concerned, G has the following outstanding property not shared by other separable p-groups.

Lemma 1.1. Let S be a subgroup of G and $\alpha : S \to G$ a homomorphism satisfying $(S \cap p^nG)\alpha \subseteq p^nG$ for all $n \in N$. Then α extends to an endomorphism of G.

Proof. By [11; Theorem 9] and [4; Proposition 40.1], α extends to a homomorphism of G into the p-adic completion $\hat{G}$ of G. The image of G must be in the torsion subgroup of $\hat{G}$, which is G.

2. Strict Ulm valuations. We say that an Ulm valuation v on G is strict if $f_v(i) \leq f(i)$ for all $i \in I$. Every endomorphism α of G

induces a valuation v_α on G by $v_\alpha x = hx\alpha$ for all $x \in G$.

<u>Lemma 2.1</u>. A valuation on G is a strict Ulm valuation if and only if it is induced by an endomorphism of G.

<u>Proof</u>. Let $\alpha \in E(G)$. Define $G_i = p^i G\alpha^{-1}$ and $G_i^* = G_i \cap p^{-1}G_{i+2}$ for all $i \in I$. Then α induces a monomorphism from G_i^*/G_{i+1} into $p^iG[p]/p^{i+1}G[p]$ for all $i \in I$. This shows that v_α is a strict Ulm valuation.

Conversely, let v be a strict Ulm valuation on G. Then we have for each $i \in N$ a monomorphism $\mu_i : G(i-1)^*/G(i) \to B_i[p]$. Combining it with the natural epimorphism $G(i-1)^* \to G(i-1)^*/G(i)$, this gives a homomorphism $\beta_i : G(i-1)^* \to B_i[p]$. This β_i does not decrease heights in G since $p^iG \subseteq G(i)$ and $B_i[p] = p^{i-1}B_i$. Therefore we can extend it by Lemma 1.1 to a homomorphism $\alpha_i : G \to B_i$ satisfying $G(i) \subseteq \operatorname{Ker} \alpha_i$. Let α be the endomorphism of G which is defined by $x\alpha = \sum_{i \in N} x\alpha_i$ for all $x \in G$. Then evidently $v_\alpha = \inf\{v_i : i \in N\}$, where v_i is the valuation induced by α_i. We claim that $v = v_\alpha$.

Let $x \in G$ and $vx = k$. First we show that $v_\alpha \geq v$. It suffices to prove $v_i x \geq vx$ for all i. If $i = 1,\ldots,k$, then $x\alpha_i = 0$, and hence $v_i x = \infty \geq vx$. If $i > k$, then $vp^{i-k}x \geq i$ implies $p^{i-k}x\alpha_i = 0$, and hence $x\alpha_i \in B_i[p^{i-k}] = p^kB_i$ and $v_i x \geq k = vx$. Thus $v_\alpha \geq v$. To prove the converse $v_\alpha \leq v$, we need only find an $i \in N$ with $v_i x = k$. Since G is torsion, there exists a smallest $n \in I$ with $p^n x \in G(k+n)^*$. Then $0 \neq p^n x\alpha_{k+n+1} \in B_{k+n+1}[p] = p^{k+n}B_{k+n+1}$. It follows that $x\alpha_{k+n+1} = p^kb + y$ with $y \in B_{k+n+1}[p^n] = p^{k+1}B_{k+n+1}$ and $hp^kb = k$. But $h(y) \geq k+1$, hence $h(p^kb+y) = k$. Therefore $v_{k+n+1}x = k$, and the proof is complete.

<u>Lemma 2.2</u>. Let $\alpha,\varphi \in E(G)$. Then $v_\alpha \leq v_\varphi$ if and only if there exists a $\lambda \in E(G)$ such that $\alpha\lambda = \varphi$.

<u>Proof</u>. If $\alpha\lambda = \varphi$, then $v_\varphi x = v_{\alpha\lambda}x = hx\alpha\lambda \geq hx\alpha = v_\alpha x$ for all $x \in G$. Conversely, assume that $v_\alpha \leq v_\varphi$. Then $\operatorname{Ker}\alpha \subseteq \operatorname{Ker}\varphi$. There-

fore we can define a homomorphism $\lambda_0 : G\alpha \to G\varphi$ by $(x\alpha)\lambda_0 = x\varphi$ for all $x \in G$. By hypothesis λ_0 doesn't decrease heights in G. Therefore we deduce from Lemma 1.1 that λ_0 extends to an endomorphism λ of G. Then $\alpha\lambda = \varphi$, as desired.

The two lemmas just presented accomplish a complete classification of strict Ulm valuations on G.

<u>Theorem 2.3</u>. $v \to v^{\perp}$ and $R \to R^{\perp}$ are inverse anti-isomorphisms between the partially ordered sets of strict Ulm valuations v on G and principal right ideals R of $E(G)$.

3. <u>n-strict Ulm valuations</u>. Let n be a natural number. We say that an Ulm valuation v on G is <u>n-strict</u> if $f_v(i) \le n \cdot f(i)$ for all $i \in I$.

<u>Lemma 3.1</u>. A valuation v on G is an n-strict Ulm valuation if and only if there exist finitely many endomorphisms $\alpha_1,\dots,\alpha_n$ of G such that $v = \inf\{v_{\alpha_1},\dots,v_{\alpha_n}\}$.

<u>Proof</u>. Let v be an n-strict Ulm valuation on G. Then $f_v(i) \le n \cdot f(i)$ for all $i \in I$. Since $G(i-1)^*/G(i)$ is a vector space over the field of p elements, we can write $G(i-1)^*/G(i) = X_1 \oplus \cdots \oplus X_n$ with $r_p(X_j) \le f(i-1) = r_p(B_i)$ for $1 \le j \le n$ and all $i \in N$. Consequently there exist homomorphisms $\mu_j : G^*(i-1)/G(i) \to B_i[p]$ for $j = 1,\dots,n$ such that μ_j is monomorph on X_j but vanishes on all X_k with $k \neq j$. As in the proof of Lemma 2.1 one concludes that $\mu_1,\dots,\mu_n$ can be extended to endomorphisms $\alpha_1,\dots,\alpha_n$ of G satisfying $v = \inf\{v_{\alpha_1},\dots,v_{\alpha_n}\}$.

Conversely, assume that $v = \inf\{v_1,\dots,v_n\}$ where v_j is the valuation induced by α_j for $j = 1,\dots,n$. Write $G_j(i) = \{x \in G : v_j x \ge i\}$. Then $G_v(i) = \bigcap_{j=1}^{n} G_j(i)$ and $G_v(i)^* = \bigcap_{j=1}^{n} G_j(i)^*$. By Lemma 2.1 we have $r_p(G_j(i)^*/G_j(i+1)) \le f(i)$. It is now routine to verify that

$$r_p(G_v(i)^*/G_v(i+1)) \le \sum_{j=1}^{n} r_p(G_j(i)^*/G_j(i+1)) \le n\cdot f(i) ,$$

finishing the proof.

Lemma 3.2. Let $\varphi, \alpha_1, \ldots, \alpha_n \in E(G)$. Then $\inf\{v_{\alpha_1}, \ldots, v_{\alpha_n}\} \le v_\varphi$ if and only if there exist $\lambda_1, \ldots, \lambda_n \in E(G)$ such that $\varphi = \alpha_1\lambda_1 + \cdots + \alpha_n\lambda_n$.

Proof. The "if" part is clear. To prove the "only if" we use induction on n . Assume our assertion true for $n-1$. Let v_j denote the valuation induced by α_j and write $G_j(i) = \{x \in G : v_j x \ge i\}$, $S_i = G_n(i)$ and $T_i = \bigcap_{j=1}^{n-1} G_j(i)$. Since $S_i \cap T_i$ is in the kernel of $\varphi\pi_i$ and since $S_i + T_i$ is modulo $S_i \cap T_i$ a direct sum, there exists a homomorphism $\xi_i : S_i + T_i \to B_i$ with $\xi_i = \varphi\pi_i$ on S_i and $\xi_i = 0$ on T_i . This ξ_i does not decrease heights in G . Let $x \in p^kG \cap (S_i + T_i)$. If $k \ge i$, then $x\xi_i = 0$ because $p^iG \subseteq S_i \cap T_i \subseteq \operatorname{Ker} \xi_i$. If $k < i$, then modulo $S_i \cap T_i$ the following is true: $p^{i-k}x = 0$ and there exist elements $y \in S_i$ and $z \in T_i$ with $x = y + z$ and $p^{i-k}y = 0 = p^{i-k}z$. It follows that $x\xi_i = y\varphi\pi_i \in B_i[p^{i-k}] = p^kB_i$. We can therefore apply Lemma 1.1 to conclude that ξ_i extends to an endomorphism of G which maps G into B_i .

Let ξ be the endomorphism of G defined by $g\xi = \sum_{i\in N} g\xi_i$ for all $g \in G$. By construction of ξ_i we have $\xi_i - \varphi\pi_i = 0$ on S_i for all i . If $k \le i$, then $S_i \subseteq S_k$ which implies $\xi_k - \varphi\pi_k = 0$ on S_i . If $k > i$, then $p^{k-i}S_i \subseteq S_i$ implies $S_i(\xi_k - \varphi\pi_k) \subseteq B_k[p^{k-i}] = p^iB_k$. In any case, we obtain $S_i(\xi_k - \varphi\pi_k) \subseteq p^iG$ for all $k \in N$ and all i . This shows that $S_i(\xi - \varphi) \subseteq p^iG$ for all i because of $\xi_k = \xi\pi_k$ and the identity $g = \sum_{k\in N} g\pi_k$ for all $g \in G$. Hence $v_n \le v_{\xi-\varphi}$, and therefore $\xi - \varphi \in \alpha_nE(G)$, by Lemma 2.2. We also know that $\xi_k = 0$ on T_k . By the same argument as above we find that $T_i\xi \subseteq p^iG$ for all i. By induction hypothesis, $\xi \in \sum_{j=1}^{n-1} \alpha_jE(G)$. Consequently $\varphi \in \sum_{j=1}^{n} \alpha_jE(G)$, and all is proved.

Lemma 3.1 shows that in the partially ordered set of n-strict Ulm valuations on G, with n ranging over all natural numbers, finite infs always exist. The following theorem is now an immediate consequence of Lemmas 3.1 and 3.2.

Theorem 3.3. $v \to v^{\perp}$ and $R \to R^{\perp}$ are inverse anti-isomorphisms between the meet-semilattice of n-strict Ulm valuations v on G and the join-semilattice of finitely generated right ideals R of $E(G)$.

4. Strict Ulm co-valuations. Let v be an Ulm co-valuation on G. Then the descending subgroup chain $G(i) \supseteq pG(i+1) \supseteq p^2G(i+2) \supseteq \cdots$ defines a topology on $G(i)$ which we call the *filtration topology* of $G(i)$. We say that v is *strict* if $f_v(i) \leq f(i)$ and $G(i)$ is complete in its filtration topology for all $i \in I$. Every endomorphism α of G induces a co-valuation v_α on G defined by $v_\alpha x = i$ if $x \in G[p^i]\alpha - G[p^{i-1}]\alpha$, $v_\alpha 0 = 0$ and $v_\alpha x = \infty$ if $x \notin G\alpha$.

Lemma 4.1. A co-valuation on G is a strict Ulm co-valuation if and only if it is induced by an endomorphism of G.

Proof. Let $\alpha \in E(G)$ and $G_i = G[p^i]\alpha$. Then α induces an epimorphism from $G[p^{i+1}]/(G[p^i] + pG[p^{i+2}])$ onto $G_{i+1}/(G_i + pG_{i+2})$. Note that $G[p^i]$ is complete in the topology inherited from the p-adic topology of G and defined by the descending chain of subgroups $G[p^i] \cap p^jG = p^jG[p^{i+j}]$ with $j \in I$. Since $\operatorname{Ker}\alpha$ is closed in the G-topology, it follows that $G[p^i]\alpha = G_i$ must be complete in the image topology defined by the descending chain of subgroups $p^jG[p^{i+j}]\alpha = p^jG_{i+j}$ with $j \in I$. Thus v_α is a strict Ulm co-valuation.

Conversely, assume that v is a strict Ulm co-valuation on G. For each $i \in N$ choose a basis for $G(i)$ modulo $G(i-1)^*$ and denote by S_i the subgroup of G spanned by these basis elements. Let $C_i = S_1 + \cdots + S_i + pS_{i+1} + p^2S_{i+2} + \cdots$. Then C_i inherits a topology from the filtration topology of $G(i)$, but it also has its own filtration topology defined by $C_i \supseteq pC_{i+1} \supseteq p^2C_{i+2} \supseteq \cdots$, which is contained in the G_i-topology. In general the two topologies don't match, but the reader will have no difficulty in verifying that $G(i)$ is the completion of C_i in both topologies. So we have $S_i \subseteq G[p^i]$ and $r_p(S_i) \leq f(i-1)$. Because of $r_p(B_i) = f(i-1)$, there exists a homomorphism $\alpha: B \to G$ with $B_i\alpha = S_i$ for all $i \in \mathbb{N}$. Then α extends to an endomorphism of G, by Lemma 1.1. Let $A_i = \sum_{j=1}^{i} B_j + \sum_{j>i} p^{j-i}B_j$. Then $A_i\alpha = C_i$. Now $G[p^i]$ is complete in the G-topology and has A_i as a dense subgroup. The purity of the B_j's in G readily implies that $(A_i \cap p^jG)\alpha = p^jC_{i+j}$. So we need only equip A_i with the G-topology and C_i with its own filtation topology to conclude from [8; Theorem 9, p.398] that α maps $G[p^i]$ onto $G(i)$. This shows that $v_\alpha = v$, completing the proof.

Lemma 4.2. Let $\alpha, \varphi \in E(G)$. Then $v_\alpha \leq v_\varphi$ if and only if there exists a $\lambda \in E(G)$ such that $\varphi = \lambda\alpha$.

Proof. Let $v_\alpha x = i < \infty$. Then $x \in G[p^i]\alpha - G[p^{i-1}]\alpha$. Hence $x \notin G[p^{i-1}]\lambda\alpha \subseteq G[p^{i-1}]\alpha$, which implies $v_\alpha x \leq v_{\lambda\alpha}x$. And if $v_\alpha x = \infty$, then $x \notin G\alpha \supseteq G\lambda\alpha$ shows $v_{\lambda\alpha}x = \infty$. Thus $v_\alpha \leq v_{\lambda\alpha}$.

Conversely, assume $v_\alpha \leq v_\varphi$. Then $G[p^i]\varphi \subseteq G[p^i]\alpha$ for all $i \in \mathbb{N}$. Pick a basis in each B_i. For each basis element b of each B_i we can write $b\varphi = x_b$ with $x \in G[p^i]$. Then there exists a $\lambda \in E(G)$ such that $b\lambda = x_b$ for all b. It follows that φ and $\lambda\alpha$ agree on each b and hence on the whole basic subgroup B of G. Consequently $\varphi = \lambda\alpha$ altogether.

If we combine Lemma 4.1 with Lemma 4.2, we derive:

Theorem 4.3. $v \to v^\perp$ and $L \to L^\perp$ are inverse anti-isomorphisms between the partially ordered sets of strict Ulm co-valuations v on G and principal left ideals L of $E(G)$.

5. n-strict Ulm co-valuations. Let n be a natural number. We say that an Ulm co-valuation v on G is n-strict if $f_v(i) \leq n \cdot f(i)$

and $G_v(i)$ is complete in its filtration topology for all $i \in I$.
Before we shall give a characterization of n-strict Ulm co-valuations on G, we would like to call the reader's attention to the fact that the inf of co-valuations is not necessarily taken pointwise. Let $\{v_j : j \in J\}$ by any set of co-valuations on G. For each $i \in I$ define $G(i)$ to be the algebraic sum (not necessarily the set-union) of the groups $G_{v_j}(i)$ for $j \in J$. Then the co-valuation $v = \inf\{v_j : j \in J\}$ is given by $vx = \inf\{i : x \in G(i)\}$ and $vx = \infty$ if x lies in no $G(i)$.

Lemma 5.1. A co-valuation v on G is an n-strict Ulm co-valuation if and only if there exist finitely many endomorphisms $\alpha_1,\ldots,\alpha_n$ of G such that $v = \inf\{v_{\alpha_1},\ldots,v_{\alpha_n}\}$.

Proof. Let v be an n-strict Ulm co-valuation on G. For each $i \in N$, we pick a basis for $G(i)$ modulo $G(i-1)^*$ and denote by S_i the subgroup of $G(i)$ spanned by these basis elements. By hypothesis we can write $S_i = S_i^1 + \ldots + S_i^n$ with $r_p(S_i^j) \le f(i-1)$ for every $i \in N$ and $j = 1,\ldots,n$. Now there exist endomorphisms $\alpha_1,\ldots,\alpha_n$ of G satisfying $B_i\alpha_j = S_i^j$ for all $i \in N$ and $j = 1,\ldots,n$. As in the proof of Lemma 4.1 we obtain $G(i) = \sum_{j=1}^{n} G[p^i]\alpha_j$. If follows that $v = \inf\{v_{\alpha_1},\ldots,v_{\alpha_n}\}$.

Conversely, suppose that $v = \inf\{v_1,\ldots,v_n\}$, where $v_j = v_{\alpha_j}$. Then each v_j is a strict Ulm co-valuation, by Lemma 4.1. Write $G_v(i) = G(i)$, $G_{v_j}(i) = G_j(i)$ and $f_{v_j}(i) = f_j(i)$. Then $G(i+1) = \sum_{j=1}^{n} G_j(i+1)$ and $G(i)^* = \sum_{j=1}^{n} G_j(i)^*$. Since each $G_j(i+1)$ is complete in its filtration topology, it is trivially true that $G(i+1)$ itself is complete in its filtration topology. It is easy to see that $r_p[G(i+1)/G(i)^*] \le \sum_{j=1}^{n} r_p[G_j(i+1)/G_j(i)^*]$. We obtain $f_v(i) \le \sum_{j=1}^{n} f_j(i) \le n\cdot f(i)$. This shows that v is indeed n-strict, and the lemma is proved.

Lemma 5.2. Let $\varphi,\alpha_1,\ldots,\alpha_n \in E(G)$. Then $\inf\{v_{\alpha_1},\ldots,v_{\alpha_n}\} \le v_\varphi$ if and only if there exist $\lambda_1,\dots,\lambda_n \in E(G)$ such that $\varphi = \lambda_1\alpha_1+\dots+\lambda_n\alpha_n$.

Proof. Only the "only if" needs to be proved. The hypothesis then reads: $G[p^i]\varphi \subseteq \sum_{j=1}^{n} G[p^i]\alpha_j$ for all $i \in \mathbb{N}$. Pick a basis $\{b_{ik} : k \in J_i\}$ in each B_i. For each b_{ik} we can write $b_{ik}\varphi = \sum_{j=1}^{n} x_{ikj}\alpha_j$ with $x_{ikj} \in G[p^i]$. Then there exist $\lambda_1,\dots,\lambda_n \in E(G)$ such that $b_{ik}\lambda_j = x_{ikj}$ for all $i \in \mathbb{N}$ and all $k \in J_i$. It follows that φ and $\sum_{j=1}^{n} \lambda_j\alpha_j$ agree on each b_{ik}, and hence on B. Therefore $\varphi = \sum_{j=1}^{n} \lambda_j\alpha_j$.

We conclude from Lemma 5.1 that the n-strict Ulm co-valuations on G form a meet-semilattice if we let n range over all natural numbers. Combining Lemmas 5.1 and 5.2 yields the following characterization theorem.

Theorem 5.3. $v \to v^{\perp}$ and $L \to L^{\perp}$ are inverse anti-isomorphisms between the meet-semilattice of n-strict Ulm co-valuations v on G and the join-semilattice of finitely generated left ideals L of $E(G)$.

Corollary 5.4. The following five statements are equivalent.

(1) Every finitely generated left ideal of $E(G)$ is principal.

(2) Every finitely generated right ideal of $E(G)$ is principal.

(3) Every n-strict Ulm valuation on G is strict.

(4) Every n-strict Ulm co-valuation on G is strict.

(5) G is either finite with all basis elements having the same order or infinite with no non-zero finite Ulm invariants.

Proof. It is a consequence of our characterization theorems 2.3, 3.3, 4.3, and 5.3 that we only have to prove the following: if G has at least two non-zero (classical) Ulm invariants $f(n)$ and $f(m)$ one of which, say $f(n)$, being finite, then there exists an n-strict Ulm valuation v and an n-strict Ulm co-valuation w on G which both are not strict. This is easily done. We may take v with $vx = \infty$ if $x \in A \oplus pC$, where $G = A \oplus C$ and $C = B_{n+1} \oplus \langle y\rangle$ with y a basis element of B_{m+1} , and $vx = n$ otherwise. And an example for w is provided by $wx = n+1$ if $x \in B_{n+1}[p] \oplus \langle z \rangle$, where $0 \neq z \in B_{m+1}[p]$, and $wx = \infty$ otherwise.

Corollary 5.5. The following three statements are equivalent.

(1) Every left ideal of $E(G)$ is principal.

(2) Every right ideal of $E(G)$ is principal.

(3) G is finite with all basis elements having the same order.

Proof. Because of Corollary 5.4 it is sufficient to show that for infinite G the ring $E(G)$ contains non-principal left and right ideals. Assume therefore that G is infinite and let $E_0(G) = \{\alpha \in E(G) : G\alpha \text{ finite}\}$. Then $E_0(G)$ is a two-sided ideal of $E(G)$. If A is any p-group without elements of infinite height, then every finite subgroup of A can be embedded in a direct summand of A [5; Proposition 65.1], and every co-finite subgroup of A contains a direct summand of A which is co-finite in A [9; Lemma 16.5]. It follows that for every $\alpha \in E(G)$ there exists a decomposition $G = H \oplus K$ such that H is finite, $G\alpha \subseteq H$, and $K \subseteq \operatorname{Ker} \alpha$. Thus, if $E_0(G)$ were a principal left ideal in $E(G)$ then it would map G into a finite direct summand of G . And if $E_0(G)$ were a principal right ideal in $E(G)$, then it would vanish on an infinite direct summand of G . Clearly neither one can happen because $E_0(G)$ would miss many projections onto cyclic direct summands of G .

6. Ulm valuations and co-valuations. Let M be a right [left] ideal of $E(G)$. Then $M^{\perp}$ is always an Ulm valuation [co-valuation] on G . But the correspondence $M \to M^{\perp}$ is in general not one-to-one between right [left] ideals of $E(G)$ and Ulm valuations [co-valuations] on G . This can be seen from the simple example $E_0(G)^{\perp} = E(G)^{\perp}$, where $E_0(G) = \{\alpha \in E(G) : G\alpha \text{ finite}\}$. Nevertheless, it is again possible to establish a Galois correspondence between Ulm valuations [co-valuations] on G and certain right [left] ideals of $E(G)$. To see which

right [left] ideals are candidates, let us introduce the <u>finite topology</u> of $E(G)$. It is defined by taking the annihilators of the finite subsets of G as a neighborhood basis at 0. In this topology $E(G)$ is a complete topological ring [5; Theorem 107.1].

Now let v be an Ulm valuation [co-valuation] on G and $\{\alpha_i\}_{i\in D}$ a Cauchy net in $v^{\perp}$, where D is partially ordered inversely to the finite subsets of G. Let $\alpha = \lim \alpha_i$. Given $x \in G$, there exists $j \in D$ such that $x(\alpha-\alpha_i) = 0$ holds for all $i \geq j$. Then $x\alpha = x\alpha_j$, so that $v_\alpha x = v_{\alpha_j} x$. Therefore $v_\alpha \geq \inf\{v_{\alpha_i} : i \in D\}$. But $v_{\alpha_i} \geq v$ for all i, since $v^{\perp} = \{\varphi \in E(G) : v_\varphi \geq v\}$. Hence $v_\alpha \geq v$ and $\alpha \in v^{\perp}$. This shows that the right [left] ideal $v^{\perp}$ is closed in the finite topology of $E(G)$.

<u>Theorem 6.1</u>. $v \to v^{\perp}$ and $M \to M^{\perp}$ are inverse anti-isomorphisms between the lattices of Ulm valuations [co-valuations] v on G and right [left] ideals M of $E(G)$ which are closed in the finite topology of $E(G)$.

If G is finite, then Theorem 6.1 follows from Theorem 3.3 [Theorem 5.3], since in that case each Ulm valuation [co-valuation] is automatically n-strict. In fact, Theorem 6.1 is true for arbitrary separable p-groups. This follows from [7; Theorems 4.9 and 5.5], if we interpret Monk's results in terms of valuated [co-valuated] groups.

7. <u>One-sided ideals of E(G)</u>. We saw in the previous section that in general not every right [left] ideal of $E(G)$ is of the form $v^{\perp}$ with v an Ulm valuation [co-valuation] on G. It is nevertheless possible to accomplish a complete classification of all one-sided ideals of $E(G)$ in terms of Ulm valuations [co-valuations] on G. Any right [left] ideal of $E(G)$ is uniquely determined by the finitely generated right [left] ideals of $E(G)$ which it contains. To each finitely generated right [left] ideal of $E(G)$, however, there corresponds a unique n-

strict Ulm valuation [co-valuation] in the sense of Theorem 3.3 [Theorem 5.3]. In this way we can associate with each right [left] ideal of E(G) a uniquely determined subset of the meet-semilattice of all n-strict Ulm valuations [co-valuations] on G . It shouldn't be a surprise now that these subsets are well-behaved.

Let $(\mathfrak{L},\leq,\wedge)$ $[(\mathfrak{L},\leq,\vee)]$ be a meet-semilattice [join-semilattice] which has a top [bottom] element. We recall that an <u>ideal</u> of $\mathfrak{L}$ is a non-empty subset A of $\mathfrak{L}$ such that (i) $a,b \in A$ implies $a \wedge b \in A$, $[a \vee b \in A]$ and (ii) $a \in A$, $x \in \mathfrak{L}$, $a \leq x$ $[x \leq a]$ imply $x \in A$. The top [bottom] element of $\mathfrak{L}$ is contained in every ideal of $\mathfrak{L}$. Clearly the intersection of any number of ideals of $\mathfrak{L}$ is again an ideal. Thus for every non-empty subset S of $\mathfrak{L}$ there is a smallest ideal of $\mathfrak{L}$ containing S . It consists of all elements of $\mathfrak{L}$ containing the meets [contained by the joins] of all finite subsets of S, and is called the ideal of $\mathfrak{L}$ generated by S . The ideals of $\mathfrak{L}$, ordered by set-inclusion, constitute a lattice. The meet of two ideals is their set-intersection, while their join is the ideal of $\mathfrak{L}$ generated by their set-union. The ideal of $\mathfrak{L}$ generated by the one-element set $\{a\}$ is called the <u>principal ideal</u> generated by a .

It is an immediate consequence of the definition of the concept of an ideal, for both rings and lattices, that the right [left] ideal lattice of a ring is naturally isomorphic with the ideal lattice of the join-semilattice of finitely generated right [left] ideals of this ring. If we now combine this fact with Theorems 2.3, 3.3, 4.3, and 5.3, then we arrive at our desired characterization of all right [left] ideals of E(G) .

<u>Theorem 7.1</u>. Let V be the meet-semilattice of n-strict Ulm valuations [co-valuations] on G . With each ideal A of V we associate the set $A^{\perp}$ of all $\alpha \in E(G)$ such that there exists a $v \in A$ with $v \leq v_{\alpha}$. And if M is a right [left] ideal of E(G) , we define $M^{\perp}$

to be the ideal in V which is generated by all n-strict Ulm valuations [co-valuations] of the form v_α for some $\alpha \in M$. Then $A \to A^\perp$ and $M \to M^\perp$ are reciprocal isomorphisms between the lattice of ideals A of V and the lattice of right [left] ideals M of $E(G)$. Under this isomorphism the principal ideals of V correspond to the finitely generated right [left] ideals of $E(G)$, and the principal ideals of V generated by strict Ulm valuations [co-valuations] correspond to the principal right [left] ideals of $E(G)$.

8. Duality. The project carried out in the preceding sections manifestly points out widespread duality between Ulm valuations and co-valuations on G, as well as between right and left ideals of $E(G)$. In this final section we shall briefly indicate to what extent the classical duality theory for finite-dimensional vector spaces and their endomorphism rings(see [1] for an extensive account) can be generalized to the present case. The following theory of duality is virtually contained in [2], [3], and [7].

The clue is of course to use dual groups. We follow [2] and [7] and define the dual G^* of G to be the maximal torsion subgroup of $\mathrm{Hom}(G, Z(p^\infty))$. Then G may be isomorphic to its dual without being finite. In fact, $G \cong G^*$ if and only if every Ulm invariant of G is finite [2; Lemma 2.2]. Better yet, in this case G is also the dual of G^* [2; Lemma 2.8]. Let us say with Faltings that G is torsion-compact if all Ulm invariants of G are finite.

Let $\alpha \in E(G)$. Then the adjoint of α is the endomorphism α^* of G^* defined by $\xi\alpha^* = \alpha\xi$ for all $\xi \in G^*$. If G is torsion-compact, then the mapping $\alpha \to \alpha^*$ is an anti-isomorphism of $E(G)$ onto $E(G^*)$ [2; Lemma 2.10]; we shall refer to it as the natural anti-isomorphism of $E(G)$ onto $E(G^*)$.

Define $L(G)$ to be the lattice of p-adically closed subgroups of G. For any $S \in L(G)$ we take the subgroup S' of G^* consisting of all characters vanishing on S. And for $T \in L(G^*)$ we let $T' = \{x \in G \mid xT = 0\}$. Then $S'' = S$ for every $S \in L(G)$ because of [2; Proposition 2.4]. Likewise $T = T''$ for every $T \in L(G^*)$ if G is torsion-compact since then T is the dual of G^*. It follows that for torsion-compact G the mapping $S \to S'$ constitutes an anti-isomorphism of $L(G)$ onto $L(G^*)$; we shall refer to it as the natural anti-isomorphism of $L(G)$ onto $L(G^*)$.

Theorem 8.1. The following four properties of G are equivalent.

(1) G is torsion-compact.

(2) $E(G)$ admits an anti-automorphism.

(3) $L(G)$ admits an anti-automorphism.

(4) $G[p^i]$ is p-adically compact for every i.

Proof. Assume first that G is torsion-compact. Then $G \cong G^*$, hence $E(G) \cong E(G^*)$. Therefore we obtain an anti-automorphism of $E(G)$ if the natural anti-isomorphism between $E(G)$ and $E(G^*)$ is followed by an isomorphism between $E(G^*)$ and $E(G)$. Assume next that $E(G)$ admits an anti-automorphism φ. Then φ induces an anti-automorphism ψ in the factor ring $E(G)/J(E(G))$, where $J(E(G))$ denotes the Jacobson radical of $E(G)$. We know from [9; p.288] that $E(G)/J(E(G))$ is isomorphic to the full direct product of the rings $M[f(i),p]$, where $M[f(i),p]$ denotes the endomorphism ring of a vector space of dimension $f(i)$ over the field of p elements. The identities of the rings $M[f(i),p]$ are the only indecomposable central idempotents of this factor ring. An anti-automorphism of a ring permutes indecomposable central idempotents. It follows that ψ permutes the rings $M[f(i),p]$. This is possible only if all dimensions $f(i)$ are finite [1; General Existence Theorem, p.193]. Thus (1) and (2) are equivalent.

If G is torsion-compact, then any isomorphism $G^* \cong G$ induces an isomorphism between $L(G^*)$ and $L(G)$. Combining it with the natural anti-isomorphism between $L(G)$ and $L(G^*)$ yields an anti-automorphism of $L(G)$. The converse is immediate from [3; Satz 1], so that (1) is also equivalent with (3).

The fact that (1) implies (4) follows from [4; exercise 4, p.70], and the reverse implication is routine. This completes the proof.

Remark. The endomorphism ring of a p-group A admits an anti-automorphism if and only if A is divisible of finite rank or torsion-compact.

This theorem provides us with a perfect duality theory for torsion-compact p-groups. It extends the duality for finite abelian groups to the infinite case. It will now be assumed for the remainder that G is torsion-compact. If v is strict Ulm valuation on G, then $p^iG \subseteq G(i)$ so that $G(i)$ is co-bounded and hence p-adically closed. Assume next that v is a strict Ulm co-valuation on G. Then $G(i) \subseteq G[p^i]$ whence $G(i)$ is bounded. We know from Lemma 4.1 that $v = v_\alpha$ and hence $G(i) = G[p^i]\alpha$ for some $\alpha \in E(G)$. By Theorem 8.1, $G[p^i]$ is p-adically compact. Then $G(i)$ must be p-adically compact also, since a continuous image of a compact space is compact. In view of this, the completeness of $G(i)$ in its filtration topology is equivalent with $G(i)$ being p-adically closed. This guarantees the crucial property that the value groups $G(i)$ in either case actually belong to $L(G)$. Now we have an isomorphism [anti-isomorphism] between the partially ordered sets (under the natural pointwise ordering) of strict Ulm valuations [co-valuations] on G and subgroup sequences $\{G_i\}_{i\in I}$ in $L(G)$ satisfying

(a) $p^iG \subseteq G_i$ $\quad$ $[G_i \subseteq G[p^i]]$

(b) $G_{i+1} \subseteq G_i$ $\quad$ $[G_i \subseteq G_{i+1}]$

(c) $pG_i \subseteq G_{i+1}$ $\quad$ $[pG_{i+1} \subseteq G_i]$

(d) $r_p(G_i \cap p^{-1}G_{i+2}/G_{i+1}) \leq f(i)$ $\quad$ $[r_p(G_{i+1}/G_i + pG_{i+2}) \leq f(i)]$

for all $i \in I$; it is given by $v \to \{G_i\}$ with $G_i = G(i)$ and has $\{G_i\} \to v$ defined by $vx = \sup\{i \in I : x \in G_i\}$ $[vx = \inf\{i \in \Gamma : x \in G_i\}]$ as its inverse. For any reduced p-group A, the subgroup pA is the intersection of all maximal subgroups of A, and the socle A[p] of A is the sum of all minimal subgroups of A. This observation readily implies that an anti-automorphism σ of L(G) interchanges the subgroups p^iS and $p^{-i}(S\sigma)$ for any $S \in L(G)$; in particular it interchanges p^iG with $G[p^i]$. Therefore σ induces an isomorphism between the partially ordered sets of strict Ulm valuations and co-valuations and an anti-isomorphism between the corresponding value group sequences. Consequently, strict Ulm valuations and co-valuations on G are dual concepts in this duality theory. It likewise reduces the study of n-strict Ulm co-valuations on G to that of n-strict Ulm valuations, and vice versa. Further details about this process of dualization are omitted.

REFERENCES

1. R. Baer, Linear Algebra and Projective Geometry, Academic Press, New York, 1952.

2. K. Faltings, On the automorphism group of a reduced primary abelian group, Trans. Amer. Math. Soc. 165 (1972), 1-25.

3. K. Faltings, Primäre abelsche Gruppen mit Semidualität, Arch. Math. 26 (1975), 14-19.

4. L. Fuchs, Infinite Abelian Groups, Volume I, Academic Press, New York, 1970.

5. L. Fuchs, Infinite Abelian Groups, Volume II, Academic Press, New York, 1973.

6. W. Liebert, One-sided ideals in the endomorphism rings of reduced complete torsion-free modules and divisible torsion modules over complete discrete valuation rings, Symposia Mathematica XIII (1974), 273-298.

7. G.S.Monk, One-sided ideals in the endomorphism ring of an abelian p-group, Acta math. Acad. Hungar. 19 (1968), 171-185.

8. D. Northcott, Lessons on Rings, Modules and Multiplicities, Cambridge University Press, 1968.

9. R.S. Pierce, Homomorphisms of primary abelian groups, Topics in Abelian Groups, Scott, Foresman and Co., Chicago (1963), 215-310.

10. R.S. Pierce, Endomorphism rings of primary abelian groups, Proc. Colloq. Abelian Groups (Tihany 1963), Budapest 1964, 125-137.

11. F. Richman and E.A. Walker, Valuated groups, to appear.

A RESULT ON PROBLEM 87 OF L. FUCHS

Warren May and Elias Toubassi

1. Introduction. Problem 87 of L. Fuchs [1, p. 248] asks the following: If G and H are groups of torsion-free rank one with $G/T(G) \cong H/T(H) \cong Q$ and if the endomorphism rings of G and H are isomorphic, then are G and H isomorphic? In [3] it is shown that the answer is no if the height matrix is allowed to contain infinite entries. In this paper we shall show that if the entries of the height matrix are restricted to be finite, then we obtain an affirmative answer. We now state the result.

Main Theorem 1. Let G and H be mixed groups of torsion-free rank one, and let $x \in G$ be a torsion-free element such that the entries of the height matrix of x are finite or the symbol ∞. Furthermore assume that $G/T(G) \cong H/T(H)$ (not necessarily isomorphic to Q). Then $\mathrm{End}\,(G) \cong \mathrm{End}\,(H)$ implies that $G \cong H$.

An analogue to this theorem is suggested by our results in [3]. It is gotten by dropping the hypothesis on the groups mod torsion while putting a further restriction on G. Before we can state it, however, we need some definitions. Let $(\sigma_{p0}, \sigma_{p1}, \sigma_{p2}, \dots)$ be the p-indicator of $x \in G$. We say that the p-indicator has a gap at σ_{pr} if $\sigma_{p,r+1} > \sigma_{pr} + 1$. We say that the p-indicator stabilizes at σ_{pr} if it has no gap at σ_{pi} for $i \geq r$, and if r is minimal with this property.

Main Theorem 2. Let G and H be mixed groups of torsion-free rank one, and let $x \in G$ be a torsion-free element such that the entries of the height matrix of x are finite or the symbol ∞. Furthermore assume that for almost all primes p, if the p-indicator of x stabilizes at $\sigma_{ps} < \infty$, then $\sigma_{ps} < \ell(T(G)_p) + s$. Then $\mathrm{End}\,(G) \cong \mathrm{End}\,(H)$ implies that $G \cong H$.

2. Review of Other Results. In [3] we give a detailed study of the behavior of the endomorphism ring for groups of torsion-free rank one having arbitrary height matrices and such that the reduced part of the torsion subgroup is totally

projective. In the proof of the Main Theorems we will need some results from [3]. This provides us with the opportunity to review the two fundamental theorems in that paper. We now set up some notation. *All groups considered in this paper are mixed groups of torsion-free rank one.* We shall let End (G) denote the endomorphism ring of G. Isomorphism of endomorphism rings will always be understood to be ring isomorphisms. Let $M = (\sigma_{pi})$ (p prime, $i \geq 0$) be a height matrix. We let $M_p = (\sigma_{p0}, \sigma_{p1}, \sigma_{p2}, \ldots)$ be the p-indicator of M. If A is p-primary, we let $\ell(A)$ denote the usual p-length in case A is reduced, otherwise we put $\ell(A) = \infty$.

Now let T be a torsion group and let $N = (\tau_{pi})$ be another height matrix. Put $\lambda_p = \ell(T_p)$. We shall define a relation between height matrices that depends upon T, and is weaker than the usual notion of equivalence. We say that M and N are T-equivalent if:

(1) For finitely many primes p, M_p and N_p are terminally equal (i.e., there exist m and n such that $\sigma_{p,m+j} = \tau_{p,n+j}$ for every $j \geq 0$).

(2) For the remaining p:

(a) If M_p contains an entry ∞, or if M_p has infinitely many gaps, then $M_p = N_p$.

(b) If M_p stabilizes at σ_{ps}, then N_p stabilizes at τ_{ps}, and for $s > 0$, $\sigma_{pi} = \tau_{pi}$ for $0 \leq i < s$. Moreover, σ_{ps} and τ_{ps} differ by a finite ordinal, and $\sigma_{ps} = \tau_{ps}$ (thus $M_p = N_p$) if either σ_{ps} or $\tau_{ps} < \lambda_p + s$.

<u>Theorem</u> (see [3]). Let G and H be groups, let $T = T(G)$, and let $x \in G$ and $y \in H$ be torsion-free elements. Assume that the reduced part of T_p is totally projective for every p. Then End (G) $\cong$ End (H) if and only if $T(H) \cong T$ and the height matrices of x and y are T-equivalent.

By a theorem of Wallace [4], we can conclude that $G \cong H$ if the height matrices are equivalent. Thus we obtain the

<u>Corollary</u> (see [3]). Let M be the height matrix of x, and suppose that for almost all primes p, if M_p stabilizes at $\sigma_{ps} < \infty$, then $\sigma_{ps} < \lambda_p + s$. Then End (G) $\cong$ End (H) implies that $G \cong H$.

The difficulty in generalizing the "if" statement of the above theorem to arbitrary torsion subgroups is demonstrated by Example 3.19 in [3]. It shows that this fails even if the entries of the height matrix are finite.

An issue raised by the above corollary is whether the isomorphism of End (G) with End (H) can be induced by an isomorphism of G with H. It is desirable to consider the related question for a single group: Is every automorphism of End (G) inner? This is true for torsion groups by the theorem of Baer and Kaplansky, and it is not hard to show that it is true for divisible groups. In the case of groups of torsion-free rank one, we have the

Theorem (see [3]). Let G be a group, let $x \in G$ be a torsion-free element, and assume that the reduced part of $T(G)_p$ is totally projective for every p. Then every automorphism of End (G) is inner if and only if there do not exist infinitely many primes p such that the p-indicator of x stabilizes at σ_{ps}, where $s \geq 1$ and $\ell(T(G)_p) < \sigma_{ps} < \infty$.

We now state the results in [3] needed in this paper. The first result we give is a combination of Corollary 3.7, Propositions 3.8 and 3.16 in [3] and relates the height matrices of G and H.

Proposition 1. Let G and H be groups, $x \in G$ a torsion-free element, and assume that the entries of the height matrix M of x are finite or the symbol ∞. Further assume that either (i) M satisfies the hypothesis of the above Corollary or else (ii) $G/T(G) \cong H/T(H)$. Then End $(G) \cong$ End (H) implies that the height matrices of G and H are equivalent.

Let G and H be groups. We give the definition of two important ideals of the endomorphism ring. Let $I(G) = \{\alpha \in \text{End}\ (G) | \alpha(G) \subseteq T(G)\}$ and $I_p(G) = \{\alpha \in I(G) | \alpha(G) \subseteq T(G)_p\}$ (p prime). The next statement shows the invariance of $I(G)$ and $I_p(G)$ under isomorphism.

Proposition 2. Let Φ:End (G) $\to$ End (H) be an isomorphism and p a prime. Then $\Phi(I(G)) = I(H)$. Consequently, $\Phi(I_p(G)) = I_p(H)$.

Suppose that $\Phi:I(G) \to I(H)$ is a (ring) isomorphism. One observes that Kaplansky's method of proof of the Baer-Kaplansky theorem (see [2, Theorem 28]) applies to show that

(3) there exists an isomorphism $\phi:T(G) \to T(H)$ such that $\Phi(\alpha)|_{T(H)} = \phi\alpha\phi^{-1}$ for every $\alpha \in I(G)$.

3. Proofs of Theorems. Throughout the rest of the paper we shall assume the following: *every entry of the height matrix is an integer or the symbol* ∞.

We begin with some preliminaries. Let K be a group. Since at times we will be interested in looking at K for a fixed prime, it will be useful to replace K by a certain subgroup. Let $g \in K$ be torsion-free and let p be a prime. Define $K^{p,g} = \{x \in K | x + \langle g\rangle$ is p-torsion in $K/\langle g\rangle\}$. Suppose that the p-indicator of g is (σ_n). Let $n(1) < n(2) < \ldots$ be a sequence, possibly finite or even empty, such that $(\sigma_{n(i)})$ is the subsequence of gaps. We wish to consider minimal pure subgroups of $K^{p,g}$ containing g which we shall denote by $K^\#$. $K^\#$ can be described by generators $\{a_i | i \geq 0\}$ satisfying relations of the form $p^{n(i)+1}g = p^{\sigma_{n(i)+1}}a_i$, where we put $n(0) = -1$. When $\sigma_{n(i)+1} = \infty$, the generator a_i and the equation $p^{n(i)+1}g = p^{\sigma_{n(i)+1}}a_i$ are symbolic and represent a countable set of generators a_{ij} $(j \geq 1)$, which satisfy the relations $p^{n(i)+1}g = pa_{i1}$, $a_{ij} = pa_{i,j+1}$ for $j \geq 1$. *We shall always consider generating sets of this type.*

Although the subgroup $K^\#$ is not unique, it contains information on the p-indicator of K. We now define two parameters which are invariant regardless of the particular choice of pure subgroup $K^\#$. We set $f_i = n(i + 1) - n(i)$ $(i \geq 0)$, $e_0 = \sigma_0$, and $e_i = \sigma_{n(i)+1} - \sigma_{n(i-1)+1} - f_{i-1}$ $(i \geq 1)$. Note that the f_i measure the spread between two consecutive gaps, while the e_i measure the size of the gap. When $e_i = \infty$, the symbolic term $p^{e_i}a_i$ will refer to that unique generator a_{ij} which satisfies the relation $p^{n(i)+1}g = p^{\sigma_{n(i-1)+1}+f_{i-1}}a_{ij}$. We observe that $T(K^\#) = \bigoplus_{i\geq 1}\langle t_i\rangle$, where $t_i = a_{i-1} - p^{e_i}a_i$ and $\exp(t_i) = \sum_{0\leq j<i}(e_j + f_j)$

$= \sigma_{n(i-1)+1} + f_{i-1} \quad (i \geq 1)$.

We adopt the following convention. *If* $K^{\#}$ *is generated by* $\{a_i\}$ *and we write* $T(K^{\#}) = \bigoplus \langle t_i \rangle$, *then it is to be understood that* $t_i = a_{i-1} - p^{e_i} a_i$.

We begin with three technical lemmas. In the first we show that under appropriate conditions we can ensure that a certain torsion subgroup arises as $T(K^{\#})$.

Lemma 4. Let K be a group and $g \in K$ a torsion-free element. Suppose that $\langle a_i | i \geq 0 \rangle$ is a minimal pure subgroup of $K^{p,g}$ containing g. Let $t_i = a_{i-1} - p^{e_i} a_i$. If $\bigoplus_{i \geq 1} \langle s_i \rangle$ is a pure subgroup of $\bigoplus_{i \geq 1} \langle t_i \rangle$ with $\langle s_i \rangle \cong \langle t_i \rangle$ for all i, then there exists a minimal pure subgroup $\langle b_i | i \geq 0 \rangle$ containing g such that $b_{i-1} - p^{e_i} b_i = k_i s_i$ and $p \nmid k_i \quad (i \geq 1)$.

Proof. Set $\sigma_{n(i)+1} = h_i$. By purity we may write $s_i = u_i + t_i + v_i$ with $u_i \in \langle t_j | 1 \leq j < i \rangle$, $v_i \in \langle t_j | j > i \rangle$, and where we are assuming, without loss of generality, that the coefficient of t_i is 1. Our approach will be to replace each generator $a_i \quad (i \geq 0)$, by a new generator $a_i + w_i \quad (= b_i)$, where w_i satisfies $p^{h_i} w_i = 0$. We note that the w_i are related recursively by

$$(5) \quad k_i s_i - t_i = w_{i-1} - p^{e_i} w_i \quad \text{for} \quad i > 0.$$

In the case where the number of generators is finite, $0 \leq i \leq n$, set $w_n = 0$ and work backwards to determine the remaining w_i.

We may assume that the number of generators is infinite. This case entails repeated changes in the generators which will be given inductively. We now describe an induction step on $r \geq 0$ to change the generators a_i. Suppose that $\{a_i'\}$ is a new choice of generators with $t_i' = a_{i-1}' - p^{e_i} a_i'$, and such that

$$(6) \quad t_i' = \ell_i t_i + (\ell_i + c_i p^r)(u_i + v_i) + p^r u_i' \quad (i \geq 1),$$

where $u_i' \in \langle t_j | 1 \leq j < i \rangle$. We claim that there exists another choice of generators $\{a_i''\}$ with $a_i'' = a_i' + w_i^{(r)}$, $\exp(w_i^{(r)}) \leq \max\{0, h_i - r\}$, and $t_i'' = a_{i-1}'' - p^{e_i} a_i''$, and such that

(7) $t_i'' = (\ell_i + c_i' p^{r+1}) t_i + \ell_i(u_i + v_i) + p^{r+1} u_i'' \quad (i \geq 1)$,

where $u_i'' \in \langle t_j | 1 \leq j < i \rangle$. First note that for $i \geq 1$, there exists v_{ij} $(j \geq 1)$, with

$$(8)\quad \begin{cases} v_i = p^{e_i}(p^{f_i} m_{i1} t_{i+1} + v_{i1}) \\ \\ v_{ij} = p^{e_{i+j}}(p^{f_{i+j}} m_{i,j+1} t_{i+j+1} + v_{i,j+1}), \end{cases}$$

where $v_{ij} \in \langle t_k | k > i + j \rangle$. This is possible since $\exp(v_i) \leq \exp(t_i)$ and $\exp(t_{i+j+1}) = e_{i+j} + f_{i+j} + \exp(t_{i+j})$. Define $w_0^{(r)} = 0$ and $w_i^{(r)} = p^r(\sum c_j(p^{f_{j+k}} m_{j,k+1} t_{j+k+1} + v_{j,k+1}) - c_{i+1} u_{i+1} - u_{i+1}')$, where $i > 0$, and the sum is taken over (j,k) with $j > 0$, $k \geq 0$, and $j + k = i$. Computing $t_i'' = a_{i-1}'' - p^{e_i} a_i''$, using (6), (8), and grouping appropriate terms, we obtain (7). One can verify that $p^{h_i - r} w_i^{(r)} = 0$ for $r \leq h_i$, otherwise $w_i^{(r)} = 0$.

One starts the induction with $\bigoplus_{i \geq 1} \langle t_i \rangle$, i.e., with $\ell_i = 1$, $c_i = -1$, $r = 0$, and $u_i' = 0$ in (6). Define $w_i = \sum_{r \geq 0} w_i^{(r)}$. This makes sense since $w_i^{(r)} = 0$ for $r \geq h_i$. Observe that the induction step is stable for $r \geq \exp(t_i)$. We now take $b_i = a_i + w_i$ $(i \geq 0)$, and note that $\langle b_i | i \geq 0 \rangle$ is a minimal pure subgroup of $K^{p,g}$ with $b_{i-1} - p^{e_i} b_i = k_i s_i$, where $p \nmid k_i$ for $i \geq 1$.

<u>Remark</u>. Let $\langle t_i' \rangle$ be a summand isomorphic to $\langle t_i \rangle$. Since $\exp(w_{i-1}) \leq h_{i-1} < \exp(t_i)$ the projection of $w_{i-1} - p^{e_i} w_i$ onto $\langle t_i' \rangle$ is not a generator. From (5) it follows that if the projection of t_i onto $\langle t_i' \rangle$ is a generator, then the same holds for s_i.

In the next two lemmas we examine special maps in $I(K)$.

<u>Lemma 9</u>. Let K be a group and $g \in K$ a torsion-free element. Suppose that $K^{\#}$ is a minimal pure subgroup of $K^{p,g}$ containing g, $T(K^{\#}) = \bigoplus_{i \geq 1} \langle t_i \rangle$, and let $\bigoplus_{i \geq 1} \langle w_i \rangle$ be pure in $T(K)$. If $\alpha \in I_p(K)$ is such that $\alpha(t_i) \in \langle w_i \rangle$ for all i, then $\exp(t_i) - \exp(\alpha(t_i)) \geq i/2$ for large i.

Proof. Since $\{\langle w_i\rangle\}$ are pure and independent, there exists a set of mutually orthogonal idempotents which correspond to $\{\langle w_i\rangle\}$. We denote the generators of $K^{\#}$ by $\{a_i\}$ and consider α on a_i $(i \geq 0)$. Let η be the projection on $\langle w_\ell\rangle$ and let $\eta\alpha(a_i) = c_i w_\ell$. Using the fact that $t_i = a_{i-1} - p^{e_i} a_i$ and that $\eta\alpha(t_i) = 0$ $(i > \ell)$, we obtain $(c_i - p^{e_{i+1}} c_{i+1}) w_\ell = 0$ for $i \geq \ell$. This implies that $c_i w_\ell = 0$ for $i \geq \ell$, and in particular $c_\ell w_\ell = 0$. Thus $\alpha(t_\ell) = c_{\ell-1} w_\ell$. But $\eta\alpha(t_i) = 0$ for $i < \ell$ implies that $c_i w_\ell - p^{e_{i+1}} c_{i+1} w_\ell = 0$ for $i < \ell - 1$. Since $\alpha \in I_p(K)$, there exists m such that $\alpha(p^m g) = 0$, and therefore $\exp(\alpha(a_i)) \leq \sigma_{n(i)+1}$ for $i \geq m$. Let $\ell \geq 2m + 2$. Then from the equations $c_i w_\ell - p^{e_{i+1}} c_{i+1} w_\ell = 0$ $(m \leq i \leq \ell - 2)$, we conclude that $c_m w_\ell - p^{\sum_{m<j<\ell} e_j} c_{\ell-1} w_\ell = 0$. Since $\exp(c_m w_\ell) \leq \sigma_{n(m)+1} = \sum_{0\leq j<m} (e_j + f_j) + e_m$, we have $p^{\sum_{0\leq j<m}(e_j+f_j)+\sum_{m\leq j<\ell} e_j} c_{\ell-1} w_\ell = 0$. Therefore $\exp(t_\ell) - \exp(\alpha(t_\ell)) = \exp(t_\ell) - \exp(c_{\ell-1} w_\ell) \geq \sum_{m\leq j<\ell} f_j \geq \ell/2$ for $\ell \geq 2m + 2$, as desired.

Note that in the lemma, the $\langle w_i\rangle$ may be allowed to be 0.

Lemma 10. Let K be a group and $g \in K$ a torsion-free element. Suppose that $K^{\#}$ is a minimal pure subgroup of $K^{p,g}$ containing g. Let $T(K^{\#}) = \bigoplus_{i\geq 1}\langle t_i\rangle$, and let $B = (\bigoplus_{i\geq 1}\langle t_i\rangle) \oplus B_1$ be a basic subgroup of the reduced part of $T(K)_p$. Then there exists an $\alpha \in I(K)$ such that $\alpha(B_1) = 0$ and

$$\alpha(t_i) = p^{f_{i-1}+[(i-1)/2]} t_i + p^{e_i+f_i+[i/2]} t_{i+1} \quad \text{for } i \geq 1.$$

Proof. Without loss of generality we may assume that $K = K^{p,g}$ and that $T(K)$ is reduced. We may regard K as a pushout of $K^{\#}$ with $T(K)$. Denote the generators of $K^{\#}$ by $\{a_i \mid i \geq 0\}$. Define α_1 on $K^{\#}$ by $\alpha_1(a_i) = p^{f_i+[i/2]} t_{i+1}$ $(i \geq 0)$, and note that $\alpha_1(g) = 0$. It follows that $\alpha_1(t_i) = p^{f_{i-1}+[(i-1)/2]} t_i - p^{e_i+f_i+[i/2]} t_{i+1}$ for $i \geq 1$, and that $\exp(t_i) - \exp(\alpha_1(t_i)) \to \infty$ as $i \to \infty$. We now define α_2 on B by $\alpha_2(B_1) = 0$ and $\alpha_2(t_i) = \alpha_1(t_i)$ $(i \geq 1)$. By [1, Section 46, Exercises 5 and 6], this extends to a map on $T(K)$. Since α_1 and α_2 agree on $T(K^{\#})$ they induce a map $\alpha \in I(K)$ with the desired property.

We now make an observation about the hypotheses of the upcoming proposition. If $I_p(K) \cong I_p(L)$, then Proposition 3.16 in [3] implies that there exist torsion-free elements $g \in K$ and $h \in H$ with equal p-indicators, except possibly in the trivial case when $T(K)_p = 0$.

Proposition 11. Let K and L be groups such that $I_p(K) \cong I_p(L)$. Let $g \in K$ and $h \in L$ be torsion-free elements such that $h_p(p^i g) = h_p(p^i h)$ for $i \geq 0$. Then $K^{p,g} \cong L^{p,h}$, and moreover, the isomorphism can be chosen such that g maps to h.

Proof. Without loss of generality we may assume that $K^{p,g}$ and $L^{p,h}$ are reduced groups with $T(K) \cong T(L)$. Let $\Phi: I_p(K) \to I_p(L)$ be the given isomorphism and let $\phi: T(K) \to T(L)$ be the isomorphism given by (3). Denote the image of t under ϕ by t^*. One of our objectives is to obtain particular minimal pure subgroups, $K^{\#}$ and $L^{\#}$, of $K^{p,g}$ and $L^{p,g}$ containing g and h respectively. We shall denote the generators of $K^{\#}$ and $L^{\#}$ by $\{a_i | i \geq 0\}$ and $\{b_i | i \geq 0\}$ respectively, and we let $T(K^{\#}) = \bigoplus_{i \geq 1} \langle t_i \rangle$ and $T(L^{\#}) = \bigoplus_{i \geq 1} \langle s_i^* \rangle$.

Although we know that the p-indicators of $K^{p,g}$ and $L^{p,h}$ are equivalent and that their torsion subgroups are isomorphic, we are still far from showing that $K^{p,g}$ and $L^{p,h}$ are isomorphic. Our aim is to show that the isomorphism between $K^{\#}$ and $L^{\#}$ which takes a_i to b_i $(i \geq 0)$, can be extended to an isomorphism of $K^{p,g}$ with $L^{p,h}$. Note that the isomorphism of $K^{\#}$ with $L^{\#}$ maps g to h. Since the groups $K^{p,g}$ and $L^{p,h}$ are pushouts of $K^{\#}$ with $T(K)_p$, and $L^{\#}$ with $T(L)_p$, respectively, the extendability of the aforementioned map is equivalent to finding an automorphism of $T(K)_p$ which takes t_i to s_i for $i \geq 1$. In the case that $T(K^{\#})$ is finite, the automorphism is easily seen to exist. In the case that $T(K^{\#})$ is unbounded, the desired automorphism has to be constructed carefully. We need some preliminaries.

Let $\langle a_i' | i \geq 0 \rangle$ and $\langle b_i' | i \geq 0 \rangle$ be minimal pure subgroups of $K^{p,g}$ and $L^{p,h}$ containing g and h respectively. Since the parameters associated with these subgroups are equal, we use the symbols σ_n, e_n, f_n to denote the parameters in either group. Put $t_i' = a_{i-1}' - p^{e_i} a_i'$ and $(s_i')^* = b_{i-1}' - p^{e_i} b_i'$ $(i \geq 1)$.

Extend $\bigoplus_{i\geq 1}\langle t_i'\rangle$ to a basic subgroup $B = (\bigoplus_{i\geq 1}\langle t_i'\rangle) \oplus (\bigoplus_{\nu}\langle u_\nu'\rangle)$ of $T(K)_p$. Consider the projection of s_i' onto $\langle t_i'\rangle$. If this projection is not a generator of $\langle t_i'\rangle$, then the projection of s_i' on $\langle u_{\nu_i}'\rangle$ will be a generator of $\langle u_{\nu_i}'\rangle$ for some ν_i. Set $u_{\nu_i} = u_{\nu_i}' - t_i'$, and $u_\nu = u_\nu'$ if $\nu \neq \nu_i$ for any i. Note that in the decomposition $B = (\bigoplus_{i\geq 1}\langle t_i'\rangle) \oplus (\bigoplus_{\nu}\langle u_\nu\rangle)$, the projection of s_i' onto $\langle t_i'\rangle$ is now a generator of $\langle t_i'\rangle$.

We are now ready to define the subgroups $K^\#$ and $L^\#$. In defining $L^\#$, $T(L^\#) = \bigoplus_{i\geq 1}\langle s_i^*\rangle$ will satisfy the property that for each i, $s_i = w_i + v_i$ where $w_i \in \bigoplus_{i\geq 1}\langle t_i'\rangle$, $v_i \in \bigoplus_{\nu}\langle u_\nu\rangle$, and the p-height of each nonzero component of w_i and v_i is less than e_i. The generators of $L^\#$, $\{b_i | i \geq 0\}$, will be chosen inductively. We start the induction with $b_0 = b_0'$. Suppose that the generators $\{b_i | 0 \leq i \leq k-1\}$ have been chosen, with s_i having the desired property for $i \leq k-1$. We want to choose b_k such that s_k has the required property. Put $s_k'' = b_{k-1} - p^{e_k}b_k'$. Since B is basic we can write $s_k'' = p^{e_k}t + s$, where $t \in T(K)_p$, $s \in B$, and $\exp(t) \leq \exp(s_k'') + e_k = \sigma_{n(k)+1}$. We may assume that the nonzero components of s have p-height less than e_k. Set $b_k = b_k' + t$. Note that $s_k^* = (s_k'')^* - p^{e_k}t^* = s^*$. Thus $s_k = s$, and hence it has the desired property. Note that by the remark following Lemma 4, the projection of s_i onto $\langle t_i'\rangle$ is a generator since that is the case for s_i'. In particular, this says that the projection of w_i onto $\langle t_i'\rangle$ is a generator. Thus $\bigoplus_{i\geq 1}\langle w_i\rangle$ is a pure subgroup of $\bigoplus_{i\geq 1}\langle t_i'\rangle$ with $\langle w_i\rangle \cong \langle t_i'\rangle$ for all i. In fact the condition on the p-height of the nonzero components of w_i implies that $w_i \in \langle t_j' | j \leq i\rangle$. Therefore $\bigoplus_{i\geq 1}\langle w_i\rangle = \bigoplus_{i\geq 1}\langle t_i'\rangle$, and we may write $B = (\bigoplus_{i\geq 1}\langle w_i\rangle) \oplus (\bigoplus_{\nu}\langle u_\nu\rangle)$. By Lemma 4 we can find generators $\{a_i | i \geq 0\}$ such that $t_i = a_{i-1} - p^{e_i}a_i = k_i w_i$, where $p \nmid k_i$ for all i. Put $K^\# = \langle a_i | i \geq 0\rangle$. Relative to this choice, the equation $s_i = w_i + v_i$ becomes $s_i = k_i^{-1}t_i + v_i$ $(i \geq 1)$.

At this point, we would like to assume two results and defer their proofs until later as they are rather technical and lengthy. The first asserts that $\exp(s_i) - \exp(v_i) \to \infty$ as $i \to \infty$. The second result which we will assume for now states that the sequence $\{k_i | i \geq 1\}$ converges to a p-adic integer π. In fact π

is a unit since $p \nmid k_i$ for all i. Moreover $\pi(s_i) = k_{j_i}s_i$, where $j_i > i$. Let $k_{j_i} - k_i = \ell_i$. Note that the power of p dividing ℓ_i tends to ∞ as $i \to \infty$. Thus

(12) $\pi(s_i) = k_{j_i}s_i = k_is_i + \ell_is_i = t_i + k_iv_i + \ell_is_i$ for $i \geq 1$.

We are now ready to define an automorphism of $T(K)_p$ which takes s_i to t_i for all i. First we define a map ψ on $B = (\bigoplus_{i\geq 1}\langle t_i\rangle) \oplus (\bigoplus_\nu \langle u_\nu\rangle)$, a basic subgroup of $T(K)_p$. Let $\psi(t_i) = -k_i^2v_i - k_i\ell_is_i$ $(i \geq 1)$, and $\psi(u_\nu) = 0$ for every ν. Note $\exp(t_i) - \exp(\psi(t_i)) \to \infty$ since $\ell_i \to 0$ p-adically, $\langle t_i\rangle \cong \langle s_i\rangle$, and $\exp(s_i) - \exp(v_i) \to \infty$. By [1, Section 46, Exercises 5 and 6], we can extend ψ to $T(K)_p$. We claim that $\pi + \psi$ is the desired automorphism. First, we produce an inverse for $\pi + \psi$. Define θ on B by $\theta(t_i) = -\pi^{-1}k_i^{-1}\psi(t_i)$ $(i \geq 1)$, and $\theta(u_\nu) = 0$ for every ν. Note that $\exp(t_i) - \exp(\theta(t_i)) \to \infty$, and so we can extend θ to all of $T(K)_p$. Before we verify that $\pi^{-1} + \theta$ is the inverse to $\pi + \psi$, we calculate that $\theta(t_i) = \pi^{-1}k_iv_i + \pi^{-1}\ell_is_i$ $(i \geq 1)$, and $\psi(s_i) = \psi(k_i^{-1}t_i + v_i) = -k_iv_i - \ell_is_i$ $(i \geq 1)$. Using (12), we have that $(\pi + \psi)(s_i) = t_i$ and $(\pi^{-1} + \theta)(t_i) = s_i$ $(i \geq 1)$. Observe that $(\pi^{-1} + \theta)(\pi + \psi)|_B = 1_B$. Since B is a basic subgroup of the reduced group $T(K)_p$, this implies that $(\pi^{-1} + \theta)(\pi + \psi) = 1_{T(K)_p}$. Similarly, $(\pi + \psi)(\pi^{-1} + \theta) = 1_{T(K)_p}$. Thus $\pi + \psi$ is an automorphism of $T(K)_p$ taking s_i to t_i for all i, as desired. The proof is complete except for the two assertions we made earlier. We now give their proofs.

Suppose, contrary to our first claim, that $\exp(s_i) - \exp(v_i) \not\to \infty$. Let d be the minimal drop in exponent which occurs infinitely often. Let $\{i(k)\}$ be the indices such that $\exp(s_{i(k)}) - \exp(v_{i(k)}) = d$. In order to ensure that $v_i \neq 0$, we may assume that $\exp(s_{i(k)}) > d$. For each k choose a generator u_{ν_k} such that the projection of $s_{i(k)}$ on $\langle u_{\nu_k}\rangle$ has an exponent drop of d. In general we shall refer to the projection of s_j on $\langle u_{\nu_\ell}\rangle$ by $c_{j\ell}u_{\nu_\ell}$. Thus $\exp(s_{i(k)}) - \exp(c_{i(k)k}u_{\nu_k}) = d$. By passing to a subsequence, we may assume the

u_{ν_k} are distinct since $\exp(s_{i(k)}) \to \infty$. We may delete the finite number of u_{ν_k} which occur nontrivially in s_i for those i where $\exp(s_i) - \exp(v_i) < d$. We now show that $c_{jk}u_{\nu_k} = 0$ for $j < i_k$. We have $h_p(c_{jk}u_{\nu_k}) = \exp(u_{\nu_k}) - \exp(c_{jk}u_{\nu_k}) \geq \exp(u_{\nu_k}) - \exp(s_j) + d > \exp(u_{\nu_k}) - \exp(s_{i(k)}) + e_j + d \geq e_j$. This implies that $c_{jk}u_{\nu_k} = 0$ since the p-heights of nonzero components of s_j are less than e_j. Our aim is to replace the subgroup $L^{\#}$ by another minimal pure subgroup $L_1^{\#}$ containing h, and such that if $T(L_1^{\#}) = \bigoplus_{i\geq 1}\langle r_i^*\rangle$, then the projection of r_n on $\langle u_{\nu_k}\rangle$ is 0 for $n \neq i(k)$, and is $c'_{nk}u_{\nu_k}$ for $n = i(k)$, where c'_{nk} and $c_{i(k)k}$ are divisible by the same power of p.

The group $L_1^{\#}$ will be described by generators $\{b''_i \mid i \geq 0\}$ which are a modification of those of $L^{\#}$. Let $b''_j = b_j$ for $j < i(1)$, and for $i(k) \leq j < i(k+1)$, let $b''_j = b_j + z_j$ where

$$z_j = -\sum_{1\leq \ell\leq k}(c_{j+1,\ell} + \sum_{m\geq 0} p^{e_{j+1}+\dots+e_{j+1+m}}c_{j+m+2,\ell})u^*_{\nu_\ell}.$$

Note that this makes sense since for large m the terms in the sum are zero. We need to check that $\exp(z_j) \leq \sigma_{n(j)+1}$. However, for some ℓ $(1 \leq \ell \leq k)$, $\exp(z_j) \leq \exp(u^*_{\nu_\ell}) \leq \exp(s^*_{i(\ell)}) + e_{i(\ell)} = \sigma_{n(i_\ell)+1} \leq \sigma_{n(j)+1}$. Thus $T(L_1^{\#}) = \bigoplus_{j\geq 1}\langle r_j^*\rangle$, where $r_j^* = b''_{j-1} - p^{e_j}b''_j = s_j^* + z_{j-1} - p^{e_j}z_j$. A direct calculation yields that

$$z_{j-1} - p^{e_j}z_j = \begin{cases} 0 \quad \text{for } j < i(1), \\ (p^{e_{i(1)}}c_{i(1)+1,1} + p^{e_{i(1)}}\sum_{m\geq 0} p^{e_{i(1)+1}+\dots+e_{i(1)+1+m}}c_{i(1)+2+m,1})u^*_{\nu_1} \\ \qquad\qquad \text{for } j = i(1), \\ -\sum_{1\leq \ell\leq k} c_{j\ell}u^*_{\nu_\ell} \qquad \text{for } i(k) < j < i(k+1), \\ -\sum_{1\leq \ell\leq k} c_{j\ell}u^*_{\nu_\ell} + p^{e_j}[c_{j+1,k+1} + \sum_{m\geq 0} p^{e_{j+1}+\dots+e_{j+1+m}}c_{j+2+m,k+1}]u^*_{\nu_{k+1}} \\ \qquad\qquad \text{for } j = i(k+1),\ k \geq 1. \end{cases}$$

Thus $r_j^* = s_j^* + z_{j-1} - p^{e_j}z_j = w_j^* + v_j^* + z_{j-1} - p^{e_j}z_j$. Reading this equation in $T(K)_p$, we have $r_j = w_j + v_j'$ where v_j' is the preimage of $v_j^* + z_{j-1} - p^{e_j}z_j$. Note that the coefficient $c'_{i(k),k}$ of the u_{ν_k}-component of $r_{i(k)}$ is the sum of $c_{i(k),k}$ and an integer which is divisible by $p^{e_{i(k)}}$. Since $c_{i(k)k}$ is not divisible by $p^{e_{i(k)}}$, we see that $c_{i(k)k}$ and $c'_{i(k)k}$ are divisible by the same p-power. Note that the u_{ν_k}-component of r_j is zero for $j \neq i(k)$ for all k, and that $\exp(r_{i(k)}) - \exp(c'_{i(k)k}u_{\nu_k}) = \exp(s_{i(k)}) - \exp(c_{i(k)k}u_{\nu_k}) = d$. Recall that $B = (\bigoplus_{i\geq 1}\langle t_i\rangle) \bigoplus (\bigoplus_\nu \langle u_\nu\rangle)$ is a basic subgroup of $T(K)_p$. Regard $K^{p,g}$ as a pushout of $K^\#$ and $T(K)_p$, where $K^\# \cap T(K)_p = \bigoplus_{i\geq 1}\langle t_i\rangle$. Define a map on B by sending t_i to zero for $i \geq 1$, u_ν to zero if $\nu \neq \nu_k$ for all k, and u_{ν_k} to $p^{[i(k)/3]}u_{\nu_k}$. By [1, Section 46, Exercises 5 and 6], this extends to a map α on $T(K)_p$. Since α and the zero map on $K^\#$ agree on $\bigoplus_{i\geq 1}\langle t_i\rangle$, we have induced a map, which we shall also call α, from $K^{p,g}$ to $T(K)_p$. Clearly $\alpha(r_j) = 0$ if $j \neq i(k)$ for all k, and $\alpha(r_{i(k)}) = c'_{i(k)k}p^{[i(k)/3]}u_{\nu_k}$. Recall that $\exp(r_{i(k)}) - \exp(c'_{i(k)k}u_{\nu_k}) = d$ and so $\exp(r_{i(k)}) - \exp(\alpha(r_{i(k)})) \leq d + [i(k)/3]$. Consider $\alpha^* \in I_p(L)$. Since $T(L_1^\#) = \bigoplus_{i\geq 1}\langle r_i^*\rangle$, $\alpha(r_{i(k)}^*) \in \langle u^*_{\nu_k}\rangle$ and $\alpha^*(r_i^*) = 0$ if $i \neq i(k)$ for all k, it follows by Lemma 9 that $\exp(r_i^*) - \exp(\alpha(r_i^*)) \geq i/2$ for large i. This is a contradiction, hence our first assertion holds.

Now we show that the sequence $\{k_i | i \geq 1\}$ converges to a p-adic integer. Let $\alpha^* \in I_p(L)$ be as in Lemma 10 relative to the torsion subgroup $T(L^\#) = \bigoplus_{i\geq 1}\langle s_i^*\rangle$ and the basic subgroup B. Then $\alpha \in I_p(K^{p,g})$, $\alpha(s_i) = p^{f_{i-1}+[(i-1)/2]}s_i - p^{e_i+f_i+[i/2]}s_{i+1}$ for $i \geq 1$, and $\alpha(u_\nu) = 0$ for all ν. Recalling that $s_i = k_i^{-1}t_i + v_i$, we have

$$(13)\quad \alpha(t_i) = p^{f_{i-1}+[(i-1)/2]}t_i - k_ik_{i+1}^{-1}p^{e_i+f_i+[i/2]}t_{i+1} + k_ip^{f_{i-1}+[(i-1)/2]}v_i - k_ip^{e_i+f_i+[i/2]}v_{i+1}.$$

Since the $\langle t_i\rangle$ and $\langle u_\nu\rangle$ are pure and independent, there are mutually orthogonal idempotents corresponding to them. In order to make conclusions about the k_i, we need to examine closely how α maps the generators $\{a_i\}$ and the torsion subgroup $\bigoplus_{i\geq 1}\langle t_i\rangle$. Since $\alpha \in I_p(K^{p,g})$, there exists an ℓ such that $\exp(\alpha(a_i)) \leq \sigma_{n(i)+1}$ for $i \geq \ell$. Let $r \geq 2\ell + 2$ and let η be the projection on $\langle t_{r+1}\rangle$.

$$\eta\alpha(a_{r+i}) = \begin{cases} p^{f_{r+i}+\sum_{i<j\leq 0}(e_{r+j}+f_{r+j})} c_{r+i}t_{r+1} & \text{for } \ell - r \leq i < 0, \\ p^{f_r}c_r t_{r+1} & \text{for } i = 0, \\ c_{r+i}t_{r+1} & \text{for } i > 0. \end{cases}$$

Using the fact that $t_j = a_{j-1} - p^{e_j}a_j$, the above implies that

$$(14)\quad \eta\alpha(t_{r+i+1}) = \begin{cases} p^{\sum_{i<j\leq 0}(e_{r+j}+f_{r+j})}(p^{f_{r+i}}c_{r+i} - c_{r+i+1})t_{r+1} & \text{for } \ell - r \leq i < 0, \\ (p^{f_r}c_r - p^{e_{r+1}}c_{r+1})t_{r+1} & \text{for } i = 0, \\ (c_{r+i} - p^{e_{r+i+1}}c_{r+i+1})t_{r+1} & \text{for } i > 0. \end{cases}$$

We now examine the divisibility of c_{r-1}. Note that from (13), $\eta\alpha(t_j) = 0$ for $\ell < j < r$. Equating that with the projection in (14), we have $p^{\sum_{i<j\leq 0}(e_{r+j}+f_{r+j})}(p^{f_{r+i}}c_{r+i} - c_{r+i+1})t_{r+1} = 0$ for $\ell - r \leq i < -1$. By working with these equations we obtain

$$(p^{f_\ell+(e_{\ell+1}+f_{\ell+1})+\dots+(e_r+f_r)}c_\ell - p^{e_{\ell+1}+\dots+e_r+f_{r-1}+f_r}c_{r-1})t_{r+1} = 0.$$

Since $\exp(t_{r+1}) = \sum_{0\leq j\leq r}(e_j + f_j)$, this implies that c_{r-1} is divisible by $p^{f_\ell+\dots+f_{r-2}}$. Since $r \geq 2\ell + 2$, it follows that c_{r-1} is divisible by $p^{[r/2]}$, hence we may write $c_{r-1} = p^{[r/2]}c'_{r-1}$. Note that $\{c'_{r-1}\} \to 0$ p-adically.

From (13) we know that $\eta\alpha(t_j) = 0$ for $j > r + 1$, hence by (14) we have $(c_j - p^{e_{j+1}}c_{j+1})t_{r+1} = 0$ for $j \geq r + 1$. This implies that

$(c_j - p^{e_{j+1}+\dots+e_{j+i}} c_{j+i})t_{r+1} = 0$ for all $i \geq 1$ and $j \geq r+1$. Thus $c_j t_{r+1} = 0$ for $j \geq r+1$, and in particular $c_{r+1}t_{r+1} = 0$. So $\eta\alpha(t_{r+1}) = p^{f_r} c_r t_{r+1}$. But from (13) $\eta\alpha(t_{r+1}) = p^{f_r+[r/2]} t_{r+1}$, thus $p^{f_r}(c_r - p^{[r/2]})t_{r+1} = 0$. Since $\exp(t_{r+1}) = \sum_{0\leq j\leq r} (e_j + f_j)$, we conclude that $p^{[r/2]} | c_r$ and we write $c_r = p^{[r/2]}c'_r$. Observe that $\{c'_r - 1\} \to 0$ p-adically. On equating $\eta\alpha(t_r)$ from (13) and (14) we get $-k_r k_{r+1}^{-1} p^{e_r+f_r+[r/2]} t_{r+1} = p^{e_r+f_r}(p^{f_{r-1}} c_{r-1} - c_r)t_{r+1}$ $= p^{e_r+f_r+[r/2]}(p^{f_{r-1}} c'_{r-1} - c'_r)t_{r+1}$. Since $\{c'_{r-1}\}$ and $\{c'_r - 1\}$ approach zero p-adically, and since $\exp(t_{r+1}) - e_r - f_r - [r/2] \to \infty$ as $r \to \infty$, we conclude that $\{k_r k_{r+1}^{-1} - 1\} \to 0$ p-adically. But this precisely means that the sequence $\{k_i\}$ converges to a p-adic integer. This completes the proof of Proposition 11.

We now give the

Proof of the Main Theorems. Let $x \in G$, $y \in H$ be torsion-free elements. By Proposition 2 the isomorphism of End (G) with End (H) implies that $I_p(G) \cong I_p(H)$ for all primes p. Applying Proposition 1 we conclude that the height matrices of G and H are equivalent. If p is a prime such that $h_p(p^i x) = h_p(p^i y)$ for $i \geq 0$ (this occurs for almost every p), then we can apply Proposition 11 to obtain $G^{p,x} \cong H^{p,y}$ with x mapping to y. If p is such that for some k and ℓ, $h_p(p^{i+k}x) = h_p(p^{i+\ell}y)$ $(i \geq 0)$, then by Proposition 11 we obtain $G^{p,x} \cong H^{p,y}$ with $p^k x$ mapping to $p^\ell y$. From these local isomorphisms we get an isomorphism of G with H (see [3, Lemma 4.2]).

REFERENCES

1. L. Fuchs, Infinite Abelian Groups, Academic Press, New York, 1970 and 1973.
2. I. Kaplansky, Infinite Abelian Groups, University of Michigan Press, Ann Arbor, 1969.
3. W. May and E. Toubassi, Endomorphism rings of mixed groups of torsion-free rank one, to appear.
4. K. Wallace, On mixed groups of torsion-free rank one with totally projective primary components, J. Algebra 17 (1971), 482-488.

LOCAL-QUASI-ENDOMORPHISM RINGS OF RANK ONE MIXED ABELIAN GROUPS

Carol L. Walker

Throughout this paper, group will mean abelian group, rank will mean torsion free rank, and Z_p will denote the ring of integers localized at a prime p. A Z_p-module which can be represented as a summand of a simply presented Z_p-module will be called a Warfield group.

A summand G of a direct sum of rank one Warfield groups need not be again a direct sum of rank one groups, but it is true that there is a totally projective p-group T for which $G \oplus T$ is a direct sum of rank one groups. A proof of this latter fact is given in [3, §8] by studying the situation in the category with objects valuated Z_p-modules and maps $\mathrm{Hom}(A,B)/\mathrm{Hom}(A,B_t)$. In this category, objects A and B are isomorphic if and only if there are torsion modules S and T with $A \oplus S$ isomorphic (as a valuated module) to $B \oplus T$.

A direct summand of a direct sum of finite rank torsion free groups need not be a direct sum of finite rank torsion free groups. An aspect of uniqueness for these decompositions for finite sums was described by Jónsson [4] with the notion of quasi-isomorphism. For infinite direct sums the situation was identified in [1] and [7]. If G is a summand of a direct sum of finite rank torsion free groups, then G has a subgroup G' which is (1) a direct sum of finite rank torsion free groups and (2) large in the sense that for each finite rank subgroup F of G there is a nonzero integer n such that $nF \subset G'$. The uniqueness statements that can be made in this case are similar to those for quasi-isomorphism. The development in [7] involves studying the situation in the category with objects all torsion free groups and maps from G to H the direct limit of the groups $\mathrm{Hom}(G', H)$ where G' ranges over the set of subgroups of G satisfying the condition in (2) above.

We combine these two situations into a single global setting. For certain rank one groups, we show that if G is a summand of a direct sum of these rank one groups, then there is a torsion group T and a subgroup G' of G

for which $G' \oplus T$ is a direct sum of rank one groups, with G' a "large" subgroup of G in the sense of (2) above. Again, uniqueness statements can be made regarding the decomposition, relative to a certain equivalence relation.

We get these results by defining a quotient category related to the two described above, which is additive, has kernels and infinite sums and satisfies a weak Grothendieck condition, and applying theorems from [8]. We show that if a rank one group has a local endomorphism ring in this category, then the height matrix of a torsion free element, with one row deleted, has no row with an infinite number of gaps and each row with an infinite ordinal also has an ∞. If the torsion subgroup is totally projective for all primes except the one corresponding to the deleted row, the converse holds also. Note that this includes all mixed groups with primary torsion subgroup. For countable groups, Megibben's theorem in [5] implies the groups with the height matrices described above are exactly those groups for which each primary subgroup is a direct summand, except possibly for one prime. Thus a countable rank one group has a local endomorphism ring in the quotient category if and only if it splits at all but at most one prime.

All finite rank "strongly indecomposable" torsion free groups have local quasi-endomorphism rings. If G/G_t is strongly indecomposable of finite rank and G quasi-splits, then G has a local endomorphism ring in the quotient category. Less trivial examples would be of interest. In this paper we are concerned primarily with rank one groups.

1. The Quotient Category $\mathcal{L}$.

Definition 1.1. A subgroup H of a group G will be called locally quasi-equal to G (denoted $H \doteq G$) if $G_t \subset H$ and if for each finite rank subgroup F of G there is a positive integer n with $nF \subset H$.

Note that if G has finite rank, then $H \doteq G$ if and only if $nG + G_t \subset H \subset G$ for some positive integer n. If G is torsion free, this reduces to the same notion of locally quasi-equal as in [7], and if G is torsion free of finite rank, this is the same as the definition of quasi-equal associated with the usual definition of quasi-isomorphism for finite rank torsion free groups.

Definition 1.2. L is the category with objects all abelian groups, maps given by

$$\mathrm{Hom}_L(G,H) = \varinjlim_{A \doteq G} \mathrm{Hom}(A,H)/\mathrm{Hom}(A,H_t)$$

and composition induced by ordinary composition of maps.

The fact that composition is well defined, and other details that need verifying to show that this is a category, follow quickly from the following proposition. The proof, which is completely routine, is omitted.

Proposition 1.3. Let $L(G)$ denote the set of subgroups of G which are locally quasi-equal to G. The following hold.

(i) $G \in L(G)$.

(ii) If $A, B \in L(G)$, then $A \cap B \in L(G)$.

(iii) If $A \in L(G)$ and $A \subset B \subset G$, then $B \in L(G)$.

(iv) If $A \in L(B)$ and $B \in L(G)$, then $A \in L(G)$.

(v) If $f : G \to H$ is a homomorphism and $A \in L(H)$, then $f^{-1}(A) \in L(G)$.

If $\alpha \in \mathrm{Hom}_L(G,H)$, then α is induced by a map $f : A \to H$ for some $A \doteq G$. We will write $\alpha = [f]$. If $[f], [g] \in \mathrm{Hom}_L(G,H)$, then $[f] = [g]$ if and only if, on the common domain of f and g, the difference $f - g$ has image contained in H_t. If $[f] \in \mathrm{Hom}_L(G,H)$, then $[f]$ is monic if and only if $f^{-1}(H_t) = G_t$. The map $[f]$ is epic if and only if $H/\mathrm{Im}\, f$ is torsion.

Theorem 1.4. The category L is additive, has kernels and infinite sums, and satisfies a weak Grothendieck condition.

Proof. It is clear that L is additive. If $[f] : G \to H$, let $K = f^{-1}(H_t)$. It is straightforward to show that $[i] : K \to G$ is a kernel for $[f]$, where $i : K \to G$ is the inclusion map.

Let $\{A_\beta\}_{\beta\in I}$ be any family of groups and let $i_\alpha : A_\alpha \to \Sigma_{\beta\in I} A_\beta = A$ be injection maps for the ordinary direct sum. Let $\{[f_\alpha] : A_\alpha \to B\}$ be a family of maps in L. Then $f_\alpha : S_\alpha \to B$ with $S_\alpha \doteq A_\alpha$ for each α. We show $\Sigma_{\alpha\in I} S_\alpha \doteq \Sigma_{\alpha\in I} A_\alpha$. Clearly this subgroup contains the torsion subgroup of A. Let $F \subset A$ of finite rank. There are only a finite number of A_α's for which the projection $\pi_\alpha : F \to A_\alpha$ does not have a torsion image. It follows that $nF \subset \Sigma_{\alpha\in I} S_\alpha$ for some positive integer n. Now the map $\Sigma f_\alpha : \Sigma S_\alpha \to B$ induces a map $[\Sigma f_\alpha] : \Sigma A_\alpha \to B$ satisfying $[\Sigma f_\alpha][i_\beta] = [f_\beta]$ for each $\beta \in I$. This map is unique, and thus $\{[i_\alpha] : A_\alpha \to \Sigma_{\beta\in I} A_\beta\}$ is a direct sum in L.

The weak Grothendieck condition referred to is the following: if $A \to \Sigma_{i\in I} B_i$ is a nonzero monic then there is a finite subset J of I and a commutative diagram

$$\begin{array}{ccc} C & \longrightarrow & \Sigma_{i\in J} B_i \\ \downarrow & & \downarrow \\ A & \longrightarrow & \Sigma_{i\in I} B_i \end{array}$$

with the right vertical map the natural injection and the composition $C \to A \to \Sigma_{i\in I} B_i$ nonzero. In this category the condition holds for any nonzero map $[f] : A \to \Sigma_{i\in I} B_i$. Since $[f] \neq 0$, there is a torsion free nonzero element $f(a) \in \Sigma_{i\in I} B_i$. Let C be the subgroup of A generated by the element a, and let J be a finite subset of I for which $f(a) \in \Sigma_{i\in J} B_i$. This gives the desired diagram.

We note that all groups of finite rank are small in L, in the sense that any map $G \to \Sigma_{i\in I} H_i$ where G has finite rank, factors through a subsum $\Sigma_{i\in J} H_i$ with J finite. This is the observation that was used in the middle of the preceding proof.

Unique decomposition and isomorphic refinement theorems applied to this category will obtain uniqueness "up to isomorphism in this category". We need to interpret this isomorphism back in the category of abelian groups. We will write $G \sim H$ if G is isomorphic to H in L.

Theorem 1.5. $G \sim H$ if and only if there are subgroups $A \doteq G$ and $B \doteq H$, and torsion groups S and T with an isomorphism $A \oplus S \doteq B \oplus T$. One may take $S = H_t$ and $T = G_t$.

Proof. The condition is clearly sufficient. Suppose $G \sim H$. Then there are homomorphisms $f : G' \to H$ and $g : H' \to G$ with $G' \doteq G$ and $H' \doteq H$, for which $\alpha = 1 - gf : f^{-1}(H') \to G_t$ and $\beta = 1 - fg : g^{-1}(G') \to H_t$.

Let $A = f^{-1}(H')$ and define the map:

$$\phi : A \oplus H_t \to H' \oplus G_t : (x, y) \mapsto (f(x) + y, \alpha(x) - g(y)).$$

For $x \in G_t$, $\phi(x, -f(x)) = (0, x)$ so $G_t \subset \operatorname{Im}\phi$. Thus $\operatorname{Im}\phi = B \oplus G_t$ for $B = \phi(A \oplus H_t) \cap H' = \{ f(x) + y \mid x \in A, y \in H_t, \alpha(x) = g(y) \}$. Note that $B \subset g^{-1}(A) \subset g^{-1}(G')$, which is the domain of β. Thus we can define another map:

$$\theta : B \oplus G_t \to G \oplus H_t : (u, v) \mapsto (g(u) + v, \beta(u) - f(v)).$$

Then $\theta\phi(x, y) = (x, y)$ for all $(x, y) \in A \oplus H_t$. Since $B \oplus G_t = \operatorname{Im}\phi$, this implies $\theta(B \oplus G_t) = A \oplus H_t$. It is then easily checked that $\phi\theta$ is the identity map also.

By Proposition 1.3, we know that $A \doteq G$. To show $B \doteq H$, let $F \subset H'$ have finite rank. Then $g(F)$ is a finite rank subgroup of G so $ng(F) \subset A$ for some $n > 0$. Note that $nF \subset g^{-1}(A)$, so β is defined on nF. Now if $x \in F$, $\phi(g(nx), \beta(nx)) = (nx, 0)$. Thus $nF \subset \operatorname{Im}\phi \cap H' = B$, and $B \doteq H'$. It follows that $B \doteq H$.

Corollary 1.6. G is isomorphic in L to a torsion-free group if and only if G has a torsion-free subgroup L with $L \oplus G_t \doteq G$.

The groups described in the Corollary will be called locally quasi-split groups. If G has finite rank, this coincides with the usual notion of a quasi-splitting group.

Lemma 1.7. If $G \doteq \Sigma_{i\in I} G_i$ and each G_i has finite rank, then $\Sigma_{i\in I} (G \cap G_i) \doteq G$.

Proof. Let $F \subset G$ of finite rank. Then $F \subset \Sigma_{i\in J} G_i + \Sigma_{i\in I}(G_i)_t$ for some finite $J \subset I$. Then for some $n \neq 0$, $n(\Sigma_{i\in J} G_i) \subset G$. Thus $n(\Sigma_{i\in J} G_i) = \Sigma_{i\in J} nG_i \subset \Sigma_{i\in J} (G_i \cap G)$. Thus $nF \subset \Sigma_{i\in J} (G_i \cap G) + \Sigma_{i\in I} (G_i)_t \subset \Sigma_{i\in I} (G_i \cap G)$.

This lemma and the preceding theorem give the following result.

Theorem 1.8. If $G \sim \Sigma_{i\in I} H_i$, where each H_i has finite rank, then there are subgroups $G' \doteq G$, $H_i' \doteq H_i$, and torsion groups T and S with

$$G' \oplus T \cong \Sigma_{i\in I} H_i' \oplus S.$$

In particular, if each H_i has rank one then $G' \oplus T$ is a direct sum of rank one groups.

2. Local Endomorphism Rings. The unique decomposition and isomorphic refinement theorems we will apply rely on certain objects having local endomorphism rings, so we will concentrate on identifying these objects. For rank one groups, the following theorem reduces this study to the case of groups with primary torsion subgroup. Let Q denote the field of rationals and Z_p the ring of integers localized at a prime p. Let $E(G) = \mathrm{Hom}_L(G, G)$, and $E_p(G) = E(G/\Sigma_{q\neq p}G_q)$.

Theorem 2.1. Let G be a rank one group. There is a natural embedding of the rings $E(G)$ and $E_p(G)$ as subrings of Q, and considered as such, $E(G) = \bigcap_p E_p(G)$. Moreover, for each prime p, $Z_p \subset E_p(G) \subset Q$, so $E(G)$ is local if and only if for all but at most one prime, $E_p(G) \cong Q$.

Lemma 2.2. If the torsion subgroup of a group G is a p-group and $(n, p) = 1$, then

(i) in G, division by n is unique (whenever possible), and

(ii) $E(G) = nE(G)$.

Proof. Part (i) is obvious. To prove (ii), let $[f] \in E(G)$, $f : A \to G$ for some $A \doteq G$. Define $g : nA + G_t \to G$ as follows: For $na + s \in nA + G_t$ there is a $k \geq 0$ with $p^k s = 0$, and there is a linear combination $mn + qp^k = 1$, so that $s = mns$. Let $g(na + s) = g(n(a + ms)) = f(a + ms)$. Then g is a homomorphism, and $[f] = n[g]$.

Lemma 2.3. Let G be any group. Then $E(G)$ is p-divisible if $E_p(G)$ is p-divisible.

Proof. Let $S = \Sigma_{q \neq p} G_q$. Then $E_p(G) = E(G/S)$. Assume $E(G/S)$ is p-divisible. Let $[f] \in E(G)$, $f : A \to G$ for some $A \doteq G$. Let $\overline{f} : A/S \to G/S$ be the map induced by f. Now $A/S \doteq G/S$ so $[\overline{f}] \in E(G/S)$ and $[\overline{f}] = p[g]$ for some $[g] \in E(G/S)$. This means, for some $B/S \doteq G/S$, $\overline{f} - pg = \alpha : B/S \to G_t/S$. We may assume $B \subset A$. Note that $B \doteq G$. The natural map $\pi : G_p \to G_t/S$ is an isomorphism, so α induces a map $h : B \to G_t$ via the composition $B \to B/S \xrightarrow{\alpha} G_t/S \xrightarrow{\pi^{-1}} G_p \to G_t$ such that the two squares

$$\begin{array}{ccc} B & \xrightarrow{h} & G_t \\ \downarrow & & \downarrow \\ B/S & \xrightarrow{\alpha} & G_t/S \end{array} \qquad \text{and} \qquad \begin{array}{ccc} B & \xrightarrow{f-h} & G \\ \downarrow & & \downarrow \\ B/S & \xrightarrow[= pg]{\overline{f}-\alpha} & G/S \end{array}$$

commute. Now we have $\overline{f} - \alpha = pg = \overline{f - h}$. So for x in the domain of $f - h$, we have $f(x) - h(x) = py + s$ where $s \in S$, and $g(x) = y + S$. Now $S = pS$, so $s = ps'$ for some s', and for $z = y + s'$, we have

(i) $pz = f(x) - h(x)$ and (ii) $z + S = g(x)$.

Suppose z' satisfies (i) and (ii) above. Then $z - z' \in S \cap G[p] = \{0\}$. Thus we can define a function $k : B \to G$ by $k(x) = z$ such that (i) and (ii) above hold. It is easy to show k is a homomorphism and $f = pk + h$. Thus $[f] = p[k]$ as desired.

Proof of Theorem 2.1. Clearly $E(G)$ is torsion-free. Let $[f], [g] \in E(G)$ and assume G has rank one. Let a be a torsion-free element of G in the common domain of f and g. Then $nf(a) = mg(a)$ for some positive integers

n and m, and $nf - mg$ has torsion image. Thus $n[f] = m[g]$. Thus $E(G)$ is a rank one torsion free group. The map $n[1_G] \mapsto n$ extends uniquely to a ring homomorphism $E(G) \to Q$ which is one-to-one. Thus $E(G)$ can be considered as a subring of Q.

The natural maps $\mathrm{Hom}(A, G) \to \mathrm{Hom}(A/S, G/S)$ induce a ring homomorphism $E(G) \to E_p(G)$. If G has rank one and these are considered as subrings of Q, this is in fact the inclusion map. Thus $E(G)$ p-divisible implies $E_p(G)$ is p-divisible also. The preceding lemma establishes the converse without the assumption G has rank one.

We now can assume G has rank one and the torsion subgroup T of G is a p-group. We attempt to determine all such groups having endomorphism ring isomorphic to Q.

First we observe that if G is torsion free, $E(G) = \varinjlim_n \mathrm{Hom}(nG, G)$ is isomorphic to Q. Thus in the case G_t is a p-group, $E_q(G) \cong Q$ for all primes $q \neq p$. This observation, along with Theorem 2.1, gives the following.

Theorem 2.4. If G has rank one and the torsion subgroup of G is a p-group, then either $E(G) \cong Q$ or $E(G) \cong Z_p$, the ring of integers localized at the prime p. In particular, $E(G)$ is local.

Lemma 2.5. If $nG + G_t \subset A \subset G$, then $p^\alpha A = p^\alpha G$ for all $\alpha \geq \omega$.

Theorem 2.6. If G has finite rank, the torsion subgroup of G is a p-group, and $E(G)$ is divisible, then the p-height sequence $H_p^G(x) = (h_p^G(x), h_p^G(px), h_p^G(p^2x), \ldots)$ of any element of G satisfies:

(i) there are only a finite number of gaps, and

(ii) if any entry is infinite, there is an ∞.

Proof. Let $x \in G$. We may as well assume x has infinite order. Since $E(G)$ is divisible, $[1_G] = p[f]$ for some $f : A \to G$, $A \doteq G$. For some $m > 0$, $mx = a \in A$. Then $H_p^G(x)$ can be obtained from $H_p^G(a)$ by a finite shift, so we will consider $H_p^G(a)$. Now $a = pf(a) + t$ for some $t \in G_t$, and for some $N \geq 0$, $p^N a = p^{N+1} f(a) = f(p^{N+1}a)$. Thus $h_p^A(p^{N+k+1}a) \leq h_p^G(p^{N+k}a)$ for all $k \geq 0$. Write $m = p^L q$ with $(q, p) = 1$. Suppose there are at least L gaps from $p^M = h_p^G(p^N a)$ to $h_p^G(p^{N+K}a)$. Then $p^{N+K}a = p^{L+M+K}b$ for some $b \in G$. Thus $p^{N+1}a = p^{L+M+1}b + t$ for some $t \in p^{M+1}G_t$. This implies $p^{N+1}a \in p^{M+1}(p^L G + T) \subset p^{M+1}A$. This is a contradiction since $p^N a \notin p^{M+1}G$. Thus there are fewer than $L + N$ gaps in the height sequence of a.

Suppose $h_p^G(p^q a) \geq \omega$. Then $h_p^G(p^{q+N+1}a) \geq \omega$ and $h_p^G(p^{N+q+1}a) = h_p^A(p^{N+q+1}a) \leq h_p^G(p^{N+q}a) < h_p^G(p^{N+q+1}a)$. This can occur only if $h_p^G(p^{N+q+1}a) = \infty$.

The following theorem gives a partial converse to the above theorem.

Theorem 2.7. If G has rank one, the torsion subgroup of G is a totally projective p-group, and the p-height sequences of elements of G satisfy conditions (i) and (ii) of Theorem 2.7, then $E(G) \cong Q$.

The proof of this theorem relies on the following facts.

(1) If A has rank one and A_p is totally projective then for any element $a \in A$, $(A/Za)_p$ is totally projective [9].

(2) A cyclic subgroup Za is nice in A [6].

(3) Suppose $S \rightarrowtail A \twoheadrightarrow A/S$ is a short exact sequence with S p-nice in A and A/S a totally projective p-group. Let $f : S \to G$ be a homomorphism satisfying $h_p^A(s) \leq h_p^G(f(s))$ for all $s \in S$. Then there is an extension of f to a homomorphism $A \to G$ [2].

Proof of Theorem 2.7. Let a be a torsion-free element of G, and suppose $h_p^G(p^k a) = \infty$ for some $k \geq 0$. Then the function

$$Zp^{k+1}a \to G : np^{k+1}a \mapsto np^k a$$

is non-decreasing on heights, so there is a homomorphism $G^p \to G$ extending this map, where $G^p/Zp^{k+1}a = (G/Zp^{K+1}a)_p$. Similarly, there is a homomorphism $G^q \to G$ extending the map for all $q \neq p$, where $G^q/Zp^{k+1}a = (G/Zp^{k+1}a)_q$. These extensions agree on the intersections of the G^q's, so induce a homomorphism $f : G \to G$ satisfying $f(p^{k+1}a) = p^k a$. Then

$1_G - pf : G \to G_t$, so $[1_G] = p[f]$, and $E(G)$ is p-divisible.

Now assume $a \in G$ is a torsion-free element with $H_p^G(a)$ having all finite entries, with no gaps. Assume $h_p^G(a) = p^N$, and suppose $p^k a \in p^{N+k}(pG + G_t)$. Then $p^k a = p^{N+k}(px + t) = p^{N+k+1}x + p^{N+k}t$ and since $t \in G_p$, for some $j \geq 0$, $p^{k+j}a = p^{N+k+j+1}x$. This cannot happen with no gaps. Thus $Zpa \to G : npa \mapsto na$ is non-decreasing on heights, with heights for npa computed in $pG + G_t$, and there is an extension $f : pG + G_t \to G$ with $f(pa) = a$. Then $[1_G] = p[f]$ and $E(G)$ is p-divisible.

To summarize, the height matrix of a group with local endomorphism ring satisfies conditions (i) and (ii) of Theorem 2.6 for all rows if $E(G) \cong Q$, and for all but the p-row if $E(G) \cong Z_p$. If G_p is totally projective for all but one prime p, then the height conditions (i) and (ii) for all but the p-row are sufficient to imply that the endomorphism ring of G is local.

If G is a countable group it follows easily from a theorem in [5] that the p-row of a height matrix of G satisfies (i) and (ii) if and only if G_p is a direct summand of G. Thus a countable rank one group has local endomorphism ring if and only if G_p is a summand of G for all but at most one prime p. Clearly this condition is sufficient to imply $E(G)$ is local even when G is not countable.

A related class of groups having local endomorphism rings is those rank one groups G for which G_p is a quasi-summand for all but at most one prime p. This is the same as saying $G/\Sigma_{q \neq p} G_q$ is a quasi-splitting group for all but at most one prime p.

3. Unique Decomposition and Isomorphic Refinement Theorems. The following theorems from [8] are valid in L, since L is additive, has kernels and infinite sums, and satisfies a weak Grothendieck condition.

Theorem 3.1. Suppose $G \sim \Sigma_{i \in I} G_i$ with $E(G_i)$ local for all $i \in I$. Then if $G \sim \Sigma_{j \in J} H_j$ with each H_j indecomposable in L, there is a bijection $\alpha : I \to J$ with $G_i \sim H_{\alpha(i)}$ for all $i \in I$.

Theorem 3.2. If $G \sim \Sigma_{i \in I} G_i \sim A \oplus B$ with $E(G_i)$ local and G_i finite rank for all $i \in I$, then $A \sim \Sigma_{i \in J} G_i$ for some $J \subset I$. Consequently, any two direct decompositions of G have isomorphic refinements.

This last theorem uses the fact that finite rank groups are small and hence countably finitely approximable. This theorem implies the following in the category of abelian groups.

Theorem 3.3. Suppose each G_i $(i \in I)$ is a finite rank group with $E_L(G_i)$ local. Then if $\Sigma_{i \in I} G_i = A \oplus B$, there are subgroups $A' \doteq A$, $G_i' \doteq G_i$, torsion groups T and S and a subset $J \subset I$ such that $A' \oplus T \cong \Sigma_{i \in J} G_i' \oplus S$.

REFERENCES

[1] L. Fuchs and G. Viljoen, On quasi-decompositions of torsion-free abelian groups of infinite rank, Math. Scand. 33(1973) 205-212.

[2] R. Hunter, Balanced subgroups of abelian groups, Ph.D. Thesis, Australian National University, Canberra (1975).

[3] R. Hunter, F. Richman and E. Walker, Warfield modules, (this publication).

[4] B. Jónsson, On unique factorization problem for torsion-free abelian groups, Bull. Amer. Math. Soc. 51(1945) 364.

[5] C. K. Megibben, On mixed groups of torsion-free rank one, Ill. Journ. of Math. 11(1967) 134-144.

[6] R. O. Stanton, The A-exchange property applied to decompositions of modules over a discrete valuation ring, Ph.D. Thesis, New Mexico State University, Las Cruces, NM (1973).

[7] C. Walker, Local quasi-isomorphisms of torsion-free abelian groups, Ill. Journ. of Math. 17(1973) 689-706.

[8] C. Walker and R. B. Warfield, Jr., Unique decomposition and isomorphic refinement theorems in additive categories, Journ. of Pure and Appl. Alg. 7(1976) 347-359.

[9] K. Wallace, On mixed groups of torsion-free rank one with totally projective primary components, Journ. of Alg. 17(1971) 482-488.

HOMOLOGICAL DIMENSION AND ABELIAN GROUPS

H. K. Farahat

1. Introduction. Let A be an abelian group and E be its endomorphism ring. Then A has a natural structure as left E-module. We are concerned with the homological dimension $d(A) = d_E(A)$ of this module, defined, as usual, by means of projective resolutions. Several questions arise: how is d(A) determined from the structure of A; what is the class of groups A with a prescribed dimension d(A); what is the range of values of d(A)? We shall give a brief survey of results and methods dealing with these and other questions.

Throughout this paper, A denotes an abelian group, E denotes the endomorphism ring of A, and $d(A) = d_E(A)$ denotes the homological dimension of A as left E-module. Generally, $d_R(M)$ denotes the homological dimension of the left R-module M, and $End_R(M)$ denotes the ring of endomorphisms of M over R.

2. The Functor Hom(-,A). For each homomorphism $f: X \to Y$ of abelian groups we have the homomorphism

$$g \in Hom(Y,A) \to gf \in Hom(X,A).$$

In particular, each endomorphism $g: A \to A$ yields an endomorphism $f \to gf$ of Hom(X,A). Thus, Hom(X,A) has a natural structure as left E-module, and Hom(-,A) determines a contravariant functor from Z-modules to E-modules. An exact sequence of Z-modules

$$0 \to X' \to X \to X'' \to 0 \qquad (*)$$

always yields an exact sequence of E-modules

$$0 \to Hom(X'',A) \to Hom(X,A) \to Hom(X',A)$$

in which the last mapping may or may not be surjective. Let us say that the sequence (*) is useful (for A) if the above mentioned mapping is surjective, and if furthermore, $Hom(X',A) \cong A$ and $Hom(X,A)$ is E-projective. Every such useful sequence then provides an exact sequence of E-modules:

$$0 \to Hom(X'',A) \to Hom(X,A) \to A \to 0,$$

which is a start on finding a projective resolution for A. In particular, we get either $d(A) = 0$ or $d(A) = 1 + d_E\, Hom(X'',A)$. Suppose, for example, that $n \geq 0$, $nA = 0$, and A has a cyclic direct summand $C \cong Z/nZ$. Then the sequence $0 \to C \to A \to A/C \to 0$ is "useful" and splits, and we conclude that $d(A) = 0$; cf. [2(2)]. Hence, if A has an infinite cyclic direct summand then $d(A) = 0$, and the same conclusion holds when A is a bounded torsion group.

These ideas may be applied in order to prove that $d(A) \leq 1$ for every torsion group A; cf. [2(15)]. The proof is firstly reduced to the case when A is a p-group, and then A is the union of the submodules $A[p^n]$ $(n = 1,2,\ldots)$, $A[p^n] = \{a \in A\colon p^n a = 0\} \cong Hom(Z/p^nZ,A)$. If A has a direct summand isomorphic to Z/p^nZ, then $A[p^n]$ is E-projective. In any case, it may be shown that each of the factor modules $A[p^{n+1}]/A[p^n]$ has dimension not exceeding 1, and hence that $d(A) \leq 1$.

Divisible groups can also be handled by these methods. If A is divisible but not torsion, then the exact sequence $0 \to Z \to Q \to Q/Z \to 0$ is "useful" for A, and yields, in fact, a projective resolution for A, namely

$$0 \to Hom(Q/Z,A) \to Hom(Q,A) \to A \to 0.$$

It follows that $d(A) \leq 1$ for every divisible group A, with strict inequality holding if and only if A is torsion-free; cf. [3].

These results have been generalized in [6], using roughly the same methods. For example, it can be shown that $d_E(H) \leq 1$ provided that H is a fully invariant torsion pure subgroup of A. Also, $d_E(H) \leq 1$ if H is a fully invariant pure subgroup of A whose reduced part is bounded. The main theorem in [6] also deals with the more general case of a pure fully invariant subgroup $H = T + D$ with divisible part D and reduced part T which is torsion. It is shown that (i) $d_E(H) \leq 2$, and more significantly that (ii) $d_E(H) \leq 1$ if and only if D is torsion or Hom(T,D) is bounded, (iii) $d_E(H) \leq 0$ if and only if D is torsion-free and T has a bounded primary component for every prime. Observe that we have here an example where a conclusion is reached on the structure of a group from a knowledge of its homological dimension. Such theorems are in short supply at present.

3. The Range of Values of d(A). We have seen that d(A) can take the values 0, 1, e.g. when A = Z, $A = Z(p^\infty)$ respectively. Initial experimentation seemed always to finish up with these two values, even when the endomorphism ring E has infinite global dimension; cf. [4,§2]. It seemed that the structure of A as an E-module remained reasonable as E was complicated, and the possibility that $d(A) \leq 1$ for all A had to be entertained. The first indication to the contrary was given in [4], where an infinite family of torsion-free groups A were constructed with $d(A) = \infty$. In fact, for each n, define

$$A_n = \{(r_1,\ldots,r_n) \in Q^n\colon r_i \in \tfrac{1}{2}Z'_{p_i} \quad \Sigma r_i \in Z_2\}$$

where $p_0 = 2$, $p_1 = 3$, $p_2 = 5$, ... is the sequence of primes, Z_p is the localization of Z at p, and Z'_p is the localization of Z at the complement of p. Then it turns out that $d(A_2) = 0$, $d(A_n) = \infty$ for $n \geq 3$.

The range of values of d(A) was finally settled brilliantly in [1], where it was proved, among other things, that for $0 \leq n \leq \infty$ there exists a countable torsion-free group A such that d(A) = n. We shall indicate the main ideas of this proof.

The first step is to choose a ring E which is reduced, torsion-free and countable as an abelian group, such that E has global dimension n + 1. For example $E = Z[x_1,\ldots,x_n]$ will do. Next, construct E-submodules $G_0,G_1,\ldots,G_n$ of the natural completion $\hat{E}$ of E such that $\mathrm{End}_Z(G_i) = E$ $(0 \leq i \leq n)$. This may be achieved by the now classical method due to A. L. S. Corner. Corner's construction admits enough freedom of choice making it possible to arrange that the following conditions hold:

$$(1) \quad d_E(G_0 + G_1 + \ldots + G_n) = n,$$

(2) $d_E(G_0 + G_1 + \dots + G_s) \leq n(1 \leq s \leq n - 1)$,

(3) $d_E(G_s \cap (G_0 + G_1 + \dots + G_{s-1})) = s - 2\ (1 \leq s \leq n)$.

These conditions imply that one of $G_0, G_1, \dots, G_{s-1}$ must have homological dimension equal to n.

It should be added that [1] contains many other interesting results involving generalizations from abelian groups to arbitrary R-modules.

Recently, H. W. K. Angad-Gaur has obtained simplifications and refinements of the above results. Specifically, he constructs, for every integer $n \geq 0$, a torsion-free group A of finite rank with $d(A) = n$. This work will appear in a dissertation being written at Tulane University.

4. Dimension of a Direct Sum. In this section we discuss two kinds of questions relating to direct sums for which some answers are known. The specific results described below without reference are among the as yet unpublished work of A. J. Douglas and the author.

Firstly, how much information is needed, preferably in terms of the dimensions of the abelian groups B, C, Hom(B,C), etc., in order to determine the dimension of their direct sum $A = B \oplus C$? The most one can hope for in this generality is an inequality involving dimensions. In [2(18)] the following is proved, on the assumption that $\mathrm{Hom}(C,B) = 0$:

$$d(B \oplus C) \leq \max\{d(C),\ d(B) + 1 + d_{\mathrm{End}(C)}\mathrm{Hom}(B,C)\}.$$

More precise information can be obtained by requiring more. For example, assume that $B = \mathrm{End}(C)$ is torsion-free, $\mathrm{Hom}(C,B) = 0$, $\mathrm{Hom}(B/B_0,B) = 0$ where B_0 is the pure subgroup of B generated by its identity element. Then it can be shown that the sequence $0 \to B_0 \to B \to B/B_0 \to 0$ is "useful" for $A = B \oplus C$, from which it follows that either $\mathrm{Hom}(B/B_0,C)$ is E-projective or $d(A) \geq 2$. In particular, we can deduce that $d(\hat{Z}_p \oplus Z(p^\infty)) = 2$, a result which was also obtained in [6] using the theorem described as the end of section 2 above.

The second question to be discussed here concerns the relation between d(A) and $d(A_I)$, where A_I is the direct sum of copies of A. This question was tackled in

[2,§3] where it was proved that $d(A_I) = d(A)$ given that $d(A) < \infty$ and that either I is finite or that A is finitely generated over the centre of E (cf. [3] for corrections to [2]). The general case is more difficult. Let Λ denote the ring of endomorphisms of A_I and set $\Gamma = \mathrm{Hom}(A,A_I)$. It is clear that Γ is (Λ,E)-bimodule and is Λ-projective. Furthermore, it can be seen that A_I is Λ-isomorphic to $\Gamma \otimes_E A$. It follows readily from this that if A is E-projective then A_I is Λ-projective, that is: $d(A) = 0 \Rightarrow d(A_I) = 0$. If we assume further that Γ is E-flat then we can show (by an induction) that for $d(A) < \infty$, $d(A) = d(A_I)$. We do not know if Γ is always E-flat, and, if not, what conditions on A are needed to make it so.

REFERENCES

1. I. V. Bobylev, Endoprojective dimension of modules, Sibirskii Mat. Zh. Vol. 16.4 (1975) pp. 663-683 (M.R. 53 #500).

2. A. J. Douglas and H. K. Farahat, The homological dimension of an abelian group as a module over its ring of endomorphisms, Monatshefte für Mathematik, 69, pp. 294-305 (1965) (M.R. 32 #2473).

3. A. J. Douglas and H. K. Farahat, The homological dimension of an abelian group as a module over its ring of endomorphisms, II, Monatshefte für Mathematik, 76, pp. 109-111 (1972) (M.R. 47 #3568).

4. A. J. Douglas and H. K. Farahat, The homological dimension of an abelian group as a module over its ring of endormorphisms, III, Monatshefte für Mathematik, 80, pp. 37-44 (1975).

5. F. Richman and E. A. Walker, Primary abelian groups as modules over their endomorphism rings, Mathematische Zeitschrift, 89, pp. 77-81 (1965) (M.R. 32 #2475).

6. F. Richman and E. A. Walker, Homological dimension of abelian groups over their endomorphism rings, Proc. Amer. Math. Soc. 54, pp. 65-68 (1976).

A GALOIS CORRESPONDENCE IN ABELIAN GROUPS

Adolf Mader

1. Introduction. We aim at describing $Hom(K,T)$ for K a p-reduced torsion-free group and T a reduced p-group. One difficulty is that practically nothing is known about the groups K. Certain well-known and easy representations of the groups K and of the groups $Hom(K,T)$ lead to the Galois Correspondence of the title. We then succeed in identifying an interesting class of groups K for which we can compute $Hom(K,T)$ when T is torsion-complete.

All facts and notations can be found in Fuchs [1]. Throughout p denotes a fixed prime and P the ring of p-adic integers. We write maps on the right. "hgt x" is the height, "exp a" is the exponential order. All topological terms refer to the p-adic topology.

2. A Galois Correspondence. Let $\#$ be an infinite cardinal. Let

$P^{\#}$ = the group of all lists $[x_i]$, $x_i \in P$, $i \in \#$,

$P^{(\#)}$ = the group of all finitely non-zero lists of $P^{\#}$,

$P^{[\#]}$ = the group of all lists $[x_i] \in P^{\#}$ such that for every natural number n, $p^n | x_i$ for all but finitely many i.

Then $P^{(\#)} \subset P^{[\#]} \subset P^{\#}$, $P^{\#}$ is the product of $\#$ copies of P, $P^{(\#)}$ is the direct sum of $\#$ copies of P, and $P^{[\#]}$ is the completion of $P^{(\#)}$ and a direct summand of $P^{\#}$. We look at P as given in the usual representation so that $\mathbb{Z} \subset P$. It is well-known that every torsion-free group K of p-rank $\#$ (i.e. $\# = \dim(K/pK)$) is embedded between $\mathbb{Z}^{(\#)}$ and $P^{[\#]}$ as a p-pure subgroup, and that $\mathbb{Z}^{(\#)}$ is a p-basic subgroup of K. For the rest of this section let T be a p-group with no elements of infinite height, and T^* its completion. Let $T^{\#}$ be the product of $\#$ copies of T with elements denoted by $[a_i]$.

2.1 Proposition. The map

$$P^{[\#]} \times T^{\#} \to T^{*} : ([x_i],[a_i]) \mapsto [x_i][a_i] = \Sigma x_i a_i$$

is a well-defined bilinear pairing with respect to the unique (unitary) P-module structures on $P^{[\#]}$, $T^{\#}$ and T^{*}.

Proof. Since $[x_i] \in P^{[\#]}$, $\Sigma x_i a_i$ has at most countably many non-zero summands and is convergent. Everything else is obvious.

2.2 Definition. For $K \subset P^{[\#]}$, let $K' = \{a \in T^{\#} : Ka \subset T\}$, and for $A \subset T^{\#}$ let $A' = \{x \in P^{[\#]} : xA \subset T\}$.

The next observation supplies the motivation for the above set-up.

2.3 Proposition. If K is a p-pure subgroup between $\mathbb{Z}^{(\#)}$ and $P^{[\#]}$ then $K' \cong \mathrm{Hom}(K,T)$. The isomorphism is given by: $f \longleftrightarrow [a_i]$ iff $[x_i]f = \Sigma x_i a_i$ for all $[x_i] \in K$.

Proof. The exact sequence $\mathbb{Z}^{(\#)} \rightarrowtail K \twoheadrightarrow K/\mathbb{Z}^{(\#)}$ with $K/\mathbb{Z}^{(\#)}$ p-divisible yields

$$(2.4) \qquad \mathrm{Hom}(K,T) \rightarrowtail \mathrm{Hom}(\mathbb{Z}^{(\#)},T) \cong T^{\#} \qquad (\mathrm{ex}).$$

The map into $T^{\#}$ is given by $f \mapsto [z_i f]$ $(f \in \mathrm{Hom}(K,T))$ where $\{z_i\}$ is the natural basis of $\mathbb{Z}^{(\#)}$, i.e. z_i is the list with i^{th} entry 1 and 0 elsewhere. Given $a_i = z_i f$ and $x = [x_i] \in K$, it follows that $xf = \Sigma x_i a_i \in T$. A list $[a_i] \in T^{\#}$ is the image of a map $f \in \mathrm{Hom}(K,T)$ iff $\Sigma x_i a_i \in T$ for all $[x_i] \in K$, and then f is the homomorphism given by $xf = \Sigma x_i a_i$. This proves the proposition.

2.5 Remark. (2.4) remains exact under the weaker hypothesis that T is reduced, but in this case it is not at all clear how $f \in \mathrm{Hom}(K,T)$ is determined by $[z_i f]$.

The following facts are immediate.

2.6 Proposition. For every $K \subset P^{[\#]}$, K' is a P-submodule of $T^{\#}$. For every $A \subset T^{\#}$, A' is a P-submodule of $P^{[\#]}$ containing $P^{(\#)}$.

2.7 Proposition. Let X, Y be subsets of either $P^{[\#]}$ or $T^{\#}$.

a) If $X \subset Y$ then $Y' \subset X'$. b) $X \subset X''$.

c) $X''' = X'$. d) $\langle X\rangle' = \cap\{\{x\}' : x \in X\}$.

($\langle X\rangle$ denotes the subgroup generated by X.)

2.8 Definition. A subset X of $P^{[\#]}$ or of $T^{\#}$ is called whole if $X'' = X$.

The Galois Correspondence now follows immediately from 2.7.

2.9 Galois Correspondence. The map $X \mapsto X'$ is a bijective inclusion reversing correspondence between whole subsets of $P^{[\#]}$ and whole subsets of $T^{\#}$.

The main and more difficult task consists in identifying whole subsets. This will be taken up in Section 3. First we give an application which illustrates that useful information is contained in the set-up just described.

2.10 Proposition. Let K be a complete torsion-free group and L a p-pure subgroup of countable index. Then every reduced p-primary epimorphic image of L is bounded.

Proof. Without loss of generality assume L is dense in K and $Z^{(\#)} \subset L \subset K = P^{[\#]}$. Let T be some unbounded torsion-complete p-group. Then $L \subset L'' \subset K'' = K$ and K/L'' is a countable, torsion-free, divisible P-module. Hence $K/L'' = 0$, i.e. $K = L''$. Furthermore $\mathrm{Hom}(K,T) \cong K' = L''' = L' \cong \mathrm{Hom}(L,T)$. Since K is complete, $\mathrm{Hom}(K,T)$ is a torsion-group and hence so is $\mathrm{Hom}(L,T)$. The claim follows easily using well-known results ([1], Vol. I, p. 152).

3. Some whole sets. In this section we assume throughout that T is torsion-complete so that T^*/T is torsion-free. This simplifies things considerably. In particular we have the following improvement of 2.6.

3.1 Proposition. For $K \subset P^{[\#]}$, K' is a P-submodule of $T^{\#}$ containing $tT^{\#}$, the maximal torsion-subgroup of $T^{\#}$, and $T^{\#}/K'$ is torsion-free. For $A \subset T^{\#}$, A' is a p-pure P-submodule between $P^{(\#)}$ and $P^{[\#]}$.

Proof. $p^n a \in T$ iff $a \in T$.

The simplest cases of whole sets are the largest and the smallest ones.

3.2 Proposition. $(tT^{\#})' = P^{[\#]}$, $P^{[\#]'} = tT^{\#}$, $(P^{(\#)})' = T^{\#}$, $(T^{\#})' = P^{(\#)}$, and these four modules are whole.

Proof. Mostly obvious. For $(P^{[\#]})' = tT^{\#}$, use that every reduced p-primary epimorphic image of the complete group $P^{[\#]}$ is bounded.

Less trivial examples of whole submodules of $P^{[\#]}$ are obtained by trying to compute $\{a\}'$ for $a \in T^{\#}$.

3.3 Proposition. Let $a = [a_i] \in T^{\#}$, $a_i = p^{n(i)}b_i$, and suppose that $\{b_i\}$ is p-independent. Then

$$\{a\}' = \{x = [x_i] \in P^{[\#]} : \overline{x} + \text{hgt } x_i \geq \exp a_i\}.$$

($\overline{x}$ is some integer depending on x. The inequality holds for all i.)

Proof. If $x = [x_i] \in \{a\}'$, then $xa = \Sigma x_i a_i = \Sigma x_i p^{n(i)} b_i \in T$ and there is an integer $\overline{x}$ such that

$$p^{\overline{x}} \Sigma x_i p^{n(i)} b_i = \Sigma x_i p^{\overline{x} + n(i)} b_i = 0. \tag{3.4}$$

Fix $j \in \#$. Choose an integer n such that $n > \overline{x} + n(j) + \text{hgt } x_j$. There is a finite subset I of $\#$ containing j such that $x_i \in p^n P$ for all $i \notin I$. It follows from (3.4) that $\Sigma_{i \in I} x_i p^{\overline{x} + n(i)} b_i \in p^n T^* \cap T = p^n T$. Since $\{b_i\}$ is p-independent, $j \in I$, and $\overline{x} + n(j) + \text{hgt } x_j < n$ it follows that $x_j p^{n(j) + \overline{x}} b_j = 0$, i.e. $\overline{x} + n(j) + \text{hgt } x_j \geq \exp b_j = n(j) + \exp a_j$ or $\overline{x} + \text{hgt } x_j \geq \exp a_j$ as claimed. The converse is trivial.

3.5 Corollary. Let $e = \{e_i\}_{i \in \#}$ be a list of non-negative integers. If $\# \leq \dim p^n T / p^{n+1} T$ for all n, then $K_e = \{x = [x_i] \in P^{[\#]} : \overline{x} + \text{hgt } x_i \geq e_i\}$ is whole.

Proof. Since $\# \leq \dim p^n T / p^{n+1} T$, it is possible to find a p-independent subset $\{b_i\}$ of T such that $\exp b_i \geq e_i$. Let $a_i = p^{(\exp b_i) - e_i} b_i$. Then $\exp a_i = e_i$, and by 3.3 $K_e = \{a\}'$ and thus is whole.

Something can be said about the structure of such groups K_e.

3.6 Proposition. Let $K_1 = \{x = [x_i] \in P^{[\#]} : \text{hgt } x_i \geq e_i\}$ and let $K_2 = K_e/K_1$. Then K_1 is complete; $\# \leq \dim(K_1/pK_1) \leq 2^{\#}$; $K_e = (K_1)_*$, the purification of K_1; and K_2 is torsion-complete with basic subgroup $\oplus \mathbb{Z}(p^{e_i})$.

Proof. a) Define

$$f : P^{\#} \to P^{\#} : [x_i]f = [p^{e_i} x_i]. \tag{3.7}$$

Clearly f is a monomorphism, so that $P^{\#}f \cong P^{\#}$ is complete. Note that $K_1 = P^{[\#]} \cap P^{\#}f$. Since $P^{\#}f/K_1 = P^{\#}f/(P^{[\#]} \cap P^{\#}f) \cong (P^{[\#]} + P^{\#}f)/P^{[\#]} \leq P^{\#}/P^{[\#]}$ and $P^{\#}/P^{[\#]}$ is torsion-free reduced, it follows that K_1 is a pure closed subgroup of the complete group $P^{\#}f$, and hence itself complete.

b) Moreover, K_1 is a direct summand of $P^{\#}f$, hence $\dim K_1/pK_1 \leq \dim P^{\#}f/p\,P^{\#}f = \dim P^{\#}/p\,P^{\#} = 2^{\#}$. On the other hand, $K_1/P^{[\#]}f \leq P^{\#}f/P^{[\#]}f \cong P^{\#}/P^{[\#]}$ is torsion free reduced, so $P^{[\#]}f$ is a direct summand of K_1 and $\dim K_1/pK_1 \geq \dim P^{[\#]}f/p\,P^{[\#]}f = \dim P^{[\#]}/p\,P^{[\#]} = \dim P^{(\#)}/p\,P^{(\#)} = \#$. We will give examples below to show that the p-rank of K_1 can be $\#$ as well as $2^{\#}$.

c) $K_e = \{x = [x_i] \in P^{[\#]} : \bar{x} + \text{hgt } x_i \geq e_i\} = \{x : p^{e_i} \mid p^{\bar{x}} x_i\} = \{x \in P^{[\#]} : p^{\bar{x}} x \in K_1\}$, so $K_e = (K_1)_*$ and K_e/K_1 is the maximal torsion subgroup of $P^{[\#]}/K_1$. It is easily checked that $P^{\#}f$ is closed in $P^{\#}$ and hence $K_1 = P^{[\#]} \cap P^{\#}f$ is closed in $P^{[\#]}$. Therefore $P^{[\#]}/K_1$ is reduced, and since $P^{[\#]}$ is complete, so is $P^{[\#]}/K_1$. Thus K_e/K_1 is torsion-complete. Let $B = (P^{(\#)} + K_1)/K_1$. We will show that B is a basic subgroup of K_e/K_1. First note that $B \cong P^{(\#)}/(P^{(\#)} \cap K_1) = P^{(\#)}/(P^{(\#)} \cap P^{\#}f) = P^{(\#)}/\oplus p^{e_i} P \cong \oplus \mathbb{Z}(p^{e_i})$. So B is a direct sum of cyclic groups. Also $(K_e/K_1)/B \cong K_e/(P^{(\#)} + K_1)$ is divisible since $K_e/P^{(\#)}$ is already divisible. Finally, suppose $p^n([x_i] + K_1) \in B$ with $[x_i] \in K_e$. Then $p^n[x_i] = [y_i] + [p^{e_i} z_i]$, $[y_i] \in P^{(\#)}$, $[p^{e_i} z_i] \in K_1$. Hence $p^n x_i = y_i + p^{e_i} z_i$ for all $i \in \#$.

Case I. i is such that $n > e_i$. Then $p^{e_i} \mid y_i$. Let $y_i = p^{e_i} y_i'$ and $y_i'' = 0$.

Case II. i is such that $n \leq e_i$. Then $p^n \mid y_i$. Let $y_i = p^n y_i''$ and $y_i' = 0$.

With these choices $[y_i'], [y_i''] \in P^{(\#)}$ since $y_i = 0$ implies $y_i' = y_i'' = 0$. Now $p^n[x_i] = p^n[y_i''] + [p^{e_i}(z_i + y_i')]$, which shows that B is pure in K_e/K_1.

3.8 Examples. (a) If $e = \{e_i\}$ is bounded, then $K_1 = P^{[\#]}f \cong P^{[\#]}$, hence $\dim(K_1/pK_1) = \#$.

(b) If, for all n, $n \le e_i$ for almost all i, then $K_1 = P^{\#}f \cong P^{\#}$ so $\dim K_1/pK_1 = 2^{\#}$.

Proof. Recall that $K_1 = P^{[\#]} \cap P^{\#}f$.

(a) For $e_i \le m$ for all i, it easily follows that $K_1 = P^{[\#]} \cap P^{\#}f \supset P^{[\#]}f \supset K_1$.

(b) Note that $P^{\#}f \subset P^{[\#]}$ in this case.

The exact sequence $K_1 \rightarrowtail K_e \twoheadrightarrow K_2$ yields information about $\mathrm{Hom}(K_e,T)$ via simple homological algebra : $\mathrm{Hom}(K_2,T) \rightarrowtail \mathrm{Hom}(K_e,T) \to \mathrm{Hom}(K_1,T)$ is exact. Moreover, $\mathrm{Hom}(K_2,T) \cong \Pi\, T[p^{e_i}]$ since T is torsion-complete, and $\mathrm{Hom}(K_1,T) \cong t(T^{\dim K_1/pK_1})$ since K_1 is complete. It is not clear what $\mathrm{Hom}(K_e,T)/\mathrm{Hom}(K_2,T)$ looks like. This question will be resolved presently.

3.9 Theorem. $\mathrm{Hom}(K_e,T) \cong K_e' = \{a = [a_i] \in T^{\#} : \exp a_i \le \bar{a} + e_i\}$.

Proof. a) $K_e' \supset \{[a_i] : \exp a_i \le \bar{a} + e_i\}$ since $\exp a_i \le \bar{a} + e_i$ and $\bar{x} + \mathrm{hgt}\, x_i \ge e_i$ imply $\exp a_i \le \bar{a} + \bar{x} + \mathrm{hgt}\, x_i$, so $\exp(xa) \le \bar{a} + \bar{x}$ and $xa \in T$.

b) For the reverse inclusion we use the function (3.7) and note that $P^{[\#]}f \subset K_e$. Since $P^{[\#]}$ is complete, $\mathrm{Hom}(P^{[\#]},T)$ is a torsion group. Now let $a = [a_i] \in K_e'$. Then $x \mapsto (xf)a$ defines a homomorphism $P^{[\#]} \to T$. Since $\mathrm{Hom}(P^{[\#]},T)$ is a torsion group, there is a positive integer $\bar{a}$ such that

(3.10) $$p^{\bar{a}}(xf)a = 0 \quad \text{for all } x \in P^{[\#]}.$$

Now fix $j \in \#$ and let $x_j = 1$, $x_i = 0$ for $i \ne j$. For the choice $x = [x_i]$, the identity (3.10) becomes $p^{\bar{a}} p^{e_j} a_j = 0$, i.e. $\exp a_j \le \bar{a} + e_j$. The proof is complete.

We now investigate the structure of K_e'.

3.11 Proposition. Let $A_1 = \{[a_i] \in T^{\#} : \exp a_i \le e_i\}$ and let $A_2 = K_e'/A_1$. Then $A_1 = \Pi\, T[p^{e_i}]$ and $A_2 \cong t\, \Pi\, p^{e_i} T$.

Proof. It is obvious that $A_1 = \prod T[p^{e_i}]$. Let $g : T^{\#} \to T^{\#} : [a_i]g = [p^{e_i} a_i]$. Then clearly $\mathrm{Ker}\ g = A_1$ and a straight forward calculation gives $K_e' g = t \prod p^{e_i} T$.

4. Remarks. This paper contains those results of an ongoing investigation which are in final form and form an entity by themselves. Further results obtained so far include an explicit description of $\{x\}'$ for an arbitrary $x \in P^{[\#]}$. A lengthy computation yields that $\{x\}''$ is the smallest pure submodule of $P^{[\#]}$ containing x and $P^{(\#)}$. This suggests that possibly K'' is the smallest pure submodule of $P^{[\#]}$ containing K and $P^{(\#)}$ for arbitrary $K \subset P^{[\#]}$. This would characterize the whole subsets of $P^{[\#]}$. A clearly interesting untouched problem is the characterization of whole subsets of $T^{\#}$.

The groups K_e have interesting properties not mentioned so far. In [3] we showed that $\mathrm{Ext}(K_e,T)[p] \neq 0$ while every p-pure subgroup of K_e of infinite p-rank has unbounded reduced p-primary epimorphic images. This is a new result on the problem of finding those torsion-free groups K for which $\mathrm{Ext}(K,T)[p] = 0$ for every p-group T. It was shown in [2] that $\mathrm{Ext}(K,T)[p] = 0$ iff $T^{\#}/K'$ is divisible. By a theorem of Nunke [4], $\mathrm{Ext}(K,T)$ is unchanged if the p-group T is replaced by a basic subgroup of T. Hence T may also be replaced by a suitable torsion-complete group. Our Galois Correspondence was discovered in connection with the $\mathrm{Ext}(K,T)[p]$-problem and for this purpose the assumption that T is torsion-complete is no loss of generality.

In [3] we also attacked the problem of the dependence on the representation of K as a p-pure subgroup of $P^{[\#]}$. We conjecture that a group K_e has the same form with respect to every representation but, of course, with a different e.

REFERENCES

1. L. Fuchs, Infinite Abelian Groups, Academic Press, Vol. I 1970, Vol. II 1973.

2. A. Mader, The group of extensions of a torsion group by a torsion free group, Arch. Math. 20(1969), 126-131.

3. _______, Finite order extensions of a primary group by a torsion free group, Acta Math. Acad. Sci. Hungar., to appear.

4. R. Nunke, On the extensions of a torsion module, Pacific J. Math. 10(1960), 597-606.

A DIFFERENT COMPLETION FUNCTOR

R. Mines

1. Introduction. For each ordinal β an idempotent endofunctor, T_β, is defined on the category of abelian groups. If β is cofinal with ω, then T_β is the usual p^β-adic completion functor. However, if β is not cofinal with ω, then T_β is a proper subfunctor of the completion functor. The fact that T_β is an idempotent functor means that it is more useful than the completion functor in studying abelian groups. This is demonstrated in section 4, where it is shown that a group G is cotorsion if and only if $T_\beta(G) = G/p^\beta G$ for each ordinal β. Section 5 outlines a development of a similar functor for the category of p-primary abelian groups. The class of fully complete p-groups is defined in terms of this functor. Finally, fully complete p-groups are characterized using their cotorsion hulls.

2. The Functor T_β. Two groups H and K containing a group G are G-isomorphic if there is an isomorphism from H onto K extending the identity on G. Define $S(G, \beta)$ to be the set of all $(G/p^\beta G)$-isomorphic equivalence classes of groups H satisfying the following.

(i) $p^\beta H = 0$.

(ii) $G/p^\beta G$ is an isotype subgroup of H.

(iii) $G/p^\beta G$ is dense in H. That is, $(G/p^\beta G) + p^{\alpha+1}H = H$ for all $\alpha < \beta$.

(iv) The cokernel of $G/p^\beta G \to H$ is a subgroup of either Q or $Z(p^\infty)$.

Note that the density of $G/p^\beta G$ in H implies that the only subgroups allowed in (iv) are Q, $Z(p^\infty)$, or 0.

We will assume that $p^\beta G = 0$. Choose a representative H from each equivalence class, and let K be the sum of the H's with G as an amalgamated subgroup. If ∇ denotes the codiagonal map, then the following diagram defines K.

$$\begin{array}{ccccccccc} 0 & \longrightarrow & \Sigma G & \longrightarrow & \Sigma H & \longrightarrow & D & \longrightarrow & 0 \\ & & \nabla\downarrow & & \downarrow & & \| & & \\ 0 & \longrightarrow & G & \longrightarrow & K & \longrightarrow & D & \longrightarrow & 0 \end{array}$$

Notice that ΣG is a p^{β}-dense isotype subgroup of ΣH. The next lemma shows that G is a p^{β}-dense isotype subgroup of K.

Lemma 2.1. If G is a p^{β}-dense isotype subgroup of H, and $f : G \to G'$ with $p^{\beta}G' = 0$, then G' is a p^{β}-dense isotype subgroup of the pushout of $G \to H$ and $G \to G'$.

Proof. Form the pushout diagram

$$\begin{array}{ccccccccc} 0 & \longrightarrow & G & \longrightarrow & H & \xrightarrow{\pi_1} & D & \longrightarrow & 0 \\ & & f\downarrow & & \downarrow f' & & \| & & \\ 0 & \longrightarrow & G' & \longrightarrow & H' & \xrightarrow{\pi_2} & D & \longrightarrow & 0 \end{array} .$$

Noticing that $H' = G' + f'(H)$, it is easy to see that G' is p^{β}-dense in H'. Let $g \in G' \cap p^{\alpha+1}H'$. Then there exists $h' \in p^{\alpha}H'$ with $ph' = g$. Using density of the top row and simple diagram chasing, there exists $h \in p^{\alpha}H$ with $h' - f'(h) \in G' \cap p^{\alpha}H' = p^{\alpha}G'$. The last equality is by induction. Therefore, $pf'(h) \in G'$. Thus $0 = \pi_2 f'(ph) = \pi_1(ph)$. That is, $ph \in G \cap p^{\alpha+1}H = p^{\alpha+1}G$. Now $g = ph' = p(h' - f(h)) + pf'(h) \in p^{\alpha+1}G'$.

Applying Lemma 2.1 to the definition of K, it follows that $G \cap p^{\beta}K = 0$. Define $T_{\beta}(G) = K/p^{\beta}K$, and $w : G \to T_{\beta}(G)$ to be the composite $G \to K \to K/p_{\beta}K = T_{\beta}(G)$. The mapping w is a monomorphism. If G is an arbitrary group, then define $T_{\beta}(G) = T_{\beta}(G/p^{\beta}G)$, and w by the composition $G \to G/p^{\beta}G \to T_{\beta}(G)$. The kernel of w is then $p^{\beta}G$. When there is no confusion, $T_{\beta}(G)$ will be written as $T(G)$. Notice that when β is not a limit ordinal, then $T_{\beta}(G) = G/p^{\beta}G$.

Theorem 2.2. The assignment $G \longmapsto T_{\beta}(G)$ defines an idempotent functor satisfying the following.

(i) The image of w is a p^{β}-dense isotype subgroup of $T_{\beta}(G)$.

(ii) If G is a p^β-dense isotype subgroup of H and $p^\beta H = 0$, then there exists a unique G-monomorphism from H into $T_\beta(G)$.

(iii) If $G \subseteq H \subseteq T_\beta(G)$, then G is p^β-dense in H if and only if H/G is p-divisible.

<u>Proof.</u> It is clear that $T^2 = T$. To show that T is a functor, let $f : G \to G'$ and form the pushout diagram

$$\begin{array}{ccccccccc}
0 & \longrightarrow & G & \longrightarrow & T(G) & \longrightarrow & D & \longrightarrow & 0 \\
 & & \downarrow & & & & & & \\
 & & G' & & \downarrow & & \| & & \\
 & & \downarrow & & & & & & \\
0 & \longrightarrow & T(G') & \longrightarrow & H & \longrightarrow & D & \longrightarrow & 0 .
\end{array}$$

Lemma 2.1 shows that $T(G') \cap p^\beta H = 0$ and thus that $T(G')$ is a p^β-dense isotype subgroup of $H/p^\beta H$. The only way that this can happen is for $T(G') \to H/p^\beta H$ to be split. Let α be the splitting and define $T(f) : T(G) \to T(G')$ to be the composition $T(G) \to H \to H/p^\beta H \xrightarrow{\alpha} T(G')$. It is clear that $T(f)|_G$ is the same as $G \to G' \to G'/p^\beta G'$. The facts that D is divisible and that $p^\beta T(G') = 0$ show that $T(f)$ is unique.

Properties (i) and (ii) follow from Lemma 2.1. The proof of property (ii) requires an argument similar to the argument showing the existence of $T(f)$. Property (iii) follows from the following more general theorem.

<u>Theorem 2.3.</u> Let β be a limit ordinal, and let G be an isotype subgroup of K. Let H be a subgroup of K satisfying $G \subseteq H \subseteq G + p^\alpha K \subseteq K$ for all $\alpha < \beta$. The following are equivalent.

(i) $G + p^\alpha H = H$ for some α, $0 < \alpha < \beta$.

(ii) $H \cap p^\alpha K = p^\alpha H$ for some α, $0 < \alpha < \beta$.

(iii) H/G is p-divisible.

(iv) Conditions (i) and (ii) hold for all $\alpha < \beta$.

<u>Proof.</u> (i) $\to$ (ii): $H \cap p^\alpha K = (G + p^\alpha H) \cap p^\alpha K = p^\alpha G + p^\alpha H = p^\alpha H$.

(ii) $\to$ (i): $H = H \cap (G + p^\alpha K) = G + H \cap (p^\alpha K) = G + p^\alpha H$.

(ii) $\to$ (iii): This is clear.

(iii) $\to$ (iv): If H/G is divisible, then we shall show by induction on

α that one of the equivalent conditions (i) or (ii) holds for α. If α is a limit ordinal, then induction shows that $H \cap p^{\alpha}K = p^{\alpha}H$. Assume that $\alpha = \gamma + 1$, and let $h \in H$. By the divisibility of H/G, there exists an $h' \in H$ and a $g \in G$ with $h = ph' + g$. Now by induction there exists an $h'' \in p^{\alpha}H$ and a $g' \in G$ with $h' = g' + h''$. Thus $h = (g + pg') + ph'' \in G + p^{\alpha+1}H$.

(iv) $\rightarrow$ (i): This is obvious.

3. The Functor T_{β} and the p^{β}-adic Completion. Let $L_{\beta}(G)$ denote the p^{β}-adic completion of the group G. It is easy to see that there exists a G-monomorphism $T_{\beta}(G) \rightarrow L_{\beta}(G)$. If β is a limit ordinal cofinal with ω, then G is a p^{β}-dense isotype subgroup of $L_{\beta}(G)$. (See [2].) Thus in this case $T_{\beta}(G) = L_{\beta}(G)$. However, if β is not cofinal with ω, then there exists a group G which is isotype but not dense in $L_{\beta}(G)$. (See [2].) In this case, one has $p^{\alpha}(L(G)/G) = (G + p^{\alpha}L(G))/G$ for all $\alpha < \beta$. Therefore, $T(G) \subseteq G + p^{\alpha}L(G)$ for all $\alpha < \beta$. Applying Theorem 2.3, it is easy to see that $T(G)$ is a p^{β}-isotype subgroup of $L(G)$, that the divisible subgroup of $L(G)/G$ is $T(G)/G$, and that $T(G)$ is the largest subgroup of $L(G)$ containing G as a p^{β}-dense isotype subgroup. An advantage of the functor T over the functor L is that T is idempotent, while L in general is not. This will be exploited in the next section to give a characterization of cotorsion groups.

4. Cotorsion Groups and the Functor T. Use will be made of the following characterization of cotorsion groups.

Theorem 4.1. A reduced group G is cotorsion if and only if whenever $G \subseteq H$ with H/G divisible, then H contains a nontrivial divisible subgroup.

Proof. Let G be cotorsion and $H \subseteq G$ with H/G divisible. Let $f : Q \rightarrow H/G$ be a nonzero homomorphism from the rationals to H/G, and form the pull back diagram

$$\begin{array}{ccccccccc} 0 & \longrightarrow & G & \longrightarrow & H' & \longrightarrow & Q & \longrightarrow & 0 \\ & & \| & & \downarrow{\scriptstyle f'} & & \downarrow{\scriptstyle f} & & \\ 0 & \longrightarrow & G & \longrightarrow & H & \longrightarrow & H/G & \longrightarrow & 0 . \end{array}$$

As G is cotorsion, $H' \cong G \oplus Q$ and $f'(Q)$ is a nonzero divisible subgroup of H.

Suppose that G satisfies the condition of the theorem, and let $G \subseteq H$ with $H/G \cong Q$. As the divisible subgroup of H is nonzero, it follows easily that $H \cong G \oplus Q$.

The following theorem characterizes cotorsion groups in terms of the functor T.

<u>Theorem 4.2</u>. A group G is cotorsion if and only if $T_\beta(G) = G/p^\beta G$ for all ordinals β.

<u>Proof</u>. It is clear using Theorem 4.1 that if G is cotorsion, then $T_\beta(G) = G/p^\beta G$ for all ordinals β. The converse will be shown using induction on β. If $\beta = \gamma + 1$ and $G/p^\gamma G$ is cotorsion, then clearly $G/p^\beta G$ is cotorsion. For β a limit ordinal, the group $L_\beta(G)$ is a cotorsion group as it is the projective limit of the cotorsion groups $G/p^\alpha G$ for $\alpha < \beta$. As $G/p^\beta G = T_\beta(G)$, it follows from section 3 that $L_\beta(G)/(G/p^\beta G)$ is reduced. Therefore $G/p^\beta G$ is cortorsion.

5. <u>The Category of p-torsion Groups</u>. If we define $T(G, \beta)$ to be all equivalence classes of p-groups $H \supseteq G$ as a p^β-dense isotype subgroup with H/G isomorphic to a subgroup of $Z(p^\infty)$, and define tT_β in a manner similar to the definition of $T_\beta(G)$, then $tT_\beta(G)$ is the torsion subgroup of $T_\beta(G)$ and is a torsion completion of the group G. A p-group G is <u>fully torsion complete</u> if $tT_\beta(G) = G/p^\beta G$ for all ordinals β. Let the cotorsion hull of G be denoted by $U(G)$.

<u>Theorem 5.1</u>. A p-group G is fully torsion complete if and only if $G/p^\beta G$ is the torsion subgroup of $U(G)/p^\beta U(G)$ for all ordinals β.

Proof. The proof is similar to the proof of Theorem 4.2.

The possibility that the fully complete p-groups might be the same as the class of p-groups satisfying the condition of Theorem 4.1 is ruled out by the following theorem.

Theorem 5.2. The class of p-groups satisfying the condition of Theorem 4.1 is the class of bounded p-groups.

Proof. Let

$$(*) \qquad 0 \longrightarrow G \longrightarrow H \xrightarrow{\pi} Z(p^{\infty}) \longrightarrow 0$$

be a non-split exact sequence with $d(G) = 0$ and $d(H) \neq 0$. Then $d(H) = Z(p^{\infty})$, and $d(H) \cap G$ is a cyclic group of order $p^n > 1$. The mapping $\pi|_{d(H)}$ is thus an automorphism of $Z(p^{\infty})$ followed by multiplication by p^n. Thus there exists a map $\alpha : Z(p^{\infty}) \to H$ so that the diagram

$$\begin{array}{ccccccccc} & & & & & & Z(p^{\infty}) & & \\ & & & & & \alpha \swarrow & \downarrow p^n & & \\ 0 & \longrightarrow & G & \longrightarrow & H & \longrightarrow & Z(p^{\infty}) & \longrightarrow & 0 \end{array}$$

commutes. This means that, considered as an element of $\mathrm{Ext}(Z(p^{\infty}), G)$, the extension (*) is a torsion element. If G is unbounded, then $\mathrm{Ext}(Z(p^{\infty}), G)$ is not a torsion group. Therefore, if G is unbounded, there exists a reduced group $H \supseteq G$ with $H/G \cong Z(p^{\infty})$. It is clear that bounded groups satisfy the condition in question.

It would be interesting to know if the group H in the above proof can be chosen to have the same length as G. The fully torsion complete groups have been studied to some extent. (See [1] and [3].) However, there remains a lot to be done with these groups. For example, it would be nice to have an idempotent functor which assigns to each p-group G a fully complete p-group containing G as an isotype subgroup. There is also a need for a set of invariants which would classify the fully torsion complete groups. The injective properties of these groups should also be investigated.

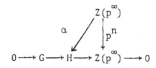

REFERENCES

[1] Cutler, D. O., Completions of topological abelian p-groups, Acta Math. Acad. Sci. Hungar., 22, (1971), 331-335.

[2] Mines, R., A family of functors defined on generalized primary groups, Pacific J. Math., 26, (1968), 349-360.

[3] Mines, R., Torsion and Cotorsion Completions, Etudes Sur les Groupes Abeliens, Paris, Dunod , (1968), 301-303.

ANALOGUES OF THE STACKED BASES THEOREM

Alfred W. Hales

1. Introduction. The so-called "Stacked Bases Theorem" was conjectured by Kaplansky in 1954 ([8], pp. 66, 80) and was first proved by Cohen and Gluck in 1970 ([1]). In this paper we consider analogous results, i.e. theorems which state that, whenever an abelian group has a presentation of a certain form, then every presentation of the group can be put in that form.

2. Preliminaries. Let G be an abelian group. We will be considering certain properties of presentations of G, so we begin with a formal definition of this concept.

Definition. A presentation of an abelian group G is a four-tuple (F,π,X,R), where F is a free abelian group, π is a homomorphism from F onto G, X is a basis for F (usually indicated by $F = F_X$), and $R \subseteq F$ is such that $\langle R \rangle = \operatorname{Ker} \pi$.

Notice in the above that we do not insist that R be a basis for $\operatorname{Ker} \pi$. In the sequel we will also be concerned with "partial" presentations (F,π) where X and R have been left unspecified.

Definition. If (F,π,X,R) is a presentation of G, then the length $\ell(R)$ of R is the supremum, over all r in R, of $\ell(r)$, where $\ell(r)$ is the number of non-zero coefficients when r is written as a linear combination of elements of X.

Hence $\ell(R)$ is either a non-negative integer or ∞, and depends of course upon the choice of X. We now focus our attention upon the following question: Given G, for which n does there exist a presentation (F,π,X,R) of G with $\ell(R) \leq n$? This gives a numerical measure of the "relational complexity" of G ([2]).

First of all, G has a presentation with $\ell(R) = 0$ if and only if G is free. For if $\ell(R) = 0$ then $\ell(r) = 0$ for all r in R, so $r = 0$. Hence $\pi : F \to G$ is an isomorphism. The converse is equally trivial.

Secondly, G has a presentation with $\ell(R) \leq 1$ if and only if G is a direct sum of cyclic groups. For if $\ell(R) \leq 1$ then $r = a_r x_r$ for all r, where a_r is in $\mathbf{Z}$ and x_r is in X. It is then immediate that $G = \Sigma_{x \in X} \pi(x)$, where $\pi(x)$ is cyclic and isomorphic to $\mathbf{Z}/I_x$, where $I_x = \langle a_r : r \in R$ and $x_r = x\rangle$. The converse is again obvious.

Next we consider groups G which have presentations with $\ell(R) \leq 2$. In the primary case the presentation can then be chosen so that each r in R is either of the form px_i or $px_i - x_j$, and we obtain the class of totally projective groups ([3], [7]). In the torsion-free case it is easy to see we obtain the class of completely decomposable groups. Following the terminology of Fuchs ([6], Vol. II), we shall call groups (torsion, torsion-free, or mixed) which have presentations with $\ell(R) \leq 2$ simply presented. The class of simply presented groups then comes close to being the largest class of groups for which a satisfactory classification theorem is available.

Finally, a group G always has a presentation with $\ell(R) \leq 3$. One way to see this is to take F to be free on a set $X = \{x_g : g \in G\}$ which is in $1 - 1$ correspondence with G itself, to define $\pi : F \to G$ so that $\pi(x_g) = g$ for all g, and to take R to be the set of all $(x_g + x_h - x_{g+h})$ as g, h run through G. We shall call this the regular presentation of G. Another way is to take any presentation (F, π, X, R) of G and modify it as follows. Suppose that $r = a_1x_1 + \cdots + a_nx_n$ is in R with $n > 3$. Adjoin to X elements $y_{r,1}$, $y_{r,2}, \ldots, y_{r,n-3}$ and replace r in R by the elements $a_1x_1 + a_2x_2 - y_{r,1}$, $y_{r,1} + a_3x_3 - y_{r,2}$, $y_{r,2} + a_4x_4 - y_{r,3}, \ldots, y_{r,n-4} + a_{n-2}x_{n-2} - y_{r,n-3}$, $y_{r,n-3} + a_{n-1}x_{n-1} + a_nx_n$. Do this for each r in R simultaneously, obtaining a larger set Y in place of X and a new $R' \subseteq F_Y$. Then (F_Y, π', Y, R') will be the desired modification, where $\pi' : F \to G$ extends π in the obvious way.

3. The Stacked Bases Theorem. If A and B are free abelian groups with $A \subseteq B$, we can ask whether there exist bases for A and B such that each basis element of A is a multiple of a basis element of B. Such bases are said to be stacked

bases for A and B. It is clear that the existence of stacked bases implies that B/A is a direct sum of cyclic groups. In 1954 ([8], pp. 66, 80) Kaplansky conjectured that the converse is true. (This is of course well-known when B is finitely generated.) In 1970 ([1]) Cohen and Gluck proved the converse, i.e., proved the

Stacked Bases Theorem. Free abelian groups $A \subseteq B$ have stacked bases if and only if B/A is a direct sum of cyclic groups.

We can restate this theorem using our terminology as follows:

SBT: If an abelian group G has a presentation (F,π,X,R) with $\ell(R) \leq 1$ then, for every partial presentation (F,π) of G, X and R can be chosen so that $\ell(R) \leq 1$.

4. Analogous Results. Our restatement of the Stacked Bases Theorem contains an obvious parameter, namely the number 1, so we now regard it as one member SBT(1) of the one-parameter family of assertions

SBT(n). If an abelian group G has a presentation (F,π,X,R) with $\ell(R) \leq n$ then, for every partial presentation (F,π) of G, X and R can be chosen so that $\ell(R) \leq n$.

From our previous remarks the only cases that are of real interest are SBT(0), SBT(2), and SBT(3).

Unfortunately the statement SBT(0) is not true as it stands, since in a partial presentation (F,π) of a free group G, π need not be an isomorphism. A slight modification of SBT(0) is, however, both true and important:

SBT'(0). If an abelian group G has a presentation (F,π,X,R) with $\ell(R) = 0$ then, for every partial presentation (F,π) of G, X and R can be chosen so that $R \subseteq X \cup \{0\}$.

This is really just the statement that free objects and projective objects coincide in the category of abelian groups.

The assertions SBT(2) and SBT(3) are more subtle. When we first considered these problems it seemed that SBT(3) ought to be false, and that an indecomposable torsion-free group of rank 3 would be a counterexample. This is far from the truth, however, and indeed the torsion-free groups are the easiest ones to handle. The explanation for this is that torsion-free groups have very few "different" types of presentations, as the following theorem indicates.

Theorem 1. If G is a torsion-free abelian group and (F,π) is a partial presentation of G, then X can be chosen so that $F = F_X$, X is the disjoint union $X_1 \dot\cup X_2 \dot\cup X_3$, $\pi(X_3) = 0$, $G = \pi(F_{X_1}) \oplus \pi(F_{X_2})$, π restricted to F_{X_2} is $1-1$, and π restricted to X_1 is a bijection onto $\pi(F_{X_1})$.

In other words, G splits as the direct sum of something and a free group, and the given presentation involves a regular presentation of the first component, a "free" presentation of the second component, and superfluous generators mapped to 0.

Proof of Theorem 1.[1] Let (F,π) be a partial presentation of G. If we could find a basis X of F such that π maps X onto G we would be finished. For then we could choose $X' \subseteq X$ so that π maps X' bijectively onto G, and each x in $X - X'$ would satisfy $\pi(x) = \pi(x')$ for some x' in X'. We could then take $X_1 = X'$, $X_2 = \varphi$, and $X_3 = \{x - x' : x \in X - X'\}$.

Next we claim that it would be sufficient to find a basis X of F such that $|X \cap \operatorname{Ker} \pi| \geq |G|$. For then, if $Y = X - (X \cap \operatorname{Ker} \pi)$, $\pi(Y)$ generates G and hence for each g in G we can write $g = \Sigma_{i=1}^{n} a_i \pi(y_i)$ where a_i is in $\mathbb{Z}$ and y_i is in Y. Associate to each g in G an x_g in $X \cap \operatorname{Ker} \pi$ (with $g \mapsto x_g$ $1-1$), and let $x'_g = x_g + \Sigma_{i=1}^{n} a_i y_i$. Then $Y \cup \{x'_g : g \in G\} \cup (X - Y - \{x'_g\})$ is a basis for F which is mapped onto G by π.

Now let m be the smallest cardinal (initial ordinal) such that F has a

[1] After the talk Professor Fuchs pointed out that Theorem 1 follows from results of J. Erdös ([4], [5], pp. 192-196).

basis $X = S \dot{\cup} T$ with $|S| = m$, $G = \pi(F_S) \oplus \pi(F_T)$, and π restricted to F_T one-to-one. Define by transfinite induction an ordinal-indexed collection S_λ of finite subsets of S as follows: if $S_{\lambda'}$ has been defined for all $\lambda' < \lambda$, then S_λ is a finite subset of $S - \bigcup_{\lambda'<\lambda} S_{\lambda'}$ such that $\pi(S_\lambda)$ is not linearly independent (if such exists). If S_λ cannot be defined for all $\lambda' < m$, then there is a subset S^* of S with $|S^*| < m$ and with $\pi(S - S^*)$ linearly independent. Since $\pi(F_{S^*}) \cap \pi(F_{S-S^*})$ has cardinality less than m, we can find S^{**} with $S^* \subseteq S^{**} \subseteq S$, $|S^{**}| < m$, and $\pi(F_{S^{**}}) \cap \pi(F_{S-S^{**}}) = 0$. This contradicts the minimality of m.

Hence we can assume S_λ has been defined for all $\lambda' < m$. Now since $\pi(S_\lambda)$ is not linearly independent, we can choose a new basis $\bar{S}_\lambda$ for F_{S_λ} so that π maps some element of $\bar{S}_\lambda$ to 0 (using the fact that $\pi(F_{S_\lambda})$ is torsion-free and hence free). Then by replacing each subset S_λ of S (for $\lambda < m$) by $\bar{S}_\lambda$, we obtain a new basis $\bar{S}$ for F_S such that $|\bar{S} \cap \mathrm{Ker}\ \pi| \geq |\pi(F_S)|$. By our previous remarks the partial presentation (F_S, π) of $\pi(F_S)$ can then be put in the desired form, and combining this with the basis T for F_T gives the final result.

Corollary 2. (SBT(3) for torsion-free groups). If G is a torsion-free abelian group then, for every partial presentation (F, π) of G, X and R can be chosen so that $\ell(R) \leq 3$.

Proof. This follows immediately from Theorem 1.

Corollary 3. (SBT(2) for torsion-free groups). If G is a completely decomposable torsion-free abelian group then, for every partial presentation (F, π) of G, X and R can be chosen so that $\ell(R) \leq 2$.

Proof. Apply Theorem 1. Then $G = \pi(F_{X_1}) \oplus \pi(F_{X_2})$. Hence $\pi(F_{X_1})$ is a direct summand of a completely decomposable group and is therefore completely decomposable itself. Hence we can choose a subset X_1' of X_1 so that $\pi(X_1')$ generates $\pi(F_{X_1})$ and so that the kernel of π restricted to $F_{X_1'}$ is generated by some R' with $\ell(R') \leq 2$. Now, for each x in $X_1 - X_1'$, write $\pi(x)$ as a linear combination $\Sigma_{i=1}^{n} a_i\pi(x_i')$ where a_i is in $\mathbb{Z}$ and x_i' is in X_1'. Let $x^* = x - \Sigma_{i=1}^{n} a_i x_i'$. Then $X_1' \cup \{x^* : x \in X_1 - X_1'\} \cup X_2 \cup X_3$ is a basis for F and $R = R' \cup \{x^* : x \in X_1 - X_1'\} \cup X_3$ generates the kernel of π.

It turns out that SBT(2) and SBT(3) are true in general. Namely we have

Theorem 4. (SBT(2)). If G is a simply presented abelian group then, for every partial presentation (F,π) of G, X and R can be chosen so that $\ell(R) \leq 2$.

Proof. The only proof we have for the general case is incredibly messy. When G is primary (i.e. totally projective) the situation is much better, and a relatively straightforward adaptation of Cohen and Gluck's proof of SBT(1) will work. Details of this will appear elsewhere.

Theorem 5. (SBT(3)). If G is an abelian group then, for every partial presentation (F,π) of G, X and R can be chosen so that $\ell(R) \leq 3$.

Proof. Here the proof involves a judicious blending of the method used in proving Theorem 1 with the second method given in section two for showing that all groups have presentations with $\ell(R) \leq 3$. Again, details will appear elsewhere.

5. Conclusions.

It seems clear that we have not found optimal proofs for SBT(2) and SBT(3) in the general case. There should be a theorem analogous to Theorem 1 which gives a "canonical form" for presentations of arbitrary abelian groups and which yields SBT(2) and SBT(3) as immediate corollaries. Indeed we feel that such theorems should exist for other algebraic systems too and that the general investigation of such problems in "presentation theory" would be well worthwhile.

REFERENCES

1. J. Cohen and H. Gluck, Stacked bases for modules over principal ideal domains, J. Algebra 14 (1970), 493-505.

2. P. Crawley and A. W. Hales, The structure of torsion abelian groups given by presentations, Bull. Amer. Math. Soc. 74 (1968), 954-956.

3. P. Crawley and A. W. Hales, The structure of abelian p-groups given by certain presentations, J. Algebra 12 (1969), 10-23.

4. J. Erdös, Torsion-free factor groups of free abelian groups and a classification of torsion-free abelian groups, Publ. Math. Debrecen 5 (1957), 172-184.

5 L. Fuchs, Abelian Groups, Publ. House of the Hungar. Acad. Sci., Budapest, 1958.

6. ______, Infinite Abelian Groups, Vols. I, II, Pure and Appl. Math., vol. 36, Academic Press, New York and London, 1970, 1973.

7. P. Hill, On the classification of abelian groups, lecture notes.

8. I. Kaplansky, Infinite Abelian Groups, University of Michigan Press, Ann Arbor, Michigan, 1954 and 1969.

COMMUTATIVE RINGS WHOSE FINITELY GENERATED MODULES ARE DIRECT SUMS OF CYCLICS

Roger Wiegand and Sylvia Wiegand

1. Introduction. The fundamental theorem on abelian groups states that principal ideal rings enjoy the property of the title, and of course this property characterizes the principal ideal rings among Noetherian rings. The literature of the last three decades contains several extensions of the fundamental theorem and its converse to special classes of non-Noetherian rings. In 1976 a complete structure theorem was finally obtained for the rings of the title, hereafter referred to as FGC-rings. This theorem is a vindication of the fundamental theorem: Every FGC-ring is a finite direct product of indecomposable rings, and the indecomposable factors look very much like principal ideal domains. (A precise statement of the structure theorem appears at the end of section 2.) The proof that finitely generated modules decompose over the rings of the structure theorem is conceptually very similar to the proof of the fundamental theorem; the converse is harder and depends on some non-trivial ideas from general topology.

In section 2 we outline a typical proof of the fundamental theorem for principal ideal domains and describe a class of non-Noetherian rings for which the "same" proof works. The third section takes care of the local case of the structure theorem. In section 4 we characterize the reduced FGC-rings, modulo two lemmas proved in sections 5 and 6. The sixth section also deals with the non-trivial problems caused by nilpotents in the ring. Finally, in section 7, we put the pieces together and present examples of FGC-rings of the various types prescribed by the structure theorem.

All rings in this paper are commutative with 1, and all modules are unital. We will use, without explicit reference, the evident facts that the class of FGC-rings is closed under the formation of finite direct products, homomorphic images, and rings of fractions. A reduced ring is a ring without non-zero nilpotent elements, and a valuation ring is a ring (possibly with zero-divisors) whose ideal lattice is totally ordered. Finally, a ring is <u>Bézout</u> in case every finitely gen-erated ideal is principal.

This paper could not have been written without the generous assistance of our colleague Tom Shores. His contributions to the solution of this problem occur hroughout the paper. Section 6 consists entirely of his unpublished results, which e presented to our ring theory seminar.

. <u>A generalization of the fundamental theorem.</u> Let M be a finitely generated odule over a principal ideal domain R. The proof that M is a direct sum of yclic modules typically proceeds in three steps: (1) Prove that M/T is free, being the torsion submodule. (2) Prove that the natural map $T \to \Sigma T_p$ (p anging over the maximal ideals of R) is an isomorphism. (3) Prove that each T_p s a direct sum of cyclic R_p-modules.

In order to apply the same three steps in a more general setting, we need some erminology. A valuation ring R is <u>maximal</u> provided every nested family of cosets f ideals $\{x_\alpha + I_\alpha\}$ has non-empty intersection, and is <u>almost maximal</u> provided R/I s maximal for every non-zero ideal I. An integral domain is <u>h-local</u> provided very non-zero ideal is contained in only finitely many maximal ideals, and every on-zero prime ideal is contained in a unique maximal ideal. Finally, an integral omain R is <u>almost maximal</u> if it is h-local and $R_{\mathfrak{m}}$ is an almost maximal valu-ation ring for every maximal ideal $\mathfrak{m}$. (See [1] for an explanation of this term-nology.)

<u>heorem 1.</u> Every almost maximal Bézout domain is an FGC-ring.

<u>roof.</u> First we need to know that a finitely generated torsion free module over a ézout domain is free. For future reference we record a more general result.

<u>emma 2.</u> Let R be a Bézout ring whose classical quotient ring is self-injective. et M be a finitely generated R-module containing an element x with annihilator . Then M has a direct summand isomorphic to R.

<u>roof.</u> An easy diagram chase shows that Q, the classical quotient ring of R, is njective as an R-module. Therefore the map $Rx \to Q$ taking x to 1 extends to

$f:M \to Q$. Since R is Bézout, $f(M)$ is cyclic, and its annihilator is 0. Thus we have a (split) surjection $M \twoheadrightarrow f(M) \cong R$.

Now let M be any finitely generated module over an almost maximal Bézout domain R, and let T be its torsion submodule. By Lemma 2, M/T is free, so $M \cong T \oplus R^m$. In particular, T is finitely generated, so $I = (0:T)$ is non-zero. Let $p_1,\ldots,p_n$ be the maximal ideals containing I, and for each i let J_i be the kernel of the natural map $R \to (R/I)_{p_i}$. Then $I \subseteq J_i \subseteq p_i$ and $I = \bigcap_i J_i$. If p is a maximal ideal distinct from p_i, then $(R_{p_i})_p$ has no non-zero prime ideal, hence is equal to the whole quotient field Q. Therefore $(J_i)_p = (I_{p_i} \cap R)_p = IQ \cap R_p = R_p$, that is, $J_i \nsubseteq p$. This means the ideals J_i are pairwise comaximal, and $R/I \cong R/J_1 \oplus \ldots \oplus R/J_n$. But each R/J_i is a proper homomorphic image of R_{p_i}, so Theorem 1 will follow from this lemma.

Lemma 3. Every maximal valuation ring is an FGC-ring.

Proof. Suppose M is an R-module with minimal generating set $\{x_1,\ldots,x_n\}$ and annihilator I. Since R is a valuation ring, $(0:x_i) = I$ for some i; thus as an R/I-module, M has an element with annihilator 0. If we can show that the classical quotient ring of R/I is self-injective, we can apply Lemma 2 to get $M \cong R/I \oplus M'$; then M' can be generated by $n-1$ elements, and so by induction is a direct sum of cyclics. For future reference, we isolate the result we need:

Lemma 4. The classical quotient ring R of a maximal valuation ring is a self-injective ring.

Proof. Since R is a valuation ring in which each non-unit is a zero-divisor, we have the annihilator relations $(0:(0:x)) = (x)$ for every $x \in R$. (See [3, Lemma 3].) Let I be an ideal in R and $f:I \to R$ an R-homomorphism. Write $I = \{a_\alpha\}$, $f(a_\alpha) = r_\alpha$. We seek an $x \in R$ such that $xa_\alpha = r_\alpha$ for all α. From the annihilator relations we deduce that $r_\alpha \in Ra_\alpha$, say $r_\alpha = x_\alpha a_\alpha$. Then the solution set of the equation $xa_\alpha = r_\alpha$ is the coset $x_\alpha + (0:a_\alpha)$. Moreover, these cosets are nested, and since R is maximal [3, Lemma 2], their intersection is non-empty, as required.

Many of the ideas in this section come from Kaplansky's 1952 paper [7]. He proved in that paper that every finitely generated torsion-free module over a Prüfer domain is a direct sum of invertible ideals, and that every almost maximal valuation ring is an FGC-ring. In the mid sixties, Matlis introduced h-local domains and characterized them as precisely those domains whose torsion modules decompose into p-components, [10,11]. He also recognized the importance of injective modules in decomposition problems. Gill generalized Matlis' techniques to allow zero-divisors, in [3]. We should also mention that an earlier paper by Klatt and Levy, [9], contains several ideas related to Lemma 3, including the proof of Lemma 4.

Theorem 1 is very nearly best possible, in the sense that the almost maximal Bezout domains are the building blocks for arbitrary FGC-rings. This point of view is partially explained by the following form of our structure theorem:

Theorem 5. A commutative ring is an FGC-ring if and only if it is a finite direct product of rings R satisfying the following four properties:

(a) R has a unique minimal prime p,

(b) $R_{\mathfrak{m}}$ is an almost maximal valuation ring for each maximal ideal $\mathfrak{m}$,

(c) R/p is an h-local Bézout domain, and

(d) the ideals of R contained in p form a chain.

The proof of this theorem will take up the rest of the paper. The role of the almost maximal Bézout domains as building blocks will be clarified in section 7, where we pin down the precise structure of the indecomposable FGC-rings.

3. The local case. The first step in the proof of Theorem 5 is to prove that the local FGC-rings are just the almost maximal valuation rings. We have already seen that almost maximal valuation domains are FGC-rings; the converse (for domains) is due to Matlis, [11, Lemma 5.6]. The general result for local rings was proved by Gill, [3]. A key step in Gill's proof was the following surprising observation:

Lemma 6. Let R be an almost maximal valuation ring that is not a domain. Then R is maximal.

Proof. Consider a nested family of cosets $\{x_\alpha + I_\alpha : \alpha \in A\}$. If $I = \bigcap_\alpha I_\alpha \neq 0$, these cosets have non-empty intersection since R/I is then maximal, so we may assume $I = 0$. Choose a non-zero element $t \in R$ such that $t^2 = 0$, and fix $\delta \in A$ such that $I_\delta \subseteq R_t$. Let $B = \{\beta \in A | I_\beta \subseteq I_\delta\}$, and write $x_\beta - x_\delta = y_\beta t$ for each $\beta \in B$. Then the cosets $y_\beta + (I_\beta : t)$ are nested, and since $0 \neq t \in \bigcap_\beta (I_\beta : t)$, there is an element $z \in \bigcap_\beta (y_\beta + (I_\beta : t))$. Clearly $x_\delta + zt \in \bigcap_\alpha (x_\alpha + I_\alpha)$, and the proof is complete.

Theorem 7. Let R be a local ring. Then R is an FGC-ring if and only if R is an almost maximal valuation ring.

Proof. By Theorem 1 every almost maximal Bézout domain is an FGC-ring, and Lemmas 3 and 6 take care of the non-domains. To prove the converse, let R be a local FGC-ring with maximal ideal $\mathfrak{m}$, and let E be the injective hull of the R-module $R/\mathfrak{m}$. We will use repeatedly the following two properties of E: (1) The R-submodule lattice of E is totally ordered. (2) If I is an ideal of R and A is its annihilator in E, then $(0:A) = I$.

To prove (1), let $x, y \in E$ and set $M = Rx + Ry$. Since E is an indecomposable injective, M, being a direct sum of cyclics, is in fact cyclic. But this forces either $Rx = M$ or $Ry = M$, because R is local. Now (2) amounts to the fact that E is an injective cogenerator. In detail, let $x \in R - I$, and let $f:(Rx+I)/I \to E$ be the map taking $x + I$ to $1 + \mathfrak{m}$. This map extends to $g: R/I \to E$; then clearly $g(1+I)$ annihilates I but not x. In other words, $x \notin (0:A)$, as desired.

Now (1) and (2) imply that R is a valuation ring. To show that R is almost maximal, consider a nest $\{x_\alpha + I_\alpha\}$ of cosets, with $\bigcap_\alpha I_\alpha \neq 0$. We may assume each $I_\alpha \neq R$. Let A_α be the annihilator, in E, of I_α. Then by (2) $E' = \bigcup_\alpha A_\alpha$ is a proper submodule of E, say $u \in E - E'$. By (1), we have $E' \subset Ru$.

There is a well-defined map $f': E' \to E$ satisfying $f'(e) = x_\alpha e$ for $e \in A_\alpha$; let $f: E \to E$ extend f'. Suppose, first of all, that $f(u) \in Ru$, say $f(u) = xu$. Then $f(e) = xe$ for $e \in E'$, that is, $x - x_\alpha \in (0:A_\alpha)$ for all α. Then by (2) we have $x \in \bigcap_\alpha (x_\alpha + I_\alpha)$ and we're done. By (1) the only other possibility is that

$u \in Rf(u)$, say $u = sf(u)$. Fix α, and note that $e = sf(e) = sx_\alpha e$ for every $e \in A_\alpha$. Then $1 - sx_\alpha \in (0:A_\alpha) = I_\alpha \subsetneq \mathfrak{m}$, so s is a unit. Therefore $f(u) \in Ru$ in this case as well.

4. Reduced FGC-rings. In this and the next section we will prove the structure theorem for reduced FGC-rings. If we restrict Theorem 5 to reduced rings it reads as follows:

Theorem 8. A reduced ring is an FGC-ring if and only if it is a finite direct product of almost maximal Bézout domains.

We will deduce this theorem from the following two lemmas about the prime ideal structure of an FGC-ring:

Lemma 9. Every FGC-ring has only finitely many minimal primes.

Lemma 10. Every non-zero prime ideal in an FGC-domain is contained in a unique maximal ideal.

The proof of Lemma 9 is very indirect and will be taken up in the next section. The idea of the proof originated in Pierce's 1967 paper, [13], where he characterized the (von Neumann) regular FGC-rings. In this context the structure theorem just says that the only regular FGC-rings are the finite direct products of fields. In 1973, Shores and R. Wiegand were able to push Pierce's idea a little further to prove Lemma 9 under the additional assumption that R has fewer than 2^c prime ideals, [16]. In 1974, S. Wiegand proved Lemma 10, [21], thereby verifying Theorem 8 for rings with fewer than 2^c prime ideals. In 1976, W. Brandal proved that every FGC-domain is an almost maximal Bézout domain, [2], and a few months later, in [19], R. Wiegand obtained the general form of Theorem 8.

We will now deduce Theorem 8 from the lemmas (which will be proved in the next two sections). One direction, of course, has already been done (Theorem 1). For the reverse implication, let R be a reduced FGC-ring with minimal primes $p_1,\ldots,p_m$. These minimal primes are pairwise comaximal; for if two minimal primes were contained in the same maximal ideal $\mathfrak{m}$, we would have a contradiction to the fact that $R_{\mathfrak{m}}$ is a valuation ring (Theorem 7). Since R is reduced, we have $R = R/p_1 \times \ldots \times R/p_m$, and we need only show that every FGC-domain D is almost maximal. (Every FGC-domain is obviously Bézout.)

Let I be a non-zero ideal of D and let $q_1,\ldots,q_n$ be the primes minimal over I. (We are applying Lemma 9 to the ring R/I). Let $\mathfrak{m}_i$ be the unique (by Lemma 10) maximal ideal containing q_i. Then by Zorn's lemma $\mathfrak{m}_1,\ldots,\mathfrak{m}_n$ are the only maximal ideals containing I, and another appeal to Lemma 10 shows that D is h-local. By Theorem 7, D is an almost maximal domain.

This completes the proof of Theorem 8, modulo proofs of Lemmas 9 and 10.

5. Proof of Lemma 9. Let spec(R) denote the set of prime ideals of R. If α is an element of R or a subset of R, let $V(\alpha)$ be the set of primes containing α, and let $D(\alpha) = \text{spec}(R) - V(\alpha)$. The sets $V(\alpha)$ are the closed sets in the Zariski topology on spec(R). We will usually work with a stronger topology, namely, the "patch" or "constructible" topology. This topology has an open base consisting of sets of the form $D(a_0) \cap V(a_1,\ldots,a_n)$, $a_i \in R$. It can be shown, [6], that with the patch topology spec(R) is Boolean, that is, compact, Hausdorff and totally disconnected. A patch is a subset of spec R that is closed in the patch topology, and a thin patch is a patch that is the closure, in the patch topology, of its set of minimal elements. (All references to ordering on spec(R) refer to set-theoretic inclusion.)

Let x be a point in a topological space X and let n be a cardinal number. Then, following Pierce, [13], we call x an n-point provided there is a family $\mathfrak{Y}$ of pairwise disjoint open subsets of X such that $|\mathfrak{Y}| = n$ and x is in the closure of each member of $\mathfrak{Y}$. We note that the compact spaces with no 2-points are just the extremally disconnected spaces (corresponding to complete Boolean algebras). Now there are well-known methods of building closed sets that are not extremally disconnected. Our task will require greater care; starting with a purported FGC-ring with infinitely many minimal primes, we will build a thin patch in spec(R) that has a 3-point relative to the patch topology. The next lemma will provide the contradiction we want. This lemma was proved by Shores and R. Wiegand, [16], but the underlying idea is Pierce's, [13].

Lemma 11. Let R be an FGC-ring. Then no thin patch in spec(R) has a 3-point in the relative patch topology.

Proof. Assume the contrary, and let T be the offending thin patch. It is harmless to assume $\bigcap T = 0$. Using compactness of T in the patch topology, one can prove that every prime ideal of R contains a member of T. (See the corollary to Theorem 1 of [6].) Therefore the set S of minimal members of T is exactly the set of minimal primes of R.

Let U_1, U_2, U_3 be pairwise disjoint sets, open in the relative patch topology on T, each having the point z in its patch closure. We claim that the sets $U_i \cap S$ are open in the relative Zariski topology on S. To prove this we will show that every patch neighborhood $V(a) \cap S$ of a point $x \in S$ contains a Zariski neighborhood of x. Form the multiplicative set $\Gamma = \{a^n b : n \geq 0, b \in R - x\}$. If $0 \notin \Gamma$, Zorn's lemma provides a prime ideal y disjoint from Γ. Then $y \subseteq x$ implies $y = x$, so $a \in y \cap \Gamma$, contradiction. Therefore $0 \in \Gamma$, and $x \in D(b) \subseteq V(a)$ as promised.

Choose ideals A_i of R such that $U_i \cap S = D(A_i) \cap S$, $i = 1,2,3$. Then for $i \neq j$ we have $A_i \cap A_j = 0$; also z is in the patch closure of each $D(A_i)$ since T is the patch closure of S. It follows that $A_1 + A_2 + A_3 \subsetneq z$. Let K be the submodule of R^2 consisting of ordered pairs (a_1+a_2, a_1+a_3), where $a_i \in A_i$. We will show that $M = R^2/K$ is not a direct sum of cyclics.

Given a prime ideal x, let R(x) be the residue field of the local ring R_x, and let M(x) be the R(x)-vector space M_x/xM_x. If $f \in R$, respectively, M, let f(x) denote the image of f in R(x), respectively M(x). Let α and β in M be the images of (1,0) and (0,1) in $R \oplus R$. Finally, let E_i be the subspace of M(z) spanned by the elements m(z), where $m \in M$ and $m(x) = 0$ for all $x \in D(A_i)$. We declare the following to be true: $\alpha(z) \in E_1 - (E_2 \cup E_3)$, $\beta(z) \in E_2 - E_1$, and $\alpha(z) + \beta(z) \in E_3$. (The proofs are not at all inspiring and may be found in [16].) From these data we deduce that E_1, E_2 and E_3 are distinct one-dimensional subspaces of the two-dimensional vector space $M(z)$.

Now suppose $M = Rm_1 \oplus \ldots \oplus Rm_k$; by renumbering we may assume $\{m_1(z), m_2(z)\}$ is a basis for $M(z)$. Fix any $i = 1,2,3$, and choose a generator of E_i of the form $m(z)$ where $m(x) = 0$ for each $x \in D(A_i)$. Write $m = r_1m_1 + \ldots + r_km_k$, and note that $r_jm_j(x) = 0$, for each $j = 1,\ldots,k$, and each $x \in D(A_i)$. In other words, $r_j(z)m_j(z) \in E_i$ for $j = 1,2$. But $r_1(z)$ and $r_2(z)$ cannot both be 0, so either $m_1(z) \in E_i$ or $m_2(z) \in E_i$. We have shown that each of the three subspaces E_1, E_2, E_3 contains either $m_1(z)$ or $m_2(z)$, which is absurd.

The rest of the proof of Lemma 9 is purely topological. Since we have no further use for the Zariski topology, all topological notions concerning spec(R) refer to the patch topology. (The partial ordering, of course, is still the usual one.) The first step in producing 3-points is to produce them "universally", that is, in $\beta N - N$. (This is the complement of the discrete space of natural numbers in its Stone-Čech compactification.) The fact that $\beta N - N$ has 3-points was deduced by Pierce, [13], from W. Rudin's theorem, [14], on the existence of P-points in $\beta N - N$. Since Rudin's theorem depended on the continuum hypothesis, this left some doubts as to the status of Pierce's result on regular FGC-rings. Fortunately, N. Hindman set things straight in 1969 with the following theorem, [5], whose proof is independent of the continuum hypothesis:

Theorem 12. $\beta N - N$ has a c-point.

It would lead us too far from the main line of reasoning to give Hindman's elegant proof here. Instead, we will indicate how Pierce uses 3-points in $\beta N - N$ to build 3-points all over the place. First, we record without proof the following well-known result ([4, 9.11]):

Lemma 13. Every infinite closed subspace of $\beta N - N$ contains a closed subspace homeomorphic to $\beta N - N$.

Here is Pierce's result ([13, 21.5]):

Lemma 14. Let Z be a Boolean space with a countably infinite dense subspace consisting of isolated points of Z. Then either Z has a 3-point, or else Z has a closed subspace homeomorphic to $\beta N - N$.

Proof (sketch). Let $Y = \{y_n : n \in N\}$ be the given dense subset. The map $n \to y_n$ induces a map ϕ from βN onto Z. If some point $z \in Z$ has three distinct preimages in βN it is easy to see that z is a 3-point. Thus we assume ϕ is at most 2 to 1.

Let K be the set of points $q \in \beta N$ for which $\phi(q)$ has only one preimage, and let $L = \beta N - K$. If $K - N$ contains an infinite compact subset we are done, by Lemma 13. If not, then L is a dense subset of $\beta N - N$. For each $q \in L$, let $\hat{q}$ be the other point of L with the same image as q. Now it is possible to build an infinite sequence, $q_1, q_2, \ldots$ in L and disjoint sets U, V, open in βN, such that each $q_i \in U$, each $\hat{q}_i \in V$, and $\phi(q_i) \neq \phi(q_j)$ if $i \neq j$. Let $W = \phi(U \cap N)^- \cap \phi(V \cap N)^-$. This set is infinite, since it contains each $\phi(q_i)$. Also, $U^- \cap V^- = \emptyset$ because $\beta(N)$ is extremally disconnected, and it follows that ϕ maps the closed set $\phi^{-1}(W) \cap U^-$ homeomorphically onto W. Another appeal to Lemma 13 completes the proof.

At last we have assembled enough information to prove Lemma 9. The proof given in [19] actually yields a more general result. By a tree we mean a partially ordered set in which $G(x) = \{y | y \leq x\}$ is a chain for every x.

Theorem 15. Suppose spec(R) is a tree and no thin patch in spec(R) has a 3-point. Then R has only finitely many minimal primes.

Proof. Suppose the set E of minimal primes is infinite. Choose a countably infinite family $\{U_n\}$ of non-empty, pairwise disjoint, relatively open subsets of E, and choose $y_n \in U_n$. Let Z be the (patch) closure of $\{y_n\}$. Since Z is a thin patch, Lemmas 11 and 14 provide a closed subspace Z_1 homeomorphic to $\beta N - N$. We will build an infinite descending chain of closed sets $Z_1 \supset Z_2 \supset \ldots$, each homeomorphic to $\beta N - N$, such that no element of Z_{n+1} is a minimal element of Z_n. Suppose, inductively, that we have built $Z_1, \ldots, Z_n$. By Lemma 11 and Theorem 12, Z_n is not a thin patch, so Z_n has a non-empty open subset U that avoids every minimal element of Z_n. Then U contains a non-empty open-and-closed set V, necessarily infinite. Now apply Lemma 13 to get a set Z_{n+1} contained in V and homeomorphic to $\beta N - N$.

By compactness there is a prime $p \in \bigcap_n Z_n$. For each n, p contains a minimal element of Z_n, say p_n. (The intersection of a chain in Z_n is again in Z_n, so Zorn's lemma applies.) Since spec(R) is a tree, the primes p_n form a chain, and it follows that p_n is properly contained in p_{n+1}. Let $q = \bigcup_n p_n$, and note that $\{q, p_1, p_2, \ldots\}$ is a countably infinite patch contained in Z_1, which is a flagrant violation of Lemma 13. This contradiction completes the proof of Theorem 15.

Lemma 9 follows instantly from what we have done: If R is an FGC-ring then spec(R) is a tree by Theorem 7, and Lemma 11 assures us that no thin patch has a 3-point. Therefore, by Theorem 15, R has only finitely many minimal primes.

6. Handling the nilpotents. Lemma 9 essentially reduces the study of FGC-rings to the case of a unique minimal prime. In this section we handle two of the remaining tasks: i) the proof of Lemma 10 (which will complete Theorem 8) and ii) verifying property (d) in the structure theorem (Theorem 5). This property of FGC-rings was established independently by T. S. Shores (unpublished) and P. Vámos, [18]. Their result conveniently provides a proof of Lemma 10 that is easier than S. Wiegand's original proof, [21]. What follows is Shores' presentation.

Theorem 16. Let R be an FGC-ring with unique minimal prime p. Then the ideals contained in p form a chain.

Proof. Suppose not, and let Rx and Ry be incomparable ideals contained in p. By factoring out $Rx \cap Ry$ we may assume $Rx \cap Ry = 0$. Let A and B be maximal proper submodules of Rx and Ry, and pass to $R/(A \oplus B)$. Thus we may as well assume Rx and Ry are minimal ideals, with annihilators $\mathfrak{m}$ and $\mathfrak{n}$ respectively. Note $\mathfrak{m} \neq \mathfrak{n}$, or else x and y would survive in $R_{\mathfrak{m}}$, violating Theorem 7. As a final reduction, we localize at the multiplicative set $R - (\mathfrak{m} \cup \mathfrak{n})$. We now have this set-up: R has exactly two maximal ideals $\mathfrak{m}$

and $\mathfrak{n}$, $(0:x) = \mathfrak{m}$ and $(0:y) = \mathfrak{n}$.

Let K and L be the kernels of the natural maps $R \to R_{\mathfrak{m}}$ and $R \to R_{\mathfrak{n}}$ respectively, and let $T = R/K \oplus R/L$. Write $m + n = 1$, with $m \in \mathfrak{m}$, $n \in \mathfrak{n}$, and let M be the submodule of T generated by the elements $(1+K, m+L)$ and $(0, n+L)$.

As in [21], the key to the proof is to examine carefully the local structure of M. When we localize M at $\mathfrak{m}$, the two generators given above split M into the direct sum of two cyclics:

$$M_{\mathfrak{m}} \cong R_{\mathfrak{m}} \oplus (R/L)_{\mathfrak{m}} . \tag{1}$$

In order to see what happens when we localize at $\mathfrak{n}$, we use a new pair of generators for M, namely, $(1+K, 1+L)$ and $u = (n+K, 0)$. Then we have

$$M_{\mathfrak{n}} \cong R_{\mathfrak{n}} \oplus (Ru)_{\mathfrak{n}} . \tag{2}$$

Next, we note that the socle S of T is generated by $(x+K, 0)$ and $(0, y+L)$, both of which lie in M. Since $R_{\mathfrak{m}}$ and $R_{\mathfrak{n}}$ are valuation rings, it is easy to see that T is an essential extension of S. Therefore M is an essential extension of its socle, which has length 2. It follows that M is the direct sum of <u>two</u> cyclic modules: $M = C \oplus D$.

Now over a valuation ring, direct sum decompositions into cyclics are unique, so by (1) we may assume $C_{\mathfrak{m}} \cong R_{\mathfrak{m}}$ and $D_{\mathfrak{m}} \cong (R/L)_{\mathfrak{m}}$. Also, by (2) we have either $C_{\mathfrak{n}} \cong R_{\mathfrak{n}}$ or $D_{\mathfrak{n}} \cong R_{\mathfrak{n}}$. The first possibility is easily ruled out: If $C_{\mathfrak{n}} \cong R_{\mathfrak{n}}$ we deduce that $C \cong R$, so the socle of C has length 2 and hence is equal to S. But since M is an essential extension of S, this forces $D = 0$, a contradiction.

Thus (2) tells us that $D_{\mathfrak{n}} \cong R_{\mathfrak{n}}$. Then we see, by checking locally, that $(0:D) = L$. Let $v = \alpha(1+K, m+L) + \beta(0, n+L)$ be a generator for D. Note that $\alpha \notin \mathfrak{n}$; otherwise we would have $yv = 0$, contradicting $(0:v) \subseteq L$. On the other hand, $Lv = 0$ implies $\alpha L \subseteq K \cap L = 0$, so α is not a unit. We conclude that $\alpha \in \mathfrak{m} - \mathfrak{n}$.

Now $\alpha \notin p$, and since $R_{\mathfrak{m}}$ and $R_{\mathfrak{n}}$ are valuation rings we deduce that

$p \subset R\alpha$. In particular, we can write $x = r\alpha$. Then $r\alpha^2 = x\alpha = 0$ forces $r \in L$. But we have already seen that $\alpha L = 0$, so $x = r\alpha = 0$, a contradiction.

Now we can prove Lemma 10. Suppose p is a non-zero prime contained in two distinct maximal ideals $\mathcal{M}, \mathcal{N}$ of the FGC-domain R. We may assume $\mathcal{M}$ and $\mathcal{N}$ are the <u>only</u> maximal ideals in R. By [8, Theorem 11] there exist prime ideals $p_1 \subset p_2 \subseteq p$, such that no prime ideal lies strictly between p_1 and p_2. Thus, by replacing p by p_2 and factoring out p_1, we may assume p has height 1. Moreover, p is the <u>only</u> height one prime since spec(R) is a tree. Write $m + n = 1$ with $m \in \mathcal{M}$ and $n \in \mathcal{N}$, and let x be any non-zero element of p. Then mx and nx are both non-zero modulo $Rmx \cap Rnx$, and Theorem 16 is violated.

The proof of Theorem 8 is now complete.

7. <u>Indecomposable FGC-rings</u>. In this section we will finish the proof of the structure theorem (Theorem 5) and then take a closer look at the indecomposable FGC-rings. One implication of Theorem 5 is very easy: Let R be any FGC-ring, and let N be the nilradical. By Theorem 8, R/N is a finite direct product of integral domains. But idempotents lift modulo N, so we deduce that R is a finite direct product of rings each of which has a unique minimal prime. To see that each of these factors satisfies (b), (c), (d) of Theorem 5, we quote Theorems 7, 8 and 16, respectively.

For the reverse implication, we need to show that every ring R satisfying (a), (b), (c), (d) is an FGC-ring. (This is essentially the content of Theorem 4.1 of [16].) We remark at the outset that p is comparable to every ideal of R.

Let M be any finitely generated R-module. If $(0:M) \supseteq p$ then M is a direct sum of cyclics, because R/p is an almost maximal Bézout domain. The other possibility is that $(0:M) \subset p$. We will show that in this case M has a cyclic direct summand whose annihilator is $(0:M)$. Assuming this for the time being, we can complete the proof as follows: Write $M = Rx_1 \oplus M_1$, where $(0:x_1) = (0:M)$. If $(0:M_1) \supseteq p$ we're done, since R/p is an almost maximal Bézout domain. If not, repeat the argument to get $M = Rx_1 \oplus Rx_2 \oplus M_2$, etc. The only worry is that we might keep splitting off cyclic summands indefinitely. But since $(0:x_1) \subseteq (0:x_2) \subseteq \ldots$ it is easy to see that $Rx_1 \oplus \ldots \oplus Rx_n$ needs n generators. Thus, if M can be generated by n elements, the process stops after at most n steps.

We still need a cyclic summand with annihilator equal to $(0:M)$. Since $(0:M) \subset p$ all hypotheses on R carry over to $R/(0:M)$, so we may as well assume M is faithful (and $p \neq 0$). Let $x_1,\ldots,x_n$ generate M. Since $(0:x_1) \cap \ldots \cap (0:x_n) = 0$, we can renumber to get $(0:x_1) \subsetneq p$. But this forces $(0:x_1)$ to be comparable with every $(0:x_i)$, which in turn forces some $(0:x_i)$ to be 0. Now R is easily seen to be Bézout, so by Lemma 2 it will suffice to prove that the classical quotient ring of R is a self-injective ring. One last lemma is in order.

Lemma 17. Let R be a ring with unique minimal prime p, and assume that the ideals of R contained in p form a chain. Then $p_{\mathcal{M}} \neq 0$ for at most one maximal ideal $\mathcal{M}$.

Proof. Suppose, first, that $x \in p$ and $(0:x)$ is contained in two distinct maximal ideals $\mathcal{M}$ and $\mathcal{N}$. We may assume $\mathcal{M}x \subseteq \mathcal{N}x$. But when we localize the inclusion at $\mathcal{N}$ we get $(Rx)_{\mathcal{N}} \subseteq \mathcal{N}(Rx)_{\mathcal{N}}$, contradicting $(Rx)_{\mathcal{N}} \neq 0$. Now fix a non-zero element $x \in p$, say $(0:x) \subseteq \mathcal{M}$. If $y \in p$ and $(0:y)$ is contained in a maximal ideal $\mathcal{N}$, either possibility $(Rx \subseteq Ry$ or $Ry \subseteq Rx)$ forces $\mathcal{N} = \mathcal{M}$. Thus $p_{\mathcal{N}} = 0$ if $\mathcal{N} \neq \mathcal{M}$.

Returning to the proof of Theorem 5, we let $\mathcal{M}$ be the unique maximal ideal for which $p_{\mathcal{M}} \neq 0$. Clearly, the set of zero-divisors of R is contained in $\mathcal{M}$, so K, the classical quotient ring of R, is a localization of $R_{\mathcal{M}}$. But this makes K an almost maximal valuation ring. (For a silly proof, use Theorem 7.) But K is certainly not a domain since it contains R, so K is maximal, by Lemma 6. Finally, Lemma 4 says K is self-injective, and the proof of Theorem 5 is complete.

The fundamental theorem of abelian groups has another aspect that we have not discussed. It provides a "canonical" decomposition for each finitely generated abelian group. In general, we define a canonical form for an R-module M to be a decomposition $M \cong R/I_1 \oplus \ldots \oplus R/I_n$, where $I_1 \subseteq \ldots \subseteq I_n \subset R$. (Such a decomposition, if it exists, is unique.) It turns out that every finitely generated module over an FGC-ring has a canonical form (although the only known proof of this fact requires the full strength of the structure theorem). Thus, the FGC-rings are exactly the "FGCF-rings" characterized in [16]. Our proof of the structure theorem almost (but not quite) produces a canonical form. The missing step is to show that a direct sum of cyclics over an h-local Bézout domain has a canonical form, [16, 1.7].

It is interesting to note that the methods of [16] sufficed to characterize FGCF-rings (rings whose finitely generated modules have canonical forms) but stopped short of characterizing FGC-rings. The reason is that it is easier to build 2-points than 3-points.

What are the indecomposable FGC-rings? There are the maximal valuation rings, the almost maximal Bézout domains, and a third type which we will describe in detail. Following Vámos, [18], we call a ring R a torch ring provided (i) R has a unique minimal prime $p \neq 0$, (ii) $R_{\mathfrak{m}}$ is a valuation ring for each maximal ideal $\mathfrak{m}$, (iii) the ideals contained in p form a chain, and (iv) R is not local. Thus, an indecomposable FGC-ring is either a maximal valuation ring, an almost maximal Bézout domain, or a torch ring. (We are using Lemma 6.) The artificial-looking condition (iv) actually contributes substantially to the structure of torch rings:

Lemma 18. Let R be a torch ring with minimal prime p. Then $p^2 = 0$, and p is a torsion, divisible R/p-module.

Proof. Let $\mathfrak{m}$ be the (unique by Lemma 17) maximal ideal for which $p_{\mathfrak{m}} \neq 0$, and let $\mathfrak{n}$ be any other maximal ideal. Then, if $x \in p$ we have $(0:x) \not\subseteq \mathfrak{n}$, so certainly $(0:x) \not\subseteq p$. But we have already noted that p is comparable to every ideal, so we have $(0:x) \supset p$. Thus $p^2 = 0$, and p is a torsion (R/p)-module. Also, if $x \in p$ and $r \in R - p$, then $(x) \subset (r)$, so p is divisible.

This lemma suggests that a torch ring might arise as a trivial extension of a suitable module over a Bézout domain. In fact, these are the only known examples of FGC-torch rings. Before describing the construction, however, we will discuss maximal valuation rings.

Given any integer n, one can "build" a maximal valuation domain of rank n as follows: Let G be the direct sum of n copies of the additive group of rational numbers, with the lexicographic ordering. Let v be the valuation on $C(x_1,\ldots,x_n)$ induced by $v(x_1^{a_1}\ldots x_n^{a_n}) = (a_1,\ldots,a_n)$. The valuation ring $\mathcal{O}$ of the maximal completion K is a maximal valuation ring of rank n. (See [15] for details on maximal completions.) Here is a fairly concrete picture of $\mathcal{O}$: If $a = (a_1,\ldots,a_n) \in G$, we abbreviate the monomial $x_1^{a_1}\ldots x_n^{a_n}$ by x^a. An element of $\mathcal{O}$ is then a "long power series" $\sum_{a\in A} c_a x^a$, where A is a well-ordered subset of G^+ and the coefficients c_a are complex numbers. (The fact that the exponents are well-ordered allows us to multiply power series in the usual way.)

Now one can define various automorphisms α_i of K and intersect the valuation rings $\alpha_i\mathcal{O}$, to obtain examples of non-local domains whose localizations are maximal valuation rings. This idea was first used by Osoksky, [12]; and later S. Wiegand, [20], adapted the method to prove the following:

Theorem 19. Let X be any finite tree with a unique minimal element. Then there is a Bezout domain R such that spec R is order-isomorphic to X, and R_p is a maximal valuation ring for every maximal ideal p.

If we restrict Theorem 19 to trees with the property that every non-minimal element is below a unique maximal element, we get a substantial collection of non-Noetherian FGC-domains. We should also mention an example due to Vâmos, [17], of a one-dimensional Bézout domain, locally maximal, and with infinitely many maximal ideals.

Finally, we indicate how to build FGC-torch rings. Start with a non-local almost maximal Bezout domain D such that $D_{\mathfrak{m}}$ is actually maximal, for some maximal ideal $\mathfrak{m}$. Let K be the quotient field of D and set $p = K/D_{\mathfrak{m}}$. As we saw in the proof of Theorem 1, $p_{\mathfrak{n}} = 0$ for every maximal ideal $\mathfrak{n} \neq \mathfrak{m}$. Let $R = D \oplus p$, with multiplication given by $(\alpha,x)(\beta,y) = (\alpha x,\alpha y+\beta x)$. Since p is a divisible D-module, every principal ideal of R not contained in p actually contains p. Also, it can be shown, [10, Corollary 8.6], that every D-submodule of p is actually a $D_{\mathfrak{m}}$-submodule, so the ideals of R contained in p form a chain. One can check directly that $R_{\mathfrak{m} \oplus p}$ is a maximal valuation ring, whereas $R_{\mathfrak{n} \oplus p} = D_{\mathfrak{n}}$ if $\mathfrak{n}$ is a maximal ideal distinct from $\mathfrak{m}$.

REFERENCES

1. W. Brandal, Almost Maximal Domains and Finitely Generated Modules, Trans. Amer. Math. Soc. 183(1973), 203-222.

2. W. Brandal, Domains Whose Finitely Generated Modules Decompose, preprint.

3. D. T. Gill, Almost Maximal Valuation Rings, J. London Math. Soc. (2), 4(1971), 140-146.

4. L. Gillman and M. Jerison, Rings of Continuous Functions, Van Nostrand, New York, 1960.

5. N. Hindman, On the Existence of c-points in βN\N, Proc. Amer. Math. Soc. 21(1969), 277-280.

6. M. Hochster, Prime Ideal Structure in Commutative Rings, Trans. Amer. Math. Soc. 142(1969), 43-60.

7. I. Kaplansky, Modules over Dedekind Rings and Valuation Rings, Trans. Amer. Math. Soc. 72(1952), 327-340.

8. I. Kaplansky, Commutative Rings, Allyn and Bacon, Inc., Boston, 1970.

9. G. Klatt and L. S. Levy, Pre-self-injective Rings, Trans. Amer. Math. Soc. 137(1969), 407-419.

10. E. Matlis, Cotorsion Modules, Memoirs Amer. Math. Soc. 49(1964).

11. E. Matlis, Decomposable Modules, Trans. Amer. Math. Soc. 125(1966), 147-179.

12. E. Matlis, Rings of Type I, J. Algebra 23(1972), 76-87.

13. R. S. Pierce, Modules over Commutative Regular Rings, Memoirs Amer. Math. Soc. 70(1967).

14. W. Rudin, Homogeneity Problems in the Theory of Čech Compactifications, Duke Math. Journal 3(1956), 409-419.

15. O. F. G. Schilling, The Theory of Valuations, Amer. Math. Soc. Surveys, No. 4, New York, 1950.

16. T. Shores and R. Wiegand, Rings Whose Finitely Generated Modules Are Direct Sums of Cyclics, J. Algebra 32(1974), 152-172.

17. P. Vámos, Multiply Maximally Complete Fields, J. London Math. Soc. 12(1975), 103-111.

18. P. Vámos, The Decomposition of Finitely Generated Modules and Fractionally Self-injective Rings, preprint.

19. R. Wiegand, Reduced Rings Whose Finitely Generated Modules Are Direct Sums of Cyclics, preprint.

20. S. Wiegand, Locally Maximal Bezout Domains, Proc. Amer. Math. Soc. 47(1975), 10-14.

21. S. Wiegand, Semilocal Domains Whose Finitely Generated Modules Are Direct Sums of Cyclics, Proc. Amer. Math. Soc. 50(1975), 73-76.